Lecture Notes in Mathematics

Edited by A. Dold and B.

T0210836

782

Bifurcation and Nonlinear Eigenvalue Problems

Proceedings, Université de Paris XIII, Villetaneuse, France, October 2–4, 1978

Edited by
C. Bardos, J. M. Lasry, and M. Schatzman

Springer-Verlag
Berlin Heidelberg New York 1980

Editors

C. Bardos
Département de Mathématiques
Université Paris-Nord
Avenue J. B. Clément
93430 Villetaneuse
France

J. M. Lasry
CEREMADE
Université de Paris 9 – Dauphine
75775 Paris Cedex 16
France

M. Schatzman
Analyse Numérique
Tour 55–65
5ème étage
Université Pierre et Marie Curie
75230 Paris Cedex 05
France

AMS Subject Classifications (1980): 35 G 20

ISBN 3-540-09758-9 Springer-Verlag Berlin Heidelberg New York
ISBN 0-387-09758-9 Springer-Verlag New York Heidelberg Berlin

Library of Congress Cataloging in Publication Data. Main entry under title: Bifurcation and nonlinear eigenvalue problems. (Lecture notes in mathematics; 782) Bibliography: p. Includes index. 1. Differential equations, Partial--Congresses. 2. Differential equations, Nonlinear--Congresses. 3. Eigenvalues--Congresses. 4. Bifurcation theory--Congresses. I. Bardos, Claude, 1940- II. Lasry, J. M. III. Schatzman, M. IV. Series: Lecture notes in mathematics (Berlin; 782. QA3.L28 no. 782 [QA374] 510s [515.3'53] 80-11358
ISBN 0-387-09758-9

Printed in Germany

Printing and binding: Beltz Offsetdruck, Hemsbach/Bergstr.
2141/3140-543210

PREFACE

This volume contains the notes of a session[+] organized on October 2,3 and 4,1978 at the Departement of Mathematics of the University Paris 13,Centre Scientifique et Polytechnique,Villetaneuse.

The aim of this session was to gather mathematicians and scientists of other fields:chemistry,biology,physics and astrophysics,and to let them exchange information and methods.

The common points to all the lectures are partial differential equations,non linear phenomena,study of the dependence with respect to a parameter,and the methods used are very diverse.

The lectures can be classified into three groups according to their relation to applied science : papers belonging to the first cluster deal with a phenomenological approach;in this case,a complete system of equations describing the experimental phenomenon is either too complicated or not entirely known and understood;therefore,a simpler system is studied which mimicks the behavior of the complete system, and one expects qualitative results.Here belong the talks of J.Heyvaerts,J.M.Lasry, M.Schatzman & P.Witomski,of G.Iooss,of J.P.Kernevez,G.Joly,D.Thomas & B.Bunow,and of P.Ortoleva.

The second group is made of mathematical and numerical studies of more complete modelizations:here,the model is better understood,and the study is more precise,so that it may give quantitative results;this group contains the contributions of C.M. Brauner & B.Nicolaenko,of C.Guillopé,of G.Iooss & R.Lozi,and of J.Mossino.

Though the papers of the third group are not directly concerned with natural phenomena,they develop theoretical tools and an understanding of non-linear phenomena,which are intended to meet the needs and preoccupations of the applied scientists. We include here the papers of H.Berestycki & P.L.Lions,of C.Bolley,M.Barnsley, F.Mignot,F.Murat & J.P.Puel,of J.C.Saut,and of D.Serre.

Striking observational data were brought by M.Dupeyrat who showed a beautiful dynamic periodic chemical phenomenon with a movie film.

We added a paper of J.P.Chollet and M.Lesieur.They show how the nonlinearity of the Navier-Stokes equation can create turbulence and give a phenomenological model that fits with the Kolmogorov law.

We thank all the participants for their active presence and interesting contributions

C.Bardos,J.M.Lasry,M.Schatzman.

[+]Supported by the University Paris 13,and the Centre National de la Recherche Scientifique,A.T.P. Mathématiques pour les Sciences de l'Ingénieur.

CONTENTS

The contributions of J.Mossino, A non Linear Eigenvalue Problem and a Variational Problem connected with Increasing Rearrangement in Plasma Physics, and of G.Iooss and R.Lozi, Convection between two Rotating Plates and Dynamo Effect occuring from a Secondary Bifurcation will appear elsewhere.

LIST OF PARTICIPANTS

ORGANIZERS

C. Bardos,Département de Mathématiques,Université Paris XIII,Av. J.B. Clément, 93430 Villetaneuse.

J.M. Lasry,CEREMADE,Université Paris IX,Place De Lattre de Tassigny, 75775 Paris CEDEX 16.

M. Schatzman,C.N.R.S.,L.A. 189 Analyse Numérique,Université Pierre et Marie Curie,2 Place Jussieu,75230 Paris CEDEX 05.

CONTRIBUTORS

M.F. Barnsley,C.E.A.,Division de la Physique,B.P.2,91190 Gif sur Yvette.

H.Berestycki,C.N.R.S.,Laboratoire d'Analyse Numérique,Université Pierre et Marie Curie,2 Place Jussieu;75230 Paris CEDEX 05.

C.Bolley,Laboratoire d'Analyse Numérique,I.N.S.A. de Rennes,B.P. 14 A, 35031 Rennes CEDEX.

C.M. Brauner,Laboratoire de Mathématiques-Informatique-Systèmes,Ecole Centrale de Lyon,69130 Ecully.

B.Bunow,National Institute of Health,Bethesda,Maryland 20014,U.S.A.

J.P. Chollet,Institut de Mécanique de Grenoble,B.P. 53 X,38041 Grenoble CEDEX.

M.Dupeyrat,Laboratoire de Chimie Physique,Université Pierre et Marie Curie, 11 Rue Pierre et Marie Curie,75230 Paris CEDEX 05.

C.Guillopé,C.N.R.S.,Analyse numérique et Fonctionnelle,Bât. 425,Université Paris-Sud,91405 Orsay.

J.Heyvaerts,Observatoire de Meudon,92190 Meudon.

G. Iooss,Institut de Mathématiques et de Sciences Physiques,Parc Valrose, 06034 Nice.

G.Joly,Université de Technologie de Compiègne,B.P. 233,60206 Compiègne.

J.P. Kernevez,Université de Technologie de Compiègne,B.P. 233, 60206 Compiègne.

P.L. Lions,C.N.R.S.,Laboratoire d'Analyse Numérique,Université Pierre et Marie Curie,2 Place Jussieu,75230 Paris CEDEX 05.

M.Lesieur,Institut de Mécanique de Grenoble,B.P. 53 X,38041 Grenoble CEDEX.

R. Lozi, Institut de Mathématiques et de Sciences Physiques,06034 Nice.

F. Mignot,U.E.R. de Mathématiques,Université de Lille I,59650 Villeneuve d'Ascq.

J.Mossino,C.N.R.S.,Analyse Numérique et Fonctionnelle,Bât. 425,Université Paris-Sud,91405 Orsay.

F. Murat,Laboratoire d'Analyse Numérique,Université Pierre et Marie Curie, 2 Place Jussieu,75230 Paris CEDEX 05.

B.Nicolaenko,University of California,Math. Anal. Group,T7,Los Alamos scientific Laboratory,Los Alamos,New Mexico 87545,U.S.A.

P.Ortoleva,Department of Chemistry,Indiana University,Bloomington,Indiana 47401,U.S.A.

J.P. Puel,Département de Mathématiques et Informatique,Université de Nancy II, 42 Avenue de la Libération,54000 Nancy.

J.C. Saut,Analyse Numérique et Fonctionnelle,Bât. 425,Université Paris-Sud, 91405 Orsay.

D. Serre,C.N.R.S.;Analyse Numérique et Fonctionnelle,Bât. 425,Université Paris-Sud,91405 Orsay.

D. Thomas,Université de Technologie de Compiègne,B.P. 233,60206 Compiègne.

P. Witomski, Laboratoire IMAG,B.P. 53 X,38041 Grenoble CEDEX.

PARAMETER DEPENDENCE OF SOLUTIONS OF CLASSES OF QUASI-LINEAR ELLIPTIC AND PARABOLIC DIFFERENTIAL EQUATIONS

by

M.F. Barnsley

Service de Physique théorique

C.E.N. Saclay

B.P. n°2

91190 Gif s/Yvette

ABSTRACT

Earlier work, on the dependence of solutions of certain classes of quasi-linear elliptic and parabolic differential equations on embedded parameters, is extended and generalized. In particular, generic classes of linearly perturbed, and inhomogeneously perturbed, quasi-linear elliptic and parabolic boundary values problems whose stable positive solutions are Laplace transforms of positive measures, are identified. For a particular class of such problems the conjecture that the solution is a Stieltjes transform of a positive measure is explored. It is shown that low order rational fraction Padé approximants provide useful bounds, independently of whether or not the conjecture itself is true.

1. INTRODUCTION

We consider some extensions and generalizations of earlier work[1,2] concerning the dependence of solutions of certain quasi-linear elliptic and parabolic differential equations on an external parameter. Our interest is in those cases where the solution, as a function of the external parameters, can be expressed as a transform of a positive measure. In such cases one can use moment theory to yield convergent sequences of upper and lower bounds on the solution throughout the range of the parameter, as described in[2]. To construct the bounds one needs to know either an initial sequence of terms in a perturbation expansion of the solution in the parameter (in some cases these can be obtained by solving a set of linear equations), or a set of experimental points corresponding to different values of the parameter. The latter possibility is attractive because the resulting bounds are to some extent "model-independent" as described in[1].

In II and III we describe two generic situations which, in the elliptic case, have positive stable solutions which are Laplace transforms of positive measures. As such, they are amenable to analysis using generalized Padé approximants.

In IV we consider a class of nonlinearly perturbed elliptic boundary value problems for which it is conjectured that the positive stable solution is a Stieltjes transform of a positive measure in the perturbation parameter. In certain cases this conjecture has been established, and then rational fraction Padé approximants provide not only convergent sequences of bounds on the solution but also they yield bounds on the associated turning point. It is shown that, for low order Padé approximants, similar results pertain in general, whether or not the conjecture itself is true.

II. LINEARLY PERTURBED NONLINEAR EQUATIONS

We consider quasi-linear differential equations of the form

$$\left.\begin{aligned} &L\phi + F(\phi) + \lambda p\phi = f \quad \text{in} \quad D, \\ &B\phi = 0 \quad \text{on } \partial D \end{aligned}\right\} \tag{2.1}$$

Here D denotes a bounded domain of real N-space $\mathbb{R}^N$ with boundary ∂D and closure $\bar{D}$. We assume that ∂D belongs to the class $C^{2+\alpha}$, where $\alpha \in (0,1)$ is fixed. L is the uniformly elliptic differential operator

$$L\phi = -\sum_{i,j=1}^{N} a_{i,j}(x) \frac{\partial^2\phi(x)}{\partial x_i \partial x_j} + \sum_{k=1}^{N} a_k(x) \frac{\partial\phi(x)}{\partial x_k} + a(x)\phi(x) \tag{2.2}$$

with real coefficients $a_{i,j} \in C^{2+\alpha}(\bar{D})$, $a_k \in C^{1+\alpha}(D)$, and $a \in C^{\alpha}(\bar{D})$ where we assume that, for all $x = (x_1, x_2, \ldots, x_N) \in \bar{D}$

$$a(x) \geq 0 \tag{2.3}$$

The matrix (a_{ij}) is supposed to be uniformly positive definite over $\bar{D}$. B is either of the boundary operators

$$B\phi \equiv \phi(x) \quad \text{on } \partial D, \tag{2.4}$$
$$B\phi \equiv \beta(x)\phi(x) + \partial\phi(x)/\partial\nu \quad \text{on } \partial D, \tag{2.5}$$

where $\beta(x) \in C^{1+\alpha}(\partial D)$ and satisfies

$$\beta(x) \geq 0 \qquad \text{for all } x \in \partial D. \tag{2.6}$$

$\partial/\partial\nu$ denotes the outward conormal derivative. In the case where (2.5) applies we assume that $a(x)$ and $\beta(x)$ do not both vanish identically. The functions p and f in (2.1) belong to $C^{\alpha}(\bar{D})$ and satisfy

$$p(x) > 0, \quad \text{and } f(x) \geq 0 \quad \text{for all } x \in \bar{D}, \tag{2.7}$$

with $f \not\equiv 0$.

The real valued function $F(\phi) = F(x,\phi)$ may depend explicitly on both $x \in \bar{D}$ and ϕ. We suppose that it has the following properties

(i) $F(0) - f \leq 0$ for all $x \in \bar{D}$.

(ii) There exists a constant $C > 0$ such that

$F(C) - f \geq 0$ for all $x \in \bar{D}$.

(iii) $F(\phi)$ is C^{∞} in ϕ for all $\phi \in [0,C]$, each of its derivatives in this range belonging to $C^{\alpha}(\bar{D})$ in x, and such that uniformly in $x \in \bar{D}$ and $n = 2,3,4,\ldots$

$$-\hat{F} \leq F^{(n)}(\phi) \leq 0 \quad \text{for all } \phi \in [0,C]$$

for some constant $\hat{F} > 0$.

(iv) $F^{(1)}(\phi) \geq 0$ for all $x \in \bar{D}$ and $\phi \in [0,C]$. This means that the linear operator in $\mathcal{L}^2(\bar{D})$ corresponding to $L + F^{(1)}(\phi)$ together with the boundary condition in (2.1) has strictly positive least eigenvalue for all smooth $\phi(x) \in [0,C]$.

The conditions above on (L,B) are such that are such that the *Maximum Principle*[3] and the *Positivity Lemma*[4] apply. Moreover, the smoothness conditions of $F(\phi)$ mean that *Amann's Theorem*[5], on the existence of solutions via sandwiching between upper and lower solutions, applies to (2.1). The key condition on $F(\phi)$ which ensures the establishment of the Laplace transform property (Proposition 1.2) is (iii).

Proposition (2.1). The problem (2.1) with $\lambda \geq 0$ possesses exactly one solution $\phi \in C^{2+\alpha}(\bar{D})$ which satisfies $0 \leq \phi(x) \leq C$ for all $x \in \bar{D}$.

Proof : The existence of at least one solution in the desired range is provided by conditions (i) and (ii), upon application of Amann's Theorem. Zero is a lower solution while the constant C is an upper solution, for all $\lambda \geq 0$.

To establish uniqueness let ϕ_1 and ϕ_2 be two solutions. Then Taylor's Theorem with remainder provides

$$F(\phi_1) - F(\phi_2) = F^{(1)}(\phi_3)(\phi_1 - \phi_2) \tag{2.8}$$

for some ϕ_3 lying between ϕ_1 and ϕ_2, so that $\phi_3 \in [0,C]$. Hence

$$\left.\begin{array}{l} [L + \lambda p + F^{(1)}(\phi_3)](\phi_1 - \phi_2) = 0 \quad \text{in } D \\ B(\phi_1 - \phi_2) = 0 \quad \text{on } \partial D \end{array}\right\}, \tag{2.9}$$

and condition (iv) now yields $\phi_1 = \phi_2$. Q.E.D.

We will denote the solution referred to above by $\phi(\lambda)$. In order to examine its analytic nature let $\lambda_0 \geq 0$ be held fixed, let $\lambda \in \mathbb{C}$ be given, and set

$$\rho = \lambda - \lambda_0 \tag{2.10}$$

Then we will investigate the formal series

$$\Psi = \Psi[\lambda_0,\rho] = \sum_{n=0}^{\infty} \frac{1}{n!}\,\psi_n[\lambda_0]\rho^n \tag{2.11}$$

where the ρ-independent functions $\psi_n = \psi_n[\lambda_0]$ are supposed to satisfy the set of equations obtained by equating the coefficients of the different powers of ρ which occur in the formal expansion of

$$\left.\begin{array}{l} L\Psi + F(\Psi) + (\lambda_0 + \rho)p\Psi = f \text{ in } D, \\ B\Psi = 0 \quad \text{on } \partial D, \end{array}\right\} \tag{2.12}$$

and where ψ_0 is constrained by

$$0 \leq \psi_0 \leq C, \qquad \text{for all } x \in \bar{D} \tag{2.13}$$

The equations to be satisfied by the ψ_n's are found to be

$$\left.\begin{array}{l} L\psi_0 + F(\psi_0) + \lambda_0 p\psi_0 = f \quad \text{in } D \\ B\psi_0 = 0 \quad \text{on } \partial D,\ \psi_0 \in [0,C] \end{array}\right\} \tag{2.14.0}$$

$$\left.\begin{array}{l} [L + F^{(1)}(\psi_0) + \lambda_0 p]\psi_1 + p\psi_0 = 0 \text{ in } D \\ B\psi_1 = 0 \quad \text{on } \partial D \end{array}\right\} \tag{2.14.1}$$

$$\left.\begin{array}{l} [L + F^{(1)}(\psi_0) + \lambda_0 p]\psi_2 + F^{(2)}(\psi_0)\psi_1^2 + 2p\psi_1 = 0 \quad \text{in } D, \\ B\psi_2 = 0 \quad \text{on } \partial D, \end{array}\right\} \tag{2.14.2}$$

$$\left.\begin{array}{l} [L + F^{(1)}(\psi_0) + \lambda_0 p]\psi_n + np\psi_{n-1} \\ \displaystyle + \sum_{m=2}^{n} F^{(m)}(\psi_0)\left\{\sum_{0<k_1<k_2<\ldots<k_{m-1}<n} C(n;k_1,\ldots,k_{m-1})\psi_{(n-k_{m-1})} \right. \\ \left. \psi_{(k_{m-1}-k_{m-2})}\cdots\psi_{(k_2-k_1)}\psi_{k_1}\right\} = 0 \quad \text{in } D \\ B\psi_n = 0 \quad \text{on } \partial D,\ n = 2,3,4,\ldots \end{array}\right\} \tag{2.14.n}$$

where $C(n,k_1,\ldots,k_{m-1}) = \binom{n-1}{k_{m-1}}\binom{k_{m-1}-1}{k_{m-2}}\cdots\binom{k_2-1}{k_1}$.

The existence of a unique set of functions $\{\psi_n \in C^{2+\alpha}(\bar{D}) : n = 0,1,2,\ldots\}$ satisfying the above set of equations is readily seen. By Proposition 1, (2.14.0) possesses a unique solution

$$\psi_0 = \phi(\lambda_0) \in C^{2+\alpha}(\bar{D}) \tag{2.15}$$

One now proceeds inductively for $n = 1,2,3,\ldots$ The nth equation is linear in ψ_n, involving it only in the term $[L + F^{(1)}(\psi_0) + \lambda_0 p]\psi_n$, and the inhomogeneous part is a function of $(\psi_0, \psi_1, \ldots, \psi_{n-1})$ which, through inductive hypothesis, belongs to $C^{\alpha}(\bar{D})$. Thus, since $(L + F^{(1)}(\psi_0) + \lambda_0 p, B)$ is positive and has smooth coefficients, the nth equation has a unique solution $\psi_n \in C^{2+\alpha}(\bar{D})$, which completes the induction. One sees, moreover, with the aid of the Positivity Lemma and by proceeding inductively using (iii), that

$$(-1)^n \psi_n(x) \geq 0 \qquad \text{for all} \quad x \in \bar{D} \tag{2.16}$$

Having shown that the formal series (2.11) is well defined, our next objective is to show that it has a finite radius of converge $R(\lambda_0)$ and that for $|\rho| \leq R(\lambda_0)$ it converges to a solution of (2.1). This solution is then readily identified to be $\phi(\lambda)$. To achieve these ends we consider the algebraic functional equation, defining $\xi(\rho)$,

$$\{\sigma(\lambda_0) + \rho\hat{p} + \hat{F}\}\xi - \hat{F}e^{\xi-\hat{\phi}_0} = \{\sigma(\lambda_0) + \hat{F}\}\hat{\phi}_0 - \hat{F} \tag{2.17}$$

Here

$$\hat{\phi}_0 = \operatorname{Max}\left\{\operatorname*{Max}_{x\in\bar{D}} \left\{\phi_0(x)\right\} ; 1\right\}, \quad \hat{p} = \operatorname*{Max}_{x\in\bar{D}}\left\{p(x)\right\}, \tag{2.18}$$

and $\sigma(\lambda_0)^{-1} = \operatorname*{Max}_{x\in\bar{D}}\{\theta(x)\} > 0$ where $\theta \in C^{2+\alpha}(\bar{D})$ is the unique positive solution of

$$\left.\begin{aligned} &\{L + F^{(1)}(\phi_0) + \lambda_0 p\}\theta = 1 \quad \text{for all} \quad x \in D, \\ &B\phi = 0 \quad \text{on } \partial D \end{aligned}\right\} \tag{2.19}$$

The motivation behind (2.17) will become clear shortly.

Equation (2.17) possesses the solution $\xi = \hat{\phi}_0$ when $\rho = 0$. Using standard analytical techniques we find that this solution $\xi(\rho)$ exists and is regular in a neighbourhood of $\rho = 0$, say for $|\rho| \leq R(\lambda_0)$ where $R(\lambda_0) > 0$. Indeed, with the aid of Bernstein's theorem[6] one finds that $\xi(\rho)$ is a regular analytic function throughout $\operatorname{Re}\rho > -R(\lambda_0)$, being expressible in the form

$$\xi(\rho) = \int_0^\infty \exp\{-(R(\lambda_0) + \rho)\gamma\}\, d\mu(\gamma) \tag{2.20}$$

where $\mu(\gamma)$ is a bounded monotone nondecreasing function over the range $0 \leq u < \infty$.

Writing the Taylor series expansion of $\xi(\rho)$ about $\rho = 0$ as

$$\xi(\rho) = \sum_{n=0}^{\infty} \frac{1}{n!}\xi_p \rho^n, \qquad |\rho| \leq R(\lambda_0) \tag{2.21}$$

we find that the ξ_n's are given recursively by

$$\xi_0 = \hat{\phi}_0, \tag{2.22.0}$$

$$\sigma(\lambda_0)\xi_1 + \hat{p}\,\xi_0 = 0, \tag{2.22.1}$$

$$\sigma(\lambda_0)\xi_2 - \hat{F}\xi_1^2 + 2\hat{p}\,\xi_1 = 0, \tag{2.22.2}$$

and

$$\sigma(\lambda_0)\xi_n + n\hat{p}\,\xi_{n-1} - \hat{F}\sum_{m=2}^{n}\Big\{\sum_{0<k_1<k_2<\ldots<k_{m-1}<n} C(n;k_1,\ldots k_{m-1})\xi_{(n-k_{m-1})} \xi_{(k_{m-1}-k_{m-2})}\cdots\xi_{(k_2-k_1)}\xi_{k_1}\Big\} = 0 \tag{2.22.n}$$

for $n = 2,3,4\ldots$

We now show that

$$0 \le (-1)^n\,\psi_n \le (-1)^n\,\xi_n \qquad \text{for all } x\in D,\ n = 0,1,2,\ldots \tag{2.23}$$

These inequalities are clearly true for $n = 0$. Now consider (2.14.1). An upper solution is provided by zero, and a lower solution is provided by $-\hat{p}\hat{\phi}_0\theta(x)$ since

$$\left.\begin{aligned}&[L + F^{(1)}(\psi_0) + \lambda_0 p](-\hat{p}\hat{\phi}_0\theta) + p\phi_0 = -p\hat{\phi}_0 + p\phi_0 \le 0 \quad \text{in } D\\ &B(-\hat{p}\hat{\phi}_0\theta) = 0 \quad \text{on } \partial D\end{aligned}\right\} \tag{2.24}$$

Hence $0 \le (-1)\psi_1 \le +\hat{p}\hat{\phi}_0\theta \le \hat{p}\hat{\phi}_0/\sigma(\lambda_0) = (-1)\xi_1$ which proves (2.23) when $n = 1$. Now assume that (2.23) is true for all $n = 0,1,2,\ldots,K$, and let us first suppose K is odd. Then it is readily seen that zero is a lower solution for (2.14. K+1), while an upper solution is

$$\Big\{-(K+1)\hat{p}\xi_K + \hat{F}\sum_{m=2}^{K+1}\Big\{\sum_{0<k_1<k_2<\ldots<k_{m-1}<K+1} C(K+1;k_1,\ldots,k_{m-1})\xi_{(K+1-k_{m-1})} \xi_{(k_{m-1}-k_{m-2})}\cdots\xi_{(k_2-k_1)}\xi_{k_1}\Big\}\theta = \bar{\psi}_{K+1}, \tag{2.25}$$

Since replacing ψ_{K+1} by $\bar{\psi}_{K+1}$ in the left-hand-side of (2.14. K+1) provides

$$\Big\{(K+1)(p\psi_K - \hat{p}\xi_K) + \sum_{m=2}^{K+1}\Big\{\sum_{0<k_1<k_2<\ldots<k_{m-1}<K+1} C(K+1;k_1,\ldots,k_K) \Big[\hat{F}\xi_{(K+1-k_{m-1})}\cdots\xi_{(k_2-k_1)}\xi_{k_1} + F^{(m)}(\psi_0)\psi_{(K+1-k_{m-1})}\cdots\psi_{(k_2-k_1)}\psi_{k_1}\Big]\Big\}\Big\} \tag{2.26}$$

which is positive via the inductive hypothesis together with assumption (iii). Moreover $B\bar{\psi}_{K+1} = 0$ on ∂D. Hence

$$0 \le \psi_{K+1} \le \bar{\psi}_{K+1} \le \xi_{K+1} \tag{2.27}$$

The induction is completed after similar treatment of the case K even, and the relations (2.23) are proved.

In particular, the series (2.21) being absolutely convergent for $|\rho| \le R(\lambda_0)$ provides, on using (2.23), that $\Psi[\lambda_0,\rho]$ is absolutely convergent for $|\rho| \le R(\lambda_0)$ uniformly for $x \in \bar{D}$. We will use the same notation $\Psi[\lambda_0,\rho]$ to denote its sum, where it converges.

Let $\Psi_N[\lambda_0,\rho]$ denote the Nth partial sum of $\Psi[\lambda_0,\rho]$ and let ρ be fixed such that $|\rho| \le R(\lambda_0)$. Then we show that $\Psi[\lambda_0,\rho] \in C^{2+\alpha}(\bar{D})$ and satisfies the differential equation (2.12) by applying the Compactness Theorem[11]. Provided with the conditions (a) through (d) which follow, the desired result is assured. We already know that (a) the sequence of functions $\{\Psi_N[\lambda_0,\rho]\}_{N=0}^{\infty}$ is uniformly convergent to $\Psi[\lambda_0,\rho]$; (b) $\Psi_N[\lambda_0,\rho] \in C^{2+\alpha}(\bar{D})$ for each N; (c) $B\Psi_N[\lambda_0,\rho] = 0$ on ∂D for each N ; and we need only to prove that (d) the sequence of functions $\{L\Psi_N[\lambda_0,\rho]\}$ is uniformly convergent to $f - \lambda p\Psi[\lambda_0,\rho] - F(\Psi[\lambda_0,\rho])$. Since F and its derivatives are continuous it suffices to prove that

$$[L + F^{(1)}(\Phi_0) + \lambda_0 p](\Psi_M - \Psi_N) \to 0 \quad \text{uniformly in D,}$$

as N tends to infinity, with M > N. But for N > 2,

$$[L + F^{(1)}(\Phi_0) + \lambda_0 p](\Psi_M - \Psi_N)| = \Big| \sum_{\ell=N+1}^{M} \rho^{\ell} \frac{1}{\ell!} \Big[\ell p \psi_{\ell-1}$$

$$+ \sum_{m=2}^{\ell} F^{(\ell)}(\psi_0) \Big\{ \sum_{0<k_1<k_2<\ldots<k_{m-1}<\ell} C(\ell;k_1,\ldots,k_{m-1}) \psi_{(\ell-k_{m-1})} \cdots \psi_{(k_2-k_1)} \psi_{k_1} \Big\}\Big]\Big|$$

$$\le \sum_{\ell=N+1}^{N} (-1)^{\ell} |\rho|^{\ell} \frac{1}{\ell!} \Big[-\ell\hat{p}\,\xi_{\ell-1} + \hat{F} \sum_{m=2}^{\ell} \Big\{ \sum_{0<k_1<k_2<\ldots<k_{m-1}<\ell} C(\ell;k_1,\ldots,k_{m-1})$$

$$\xi_{(\ell-k_{m-1})} \xi_{(k_{m-1}-k_{m-2})} \cdots \xi_{(k_2-k_1)} \xi_{k_1} \Big\}\Big] = \sigma(\lambda_0) \sum_{\ell=N+1}^{M} (-1)^{\ell} |\rho|^{\ell} \xi_{\ell} \frac{1}{\ell!} \tag{2.29}$$

where we have used the inequalities (2.23) and the definitive equations (2.22.n) (d) is now proved because the last expression in (2.29) tends to zero as N tends to infinity, uniformly for M > N, and independently of $x \in \bar{D}$, because the series $\sum_{n=0}^{\infty} \rho^n \frac{1}{n!} \xi_n$ is absolutely convergent for $|\rho| \le R(\lambda_0)$.

The identification of $\Psi[\lambda_0,\rho]$ with $\phi(\lambda) = \phi(\lambda_0 + \rho)$ for λ_0 and ρ real with $|\rho| \le R(\lambda_0)$ follows immediately from the fact that $\Psi[\lambda_0,\rho]$ is positive in some neighbourhood of λ_0, together with the uniqueness part of proposition (2.1). We have in particular that $\phi(\lambda)$ is analytic and regular in some neighbourhood of the real axis $0 \le \lambda \le \infty$, and that

$$(-1)^n \frac{d^n\phi}{d\lambda^n} \ge 0 \quad \text{for all } \lambda \in [0,\infty] \tag{2.30}$$

Thus, Bernstein's Theorem[6] completes the proof of the following proposition.

Proposition (2.2). The function $\phi(\lambda)$ *defined by Proposition (2.1) can be analytically continued throughout* $\mathrm{Re}\,\lambda \geq 0$, *where it can be expressed in the form*

$$\phi(\lambda) = \int_0^\infty e^{-\lambda s}\, d\mu_x(s), \tag{2.31}$$

the function $\mu_x(s)$ *being uniformly bounded and monotone non-decreasing for* $0 \leq s < \infty$, *for all* $x \in \bar{D}$. $\phi(\lambda)$ *satisfies (2.1) at least throughout some open neighbourhood of the real axis* $0 \leq \lambda < \infty$.

A similar result also applies in the case of the associated parabolic differential equation obtained by the adjunction of $\partial/\partial t$ to L, and adjoining an initial positive boundary condition to B, as in[2].

3. INHOMOGENEOUSLY PERTURBED NON LINEAR EQUATIONS

We consider equations of the form

$$\left.\begin{array}{l} [L - \lambda p]\phi + G(\phi) = \gamma f \quad \text{in } D \\ B\phi = 0 \quad \text{on } \partial D \end{array}\right\} \tag{3.1}$$

Everything here is defined as in §II except that now $\lambda \in (-\infty + \infty)$ and γ is the parameter of interest. The real valued function $G(\phi) = G(x,\phi)$ is assumed to have the following properties :

(i) $\min_{x \in \bar{D}} \{G(u)/u\} \to \infty$ as $u \to \infty$

(ii) $G(0) \equiv 0$

(iii) $G(u)$ and all of its derivatives exist and belong to $C^{\alpha}(\bar{D})$ for all $u \in [0,\infty)$.

(iv) $[G(u)/u]^{(1)}$ exists and is ≥ 0 for all $u \in [0,\infty)$, all $x \in \bar{D}$.

(v) $\exists$ a constant $\hat{G}$ such that $\hat{G} \geq (-1)^n G^{(n)}(u) \geq 0$ for all $u \in [0,\infty)$, $x \in \bar{D}$, $n = 2,3,\ldots$

Using Bernstein's Theorem we find that the most general form of G is

$$G(x,\phi) = A(x)\phi^2 + B(x)\phi + \int_0^\infty \left\{\exp\{-\gamma\phi\} - 1\right\} d\rho_x(\gamma) \tag{3.2}$$

where $A(x) > 0$ for all $x \in \bar{D}$, and $\rho_x(\gamma)$ is a bounded monotone non-decreasing function on $0 \leq \gamma < \infty$, for each $x \in D$.

Proposition 3.1. For each $\gamma \geq 0$ *and* $\lambda \in (-\infty,+\infty)$ *the problem* (3.1) *possesses exactly one non negative solution.* $\phi \in C^{2+\alpha}(\bar{D})$. *This solution is regular in* γ *and can be analytically continued throughout* $\mathrm{Re}\,\gamma \geq 0$, *where it satisfies a representation of the form*

$$\frac{\partial\phi}{\partial\gamma}[\gamma] = \int_0^\infty e^{-\gamma s}\, d\Pi_{x,\lambda}(s) \quad , \quad \textit{for each fixed } x,\ \lambda \tag{3.2}$$

where $\Pi_{x,\lambda}$ *is bounded monotone nondecreasing function on* $0 \le s < \infty$. *The continued solution satisfied* (3.1) *for all* γ *in some neighbourhood of* $[0,\infty)$.

We omit the proof of this proposition as it follows somewhat similar lines to the demonstrations en §2. A key point is that the operator $([L-\lambda p + G^{(1)}(\phi[\gamma])],B)$ has strictly positive least eigenvalue for all $\gamma \ge 0$.

4. PADE APPROXIMANTS AND THE PROBLEM $[L-\lambda p]\phi + \gamma q\,\phi^N = f$

We consider the γ-dependence of the positive stable solution $\phi[\gamma]$ of the problem

$$\left.\begin{array}{ll} L\phi + \gamma q\phi^N = f & \text{in } D, \\ B\phi = 0 & \text{on } \partial D;\ N = 2,3,\ldots \end{array}\right\} \tag{4.1}$$

where $q \in C^{\alpha}(\bar{D})$, $q(x) > 0$ for all $x \in \bar{D}$, and all other quantities are defined as in §2. In particular, we are interested in the location of the *turning point* γ^*, which corresponds to the first singularity in $\phi[\gamma]$ on the real axis as γ goes towards minus infinity starting from zero $\phi[\gamma]$ is a positive stable solution of (4.1) for all $\gamma > \gamma^*$.

We are also interested in the possibility of using classical rational fraction Padé approximants (P.A.'s) to provide upper and lower bounds on $\phi[\gamma]$ for all $\gamma > \gamma^*$, starting from the Taylor series expansion of $\phi[\gamma]$ about $\gamma = 0$ which we write

$$\phi[\gamma] = \sum_{n=0}^{\infty} \frac{1}{n!}\gamma^n \phi_n \tag{4.2}$$

Over and above the practical utility of such bounds, the motivation here is the *conjecture* that $\phi[\gamma]$ is a Stieltjes transform of positive measure, namely

$$\phi[\gamma] = \int_0^{-1/\gamma^*} \frac{d\mu(s)}{1+s\gamma} \quad \text{for all } \gamma \in \mathbb{C}-(-\infty,\gamma^*] \quad , \tag{4.3}$$

where $\mu(s)$ is a bounded monotone nondecreasing function on $-1/\gamma^* \le s < \infty$. This conjecture was shown indeed to be true in the case $N = 2$, in dimension ≤ 5, when $L = -\Delta$, $q = 1$, and f is sufficiently smooth,[5].

We begin by saying why a representation of the form (4.1) is suggested in the first place. In the case where the boundary condition in (4.1) is replaced by $\partial\phi/\partial\nu = 0$ on ∂D, and a, q, and f, are constants we have that $\phi[\gamma]$ is itself a constant, being the positive solution of the algebraic equation

$$a\phi[\gamma] + \gamma q\phi[\gamma]^N - f = 0 \tag{4.4}$$

The solution of the latter which is regular around $\gamma = 0$ is expressible in the form (4.3) with

$$\gamma^* = -\left[\frac{a(N-1)}{f\,N}\right]^N \frac{f}{(N-1)} , \tag{4.5}$$

see[8], example 1. In view of the close relationship which often exists between solutions of uniform elliptic equations and the analogous algebraic equations, the conjecture (4.3) is suggested.

Let us now examine some consequences of (4.3) when we suppose that it *is* true [e.g. $N = 2$]. The first and perhaps most important consequence is that the [M/M] and [(M-1)/M] sequences of PA.'s, constructed from initial sets of coefficients occuring in the expansion (4.2), provide convergent bounds on $\phi[\gamma]$ according to

$$\begin{aligned} &[M/M] \geq [(M+1)/(M+1)] \geq \ldots \geq \phi[\gamma] \geq \ldots \geq [M/(M+1)] \geq [(M-1)/M] \\ &\qquad\qquad \text{for all } \gamma > 0, \\ &\phi[\gamma] \geq \ldots \geq [(M+1)/(M+1)] \geq [M/(M+1)] \geq [M/M] \geq [(M-1)/M] \\ &\qquad\qquad \text{for } \gamma^{*} < \gamma \leq 0 \ ; \quad M = 1,2,3,\ldots \end{aligned} \tag{4.6}$$

The [R/S] P.A. is defined as follows :

$$[R/S] = \frac{P_R(\gamma)}{Q_S(\gamma)} = \frac{P_0 + P_1\gamma + \ldots + P_R\gamma^R}{1 + q_1\gamma + \ldots + q_S\gamma^S} \tag{4.7}$$

where the (R+S+1) unknowns, the p's and q's, are determined by the requirement

$$Q_S(\gamma)\left(\sum_{n=0}^{R+S} \frac{1}{n!}\gamma^n\phi_n\right) - P_R(\gamma) = \text{terms of order } \gamma^{R+S+1} \text{ and higher.} \tag{4.8}$$

Certain subtle modifications of this definition are needed in special cases, but the above suffices in general. For full details see[9]. There also exist complementary P.A.'s, denoted $[R/S]^c$, whose bounding properties complement those in (4.6). These are defined similarly to the above except that the polynomial occuring in the denominator is required to have a zero at a point $\bar{\gamma}^{*} \geq \gamma^{*}$, while the number of agreements demanded in (4.8) is decreased by one see[10]. An elementary pair of approximants is

$$[0/1] = \phi_0/(1-\gamma\phi_1/\phi_0) \quad \text{and} \quad [0/1]^c = \phi_0/(1-\gamma/\bar{\gamma}^{*}) \tag{4.9}$$

and these display in particular the bounds

$$[0/1] \leq \phi[\gamma] \leq [0/1]^c \quad \text{for} \quad \bar{\gamma}^{*} < \gamma \leq 0 \tag{4.10}$$

provided that (4.3) is true.

We note that the ϕ_n's needed for the construction of the P.A.'s can be obtained by successive solution of the set of linear equations

$$\left.\begin{aligned} L\phi_0 &= f \text{ in } D, \\ B\phi_0 &= 0 \text{ on } \partial D; \end{aligned}\right\} \tag{4.11.0}$$

$$\left.\begin{aligned} L\phi_1 + q\phi_0^2 &= 0 \text{ in } D, \\ B\phi_1 &= 0 \text{ on } \partial D; \end{aligned}\right\} \tag{4.11.1}$$

$$L\phi_{n+1} + (n+1)qN\phi_0^{N-1}\phi_n + (n+1)q\sum_{m=2}^{\mathrm{Min}\{N,n\}}\frac{N!}{(N-m)!}\phi_0^{(N-m)}\Big\{\sum_{0<k_1<\ldots<k_{m-1}\leq n} C(n;k_1,\ldots,k_{m-1})$$
$$\phi_{(n-k_{m-1})}\phi_{(k_{m-1}-k_{m-2})}\cdots\phi_{(k_2-k_1)}\phi_{k_1}\Big\} = 0 \quad \text{in } D,$$
$$B\phi_{n+1} = 0 \quad \text{on } \partial D, \quad n=1,2,\ldots \qquad (4.11.n+1)$$

The second consequence of (4.3) is that we are able to obtain a convergent sequence of lower bounds to γ^* with the aid of the P.A.s' described above. The bounds are obtained as follows. Provided with (4.3) we have that each [M/M] and [(M-1)/M] P.A. possesses a partial fractions expansion of the form

$$\sum_{m=1}^{M}\frac{V_m}{(1-\gamma/\gamma_m)} + V_0 \qquad (4.12)$$

where the Vm's and γm's are real constants, the Vm's being non negative with $V_0 = 0$ in the case of [(M-1)/M] P.A.'s, and the γm's are ordered so that

$$\gamma_M < \gamma_{M-1} < \ldots < \gamma_1 \leq \gamma^* \qquad (4.13)$$

Writing $\gamma_1 = \gamma_1^M$, we have $\{\gamma_1^M\}_{M=1}^{\infty}$ is a non decreasing sequence which converges to γ^* as $M \to \infty$. Further details, and proofs of these properties can be found in[9]. Such results apply directly to (4.1) with $N=2$ when (4.3) is known to be true, and we have for example the simple lower bound $\underset{x\in\bar{D}}{\mathrm{Max}}\ \{\phi_0/\phi_1\} \leq \gamma^*$.

Since (4.3) has not been established in general we now adopt a different point of view. *Can we prove directly results along the lines of the above consequences of (4.3), without assuming that (4.3) is itself true ?* Below we present some initial propositions in this direction.

Proposition 4.1 : *For $\gamma \geq 0$, a lower solution of (4.1) is provided by* [0/1] *corresponding to the series (4.2), while upper solutions are provided by* [0/0] *and* [1/1]. *We have*

$$[0/0] \geq [1/1] \geq \phi[\gamma] \geq [0/1] \quad \text{for all } \gamma \geq 0. \qquad (4.14)$$

For $\gamma \leq 0$, each of these three approximants is a lower solution of (4.1), provided that $\gamma \geq$ location of the pole of the approximant in the case of [0/1] *and* [1/1].

Proof : We prove the statements concerning [0/1] ; the proofs of the other statements are similar. Notice that we have

$$(-1)^n\phi_n(x) \geq 0 \quad \text{for all } x\in\bar{D}\ ;\ n=0,1,2,\ldots \qquad (4.15)$$

This follows inductively from the equations (4.11), upon making repeated use of the Maximum Principle. We have

$$[0/1] = \phi_0^2/(\phi_0 - \gamma\phi_1), \qquad (4.16)$$

so that its pole is located on the negative real axis, and

$$B[0/1] = 0 \quad \text{on } \partial D . \tag{4.17}$$

Thus [0/1] is a lower solution as claimed provided that

$$L[0/1] + \gamma q[0/1]^N - f \le 0 \text{ for all } \max_{x\in\bar{D}} \{\phi_0/\phi_1\} < \gamma < \infty \tag{4.18}$$

$$\begin{aligned} \text{But } \{L[0/1] + \gamma q[0/1]^N - f\}(\phi_0 - \gamma\phi_1)^N &= \\ (\phi_0 - \gamma\phi_1)^{N-3}\{\gamma\phi_0^2(\phi_0 - \gamma\phi_1)L\phi_1 &+ (\phi_0 - 2\gamma\phi_1)\phi_0(\phi_0 - \gamma\phi_1)L\phi_0 - 2\gamma^2\Theta\} \\ + \gamma q\phi_0^{2N} - f(\phi_0 - \gamma\phi_1)^N &= -\phi_1^2(\phi_0 - \gamma\phi_1)^{N-2}\phi_1^2 f - 2\gamma^2\Theta(\phi_0 - \gamma\phi_1)^{N-3} \\ &- \gamma q\,\phi_0^{N+2}\{(\phi_0 - \gamma\phi_1)^{N-2} - \phi_0^{N-2}\} \end{aligned} \tag{4.19}$$

Here, to obtain the second equality from the first we have substituted directly from (4.11.0) and (4.11.1) for the quantities $L\phi_0$ and $L\phi_1$, and simplified. Also,

$$\Theta = (\phi_1\underline{\nabla}\phi_0)^{\dagger}\,\underline{\underline{A}}(\phi_1\underline{\nabla}\,\phi_0) \tag{4.20}$$

where $\underline{\underline{A}}$ is the matrix with $A_{ij} = \delta_{ij} + \frac{1}{2}(1-\delta_{ij})$. $\underline{\underline{A}}$ is self-adjoint with strictly positive least eigenvalue so that $\Theta \ge 0$. (4.18) now follows, as desired. In a similar way the upper/lower solution properties of [0/0] and [1/1] are proved. In particular we deduce (4.14) by using Amann's Theorem. Q.E.D.

We next give a result on the relation between an approximant of the complementary PA type and $\phi[\gamma]$. Again we see that the situation is just as it would be if (4.3) were true.

Proposition 4.2 *When* $\alpha = \min_{x\in\bar{D}} \{f/q\,\phi_0^N\} \ne 0$, *an upper solution of (4.1) is the approximant* $\phi_0/(1+N^{N-1}\gamma/(N-1)^{N-1}\alpha) = \eta(\gamma)$, *for all* $-(N-1)^{N-1}\alpha/N^N < \gamma \le 0$, *and we have*

$$\gamma^* \le -(N-1)^{(N-1)}\alpha/N^N \tag{4.21}$$

Moreover

$$\eta(\gamma) \ge \phi[\gamma] \ge \text{both } [0/1] \text{ and } [1/1] \text{ for } -(N-1)^{N-1}\alpha/N^N < \gamma \le 0 \tag{4.22}$$

For $\gamma \ge 0$ *we have that* $\nu(\gamma) = \phi_0/(1+\gamma/\alpha)$ *is a lower solution and*

$$\phi[\gamma] \ge \nu(\gamma), \qquad \text{for all } \gamma \ge 0 . \tag{4.23}$$

Proof. We note that $B\eta(\gamma) = 0$ on ∂D, and, for all $-(N-1)^{N-1}\alpha/N^N < \gamma \le 0$ we have

$$\begin{aligned} &(1 + N^{N-1}\gamma/(N-1)^{N-1}\alpha)^N[L\eta(\gamma) + \gamma q\eta(\gamma)^N - f] \\ &= (q\phi_0^N - (1 + N^{N-1}\gamma/(N-1)^{N-1}\alpha)^{N-1}(N^{N-1}/(N-1)^{N-1}\alpha)f\} \\ &\ge \{q\phi_0^N - f/\alpha\}\,\gamma \ge 0 \qquad \text{for all } x \in D, \end{aligned} \tag{4.24}$$

so that $\eta(\gamma)$ is an upper solution as claimed. Using the fact that, for $-(N-1)^{N-1}\alpha/N^N < \gamma \leq 0$, $\eta(\gamma)$ is a positive upper solution while zero is a lower solution, we have through Amann's theorem that (4.1) has a positive stable solution over the range of γ, whence (4.21) is obtained.

(4.22) comes from the fact that the poles of [0/1] and [1/1] lie to the left of the point $-(N-1)^{N-1}\alpha/N^N$, see Proposition 4.3. Using Proposition (4.1) we now have that [0/1] and [1/1] are both lower solutions while $\eta(\gamma)$ is a corresponding upper solution for $-(N-1)^{N-1}\alpha/N^N < \gamma \leq 0$, yielding (4.22).

The proof of the last statement in the proposition follows similar lines to the above.

Finally we provide a result of the type given in the second consequence of (2.3). This yields lower bounds to γ^* and can surely be generalized.

Proposition 4.3 *Corresponding to the pole locations of the* [0/1] *and* [1/1] P.A.'s, *we have inequalities*

$$\underset{x\in\bar{D}}{\mathrm{Max}} \{\phi_0/\phi_1\} \leq \gamma^* , \qquad \underset{x\in\bar{D}}{\mathrm{Max}} \{\phi_1/\phi_2\} \leq \gamma^* \tag{4.25}$$

Proof : We will prove only the first inequality ; that of the second follows similar lines. Since the series (4.2) has coefficients of alternating signs, $-\gamma^*$ is equal to its radius of convergence. The first inequality in (4.25) is certainly true if

$$(-1)^n\phi_n \geq \phi_0(-\phi_1/\phi_0)^n \quad \text{for all } n = 0,1,2,\ldots \tag{4.26}$$

The latter can be established by induction on n, using the equations (4.11). For brevity, we show how this is done in the case dim=1 and $L = -d^2/dx^2$. First note that

$$-\frac{d^2}{dx^2}\left(\frac{\phi_1^n}{\phi_0^{n-1}}\right) = -n(n-1)\frac{\phi_1^{n-2}}{\phi_0^{n+1}}\left(\phi_0\frac{d\phi_0}{dx} - \phi_1\frac{d\phi_1}{dx}\right)^2 - \frac{n\phi_1^{n-1}}{\phi_0^{n-1}}\frac{d^2\phi_1}{dx^2} + (n-1)\frac{\phi_1^n}{\phi_0^n}\frac{d^2\phi_0}{dx^2}$$

$$\text{for } n = 0,1,2,\ldots \tag{4.27}$$

The inequalities (4.26) are trivially true for n = 0,1, and for n = 2 we have

$$\begin{aligned} -\frac{d^2}{dx^2}(\phi_2 - \phi_1^2/\phi_0) &= -\frac{d^2\phi_2}{dx^2} + \frac{2}{\phi_0^3}\left(\phi_0\frac{d\phi_0}{dx} - \phi_1\frac{d\phi_1}{dx}\right)^2 + 2\frac{\phi_1}{\phi_0}\frac{d^2\phi_1}{dx^2} - \frac{\phi_1^2}{\phi_0^2}\frac{d^2\phi_0}{dx^2} \\ &= 2q(1-N)\phi_0^{N-1}\phi_1 + \frac{2}{\phi_0^3}(\phi_0\phi_0' - \phi_1\phi_1')^2 + \frac{\phi_1^2 f}{\phi_0^2} \geq 0, \text{ for all } x\in D, \end{aligned} \tag{4.28}$$

where to obtain the second equality we have substituted from (11.0), (11.1), and (11.2). We also have $B(\phi_2 - \phi_1^2/\phi_0) = 0$ on ∂D. The Maximum Principle gives (4.26) with n = 2. We now assume the inequalities (4.26) are true for all n = 0,1,2,...,K.

Then we have

$$-\frac{d^2}{dx^2}\left((-1)^{K+1}\phi_{K+1} - \frac{(-\phi_1)^{K+1}}{\phi_0^K}\right) = (-1)^K(K+1)Nq\,\phi_0^{N-1}\,\phi_K$$

$$(-1)^K(K+1)q \sum_{m=2}^{\mathrm{Min}\{N,K\}} \frac{N!}{(N-m)!}\,\phi_0^{(N-m)}$$

$$\left\{\sum_{0<k_1<\ldots<k_{m-1}<K} C(K;k_1,\ldots,k_{m-1})\phi_{(K-k_{m-1})}\phi_{(k_{m-1}-k_{m-2})}\cdots\phi_{(k_2-k_1)}\phi_{k_1}\right\}$$

$$+K(K+1)\frac{(-\phi_1)^{K-1}}{\phi_0^{K+2}}\left(\phi_0\frac{d\phi_0}{dx} - \phi_1\frac{d\phi_1}{dx}\right)^2 - (K+1)\frac{(-\phi_1)^K}{\phi_0^K}\frac{d^2\phi_1}{dx^2} - K\frac{(-\phi_1)^{K+1}}{\phi_0^{K+1}}\frac{d^2\phi_0}{dx^2}$$

$$\geq (-1)^K(K+1)Nq\,\phi_0^{N-1}\,\phi_K - (K+1)\frac{(-\phi_1)^K}{\phi_0^K}\frac{d^2\phi_1}{dx^2} - \frac{K(-\phi_1)^{K+1}}{\phi_0^{K+1}}\frac{d^2\phi_0}{dx^2}\,, \qquad (4.29)$$

where we have used (4.11.K+1) and (4.27), and the fact that $(-1)^n\phi_n \geq 0$. Now using (4.11.0), (4.11.1), and the inductive hypothesis, we have

$$-\frac{d^2}{dx^2}\left((-1)^{K+1}\phi_{K+1} - \frac{(-\phi_1)^{K+1}}{\phi_0^K}\right) \geq (K+1)(N-1)q\,\phi_0^N\,\frac{(-\phi_1)^K}{\phi_0^K} + \frac{K(-\phi_1)^{K+1}}{\phi_0^{K+1}}\,f \geq 0 \text{ in } D. \qquad (4.30)$$

Since we also have $B((-1)^{K+1}\phi_{K+1} - (-\phi_1)^{K+1}/\phi_0^K) = 0$ on ∂D, the Maximum Principle yields the inequality (4.26) for $n = K+1$, and induction completes the establishement of these inequalities Q.E.D.

The author thanks D. Bessis for help and encouragement.

REFERENCES

[1] Barnsley M.F., Bessis D; [1978] Padé approximant bounds on the positive solutions of some nonlinear elliptic equations, Proc. Royal. Soc. Edim. (to appear)

[2] Barnsley M.F., Bessis D, [1978] Constructive methods based on analytic characterizations and their application to nonlinear elliptic and parabolic differential equations, J. Math. Phys. (to appear)

[3] Protter M.H., Weinberger H.F. *Maximum Principles in Differential Equations* (Englewood Cliffs, N.J. : Prentice Hall, 1967)

[4] Keller H.B., Cohen D.S., [1967] Some positone problems suggested by nonlinear heat generation. J. Math. Mech. 16, 1361-1376

[5] Amann H. [1971] On the existence of positive solutions of nonlinear elliptic and parabolic boundary value problems. Indiana Univ. Math. J. 21, 125-146

[6] Widder D. V., *The Laplace Transform* (Princeton : Univ. Press. 1941)

[7] Serre D, Sur une équation quasilinéaire elliptique d'ordre 2. Thèse 3ème cycle Université de Paris-sud, Orsay 1978

[8] Barnsley M.F., Bessis D., Moussa P., [1978] The Diophantine moment problem and the analytic structure in the Activity of the ferromagnetic Ising model. J. Math. Phys. (to appear)

[9] Baker G.A. Jr., *The Essentials of Padé Approximants* (New-York ; Academic Press 1975)

[10] Barnsley M.F. [1974] The bounding properties of the multipoint Padé approximant to a series of Stieltjes on the real line. J. Math. Phys. 14, 299-313
[11] Agmon S., Douglis A., Nirenberg [1959] Estimates near the boundary of elliptic partial differential equations satisfying general boundary conditions.

SOME APPLICATIONS OF THE METHOD OF SUPER AND SUBSOLUTIONS

H. Berestycki and P.L. Lions

Introduction :

The goal of this paper is to present several applications (in general simple) of order methods and of combined order and topological methods in non-linear problems. In the first section we give an elementary principle on linking local super and subsolutions defined on different subdomains. This result will be used in most of the applications: simple ones are given in section 2. Then, in section 3, we recall and extend a method due to P.H. Rabinowitz [17] which combines order and topological arguments. Finally we apply these results in section 4 to derive a new existence result for stationary waves in nonlinear classical scalar fiels equations of the type Klein-Gordon (in short Non Linear Klein-Gordon : NLKG - see [4-7-8]).

Let us give an example of the results presented in the last section: let $1 < p < q < +\infty$, let λ,μ,m be positive constants, then a necessary and sufficient condition for the existence of a solution u of

(1) $$-\Delta u = \lambda |u|^{p-1}u - \mu |u|^{q-1}u - mu \ , \ u \in H^1(\mathbb{R}^n) \ , \ u \not\equiv 0$$

is

(2) $$\exists t > 0 \text{ such that } \lambda \frac{t^{p+1}}{p+1} - \mu \frac{t^{q+1}}{q+1} - m \frac{t^2}{2} > 0$$

Moreover if (2) is satisfied, there exists a smooth solution u of (1) which is positive, radial ($u(x) = u(|x|)$), decreasing i.e. $u(x) < u(y)$ if $|x| > |y|$, satisfying :

$$\forall \alpha, \exists C_\alpha , \delta_\alpha > 0 \text{ such that } |D^\alpha u(x)| \leq C_\alpha e^{-\delta_\alpha r} \quad (\text{where } r = |x|).$$

I. Linking local super and subsolutions :

Let Ω be a regular domain in $\mathbb{R}^n$ and let Ω_1 be a subdomain of Ω such that $\partial\Omega_1$ is regular and $\overline{\Omega}_1 \subset \Omega$. We denote by $\Omega_2 = \Omega - \Omega_1$ ($\partial\Omega_2 = \partial\Omega \cup \partial\Omega_1$) and by ν the unit outward normal to Ω_1. Let A be the operator defined by

$$A = -\frac{\partial}{\partial x_i}\left(a_{ij} \frac{\partial}{\partial x_j}\right) + b_i \frac{\partial}{\partial x_i} + c$$

where $a_{ij} = a_{ji} \in L^\infty(\Omega)$, $b_i \in L^\infty(\Omega)$, $c \in L^\infty(\Omega)$.

We denote by $\nu_A^i = \sum_j a_{ij}(x)\, \nu_j$ the conormal associated with A.

We start with a simple lemma (*) :

<u>Lemma I.1.</u> : Let $u_i \in H^2(\Omega_i)$, $f_i \in L^1(\Omega_i)$ $(i = 1,2)$. We assume that we have

(3) $Au_i \le f_i$ a.e. in Ω_i, $u_1 = u_2$ on $\partial\Omega_i$, $\dfrac{\partial u_1}{\partial \nu_A} \le \dfrac{\partial u_2}{\partial \nu_A}$ on $\partial\Omega_1$ then

we have

(4) $Au \le f$ in $\mathcal{D}'(\Omega)$

where $u \in H^1(\Omega)$ is defined by $u = u^i$ on Ω_i, and $f = f_i$ on Ω_i.

<u>Proof of the lemma</u> : let $\Phi \in \mathcal{D}_+(\Omega)$, we have

$$\langle Au, \Phi \rangle = \int_\Omega \sum_{i,j} a_{ij} \frac{\partial u}{\partial x_i} \frac{\partial \Phi}{\partial x_j} + \sum_i b_i \frac{\partial u}{\partial x_i} \Phi + c\, u\, \Phi \, dx$$

$$= \int_{\Omega_1} (Au_1)\, \Phi \, dx + \int_{\Omega_2} (Au_2) \Phi \, dx + \int_{\partial\Omega_1} \left(\frac{\partial u_1}{\partial \nu_A} - \frac{\partial u_2}{\partial \nu_A} \right) \Phi \, ds,$$

then because of (3) we have obviously $\langle Au, \Phi \rangle \le \int_\Omega f\, \Phi \, dx$. ∎

<u>Corollary I.1.</u> : Let us assume that f is a measurable function on $\Omega \times \mathbb{R} \times \mathbb{R}^n$ such that if $v \in H^1(\Omega)$, $f(x,v,\nabla v) \in L^1(\Omega)$. We furthermore assume the existence of $u_i \in H^2(\Omega_i)$ satisfying (3) with $f_i = f(x,u_i, \nabla u_i)$, then u defined by $u = u^i$ on Ω_i belongs to $H^1(\Omega)$ and satisfies

(4') $Au \le f(x,u,\nabla u)$ in $\mathcal{D}'(\Omega)$.

<u>Remark I.1.</u> : A similar result obviously holds for supersolutions (just change u in $-u$). ∎

<u>Remark I.2.</u> : Let us give an elementary application of corollary I.1. : we assume that $f(x,0,p) \ge 0$ $\forall x \in \bar\Omega$ $\forall p \in \mathbb{R}^n$, that if $v \in H^1$ then $f(x,v,\nabla v) \in L^1$ and that there exists $u > 0$, $u \in H^2 \cap H^1_0(\Omega_1)$ such that $-\Delta u \le f(x,u, \nabla u)$ on Ω_1.

(*) Prof. L.A. Peletier brought our attention on the fact that a similar observation had already been made by Il'in-Kalashnikov-Oleinik [11] .

Then, if u is defined by $u = \underline{u}$ on Ω_1, $u = 0$ on $\bar{\Omega}_2$, one has

(4') $\quad - \Delta u \leq f(x,u, \nabla u)$ in $\mathcal{D}'(\Omega)$.

Indeed, notice that $\frac{\partial \underline{u}}{\partial \nu_A} \leq 0$ on $\partial\Omega_1$, and choose $u_2 = 0$. ∎

We give in the following section some simple applications of this elementary principle. We also mention that this remark is being used by C.M. Brauner and B. Nicolaenko [9] . Applying this principle, they extend results previously obtained in the case of spheres to the general case of bounded domains. Indeed, in view of remark I.1., if we have a positive solution of a non linear Dirichlet problem in some domain, this solution extended by zero is a non trivial subsolution in all larger domains for the same non linear problem.

II. Some simple applications.

II.1. Existence theorems with local conditions

Let Ω be a bounded regular domain in $\mathbb{R}^n$ and $f(x,u)$ be a function which satisfies Caratheodory conditions ($f(x,u)$ is continuous in u for almost every x in $\bar{\Omega}$ and measurable in x for all u). Actually the method developped below applies clearly in the more general case where f depends weakly on ∇u.

Theorem II.1. : Let Ω_1 be a subdomain of Ω (connected) - We assume that

$f(x,0) = 0$,

(5) $\quad \varliminf_{u \to 0_+} \frac{f(x,u)}{u} > \lambda_1$, uniformly in $x \in \bar{\Omega}_1$,

where λ_1 is the first eigenvalue of $-\Delta$ with Dirichlet conditions on Ω_1;

and

(6) $\quad \overline{\lim_{u \to +\infty}} \frac{f(x,u)}{u} < \mu_1$, uniformly in $x \in \bar{\Omega}$,

where μ_1 is the first eigenvalue of $-\Delta$ with Dirichlet conditions on Ω_1;

Then, there exists a positive solution u in $W^{2,p}(\Omega)$ ($\forall p < +\infty$) of

(7) $\quad - \Delta u = f(x,u)$ in Ω, $u|_{\partial\Omega} = 0$, $u > 0$ in Ω.

Let us remark that the preceding result gives immediately the following "classical result" (*)

Corollary II.1. : We assume that $f(x,0) = 0$ and

(5') $\displaystyle \varliminf_{u \to 0_+} \frac{f(x,u)}{u} > \mu_1$, uniformly in $x \in \bar\Omega$,

(6) $\displaystyle \varlimsup_{u \to +\infty} \frac{f(x,u)}{u} < \mu_1$, uniformly in $x \in \bar\Omega$,

where μ_1 is the first eigenvalue of $-\Delta$ with Dirichlet condition on Ω.

Then, there exists a positive solution u in $W^{2,p}(\Omega)$ ($\forall p < +\infty$) of

(7) $-\Delta u = f(x,u)$ in Ω, $u_{|\partial\Omega} = 0$, $u > 0$ in Ω.

Remark II.1. : These two results (with the same method of proof) hold in the more general case where $-\Delta$ is replaced by

$$A = -\sum_{i,j} \frac{\partial}{\partial x_i}\left(a_{ij} \frac{\partial}{\partial x_j}\right) + \sum_i b_i \frac{\partial}{\partial x_i} + c$$

where $a_{ij} = a_{ji}$, b_i, $c \in L^\infty(\Omega)$ and

i/ $\exists \nu > 0$ a.e. $x \in \bar\Omega$ $\forall \xi \in \mathbb{R}^n$ $\sum_{i,j} a_{ij} \xi_i \xi_j \geq \nu |\xi|^2$,

ii/ $c \geq 0$ a.e.,

in this case the solution u belongs to $H_0^1(\Omega) \cap C(\bar\Omega)$ (in fact $C^{0,\alpha}(\bar\Omega)$ for some $\alpha < 1$).

It is also worth noting that Corollary II.1. holds without any change (same method of proof, using [15]) for an operator A which is not under divergence form i.e.

$$A = -\sum_{i,j} a_{ij} \frac{\partial^2}{\partial x_i \partial x_j} + b_i \frac{\partial}{\partial x_i} + c$$

where $a_{ij} = a_{ji} \in C(\bar\Omega)$, $b_i \in L^\infty(\Omega)$, $c \in L^\infty_+(\Omega)$ and

(*) We have not found any reference for this result. However, existence results for particular classes of non linearities f are well known (cf. e.g. D.G. Aronson-L.A. Peletier [3] for the case where $f(x,u) = u^\beta$ with $0 < \beta < 1$). In the non-variational case (i.e. when A is not self-adjoint), this existence result seems to be new.

$\exists \nu > 0 \quad \forall x \in \bar{\Omega} \; \forall \xi \in \mathbb{R}^n \qquad \sum_{i,j} a_{ij} \xi_i \xi_j \geq \nu |\xi|^2$. ∎

Proof of Theorem II.1. : We shall give a proof which does not rely on any variational argument and which thus can easily be extended to the case of any uniformly elliptic operator (non necessarily self-adjoint, see remark II.1. above).

(In remark II.2., an easier proof is given in the variational case).

1/ We first build a subsolution - Considerer v_1, solution of

$$- \Delta v_1 = \lambda_1 v_1 \quad \text{in} \quad \Omega_1 , \quad v_{1|\partial\Omega_1} = 0, \quad v_1 > 0 \quad \text{in} \quad \Omega_1 .$$

Such a solution exists and belongs (at least) to $W^{2,p}(\Omega_1)$ $(\forall p < +\infty)$.

Then by assumption (5) we have for ε small enough

$$- \Delta \varepsilon v_1 = \lambda_1 \varepsilon v_1 \leq f(x, \varepsilon v_1) \quad \text{in} \quad \Omega_1 .$$

Thus εv_1 extended by 0, in view of corollary I.1. is a weak, non negative, non trivial subsolution of (7) which will be denoted by $\underline{u}$ (in the case $\Omega_1 = \Omega$, one does not need to use Corollary I.1.).

2/ Next, by a simple regularization process, one introduces $f_\varepsilon(x,u)$ such that :

$f_\varepsilon(x,u) \xrightarrow[\varepsilon \to 0]{} f(x,u)$ uniformly in u bounded for a.e. $x \in \bar{\Omega}$;

$\exists C_\varepsilon \; \forall u \; \forall x \; |f_\varepsilon(x,u)| \leq C_\varepsilon$; (f_ε) satisfy (5) (6) uniformly in $\varepsilon > 0$;

$f_\varepsilon(x,u) \in C^1(\bar{\Omega} \times \mathbb{R})$; $f_\varepsilon(x,0) = 0$.

Let us assume for the moment that we have proved the existence of u_ε solution of

$$(7-\varepsilon) \qquad - \Delta u_\varepsilon = f_\varepsilon(x,u_\varepsilon) \quad \text{in} \quad \Omega, \quad u_{\varepsilon|\partial\Omega} = 0, \quad u_\varepsilon > \underline{u} \quad \text{in} \quad \Omega.$$

Then by assumption (6) : $\exists k < \mu_1 \; \exists C \quad \forall x \; \forall u \geq 0 \quad f(x,u) \leq ku + C$ and we have

$$- \Delta u_\varepsilon - ku_\varepsilon \leq C \quad \text{in} \quad \Omega$$

Since $k < \mu_1$ this implies $u_\varepsilon \leq v$ where v is the solution of

$$- \Delta v - kv = C \quad \text{in} \quad \Omega, \quad v_{|\partial\Omega} = 0.$$

We deduce that $\|u_\varepsilon\|_{L^\infty(\Omega)}$ is bounded and taking $\varepsilon \to 0$ we conclude.

Thus, it remains to prove the existence of (u_ε) solution of (7-ε); remark that because of the choice of f_ε, $\underline{u}$ may be chosen as a weak non negative, non trivial subsolution of (7-ε) for every $\varepsilon > 0$.

3/ Since for any positive solution of (7-ε) (or(7)), as showed above, we have $\|u_\varepsilon\|_{L^\infty(\Omega)} \le$ const, we can replace $f_\varepsilon(x,t)$ by $f_\varepsilon(x, t \wedge C)$ for some positive C. Then there exists some constant K such that $f_\varepsilon(x,t) + Kt$ is non decreasing in $t \ge 0$ (for $x \in \bar{\Omega}$).

Finally consider the following iterative scheme

$$\begin{cases} -\Delta u^{n+1} + K u^{n+1} = f_\varepsilon(x, u^n) + K u^n & \text{in } \Omega \\ u^{n+1}\big|_{\partial\Omega} = 0 \end{cases}$$

with $u^0 = \underline{u}$. We observe that $u^n \nearrow$ as $n \nearrow$ and that by a form of the strong maximum principle, (see for example H. Brézis and P.L.Lions [10])

$$\|u^n\|_{L^\infty} \le \|u^{n+1}\|_{L^\infty} \le \lambda[C + K \|u^n\|_{L^\infty}] \quad \text{where } \lambda < \frac{1}{K}$$

Thus $\|u^n\|_{L^\infty(\Omega)} \le$ Const., and we conclude. ∎

Remark II.2. : In the variational case i.e. A self adjoint (for example $A = -\Delta$) and $f = f(x,u)$ (not depending on ∇u), a simpler proof is obtained by a direct minimization method. Consider the problem : find $u \in H_0^1$ minimizing the functional:

$$S(v) = \int_\Omega \{\tfrac{1}{2}|\nabla v|^2 - F(x,v)\}\, dx \quad \text{over } v \in H_0^1(\Omega),$$

where $F(x,t) = \int_0^t f(x,s)ds$ (and $f(x,s) = 0$ if $s \le 0$).

By assumption (6), $S(u)$ is bounded below and by hypothesis (5'), there exists some $v \in H_0^1(\Omega)$ (the same $v = \underline{u}$ used in the preceding proof) such that $S(v) < 0$. It is then easy to conclude. ∎

II.2. Existence theorems for degenerate operators.

Let Ω be as in section II.1. and let f be a bounded measurable function of $(x,u) \in \bar{\Omega} \times \mathbb{R}$ such that f is continuous in u for all x in $\bar{\Omega}$.

We consider a degenerate elliptic operator :

$$A = - a_{ij} \frac{\partial^2}{\partial x_i \partial x_j} + b_i \frac{\partial}{\partial x_i} + c$$

where $a_{ij} = a_{ji}$, b_i, $c \in C^2(\bar{\Omega})$ and satisfy :

(8) $\forall x \in \bar{\Omega}, \forall \xi \in \mathbb{R}^n, \quad \sum_{i,j} a_{ij} \xi_i \xi_j \geq 0$

(9) $\forall x \in \bar{\Omega} \quad c(x) \geq \lambda > 0$.

Theorem II.2. : We assume that the coefficients of A satisfy (8) - (9) and (8') $\exists U$ neighbourhood of $\partial\Omega$, $\exists \nu > 0$, $\forall x \in \bar{U}, \forall \xi \in \mathbb{R}^n$, $\sum_{i,j} a_{ij}(x) \xi_i \xi_j \geq \nu |\xi|^2$. We assume also that $f(x,o) = 0 \quad \forall x \in \bar{\Omega}$ and

(5') $\underline{\lim}_{u \to 0_+} \frac{f(x,u)}{u} > \mu$, uniformly $x \in \bar{U}$.

(10) $\exists M > 0$ a.e. $x \in \bar{\Omega}$, $\forall u \in \mathbb{R}$ $f(x,u) \leq M$

Then if μ is large enough (depending only on U and the coefficients of A) there exists a positive solution of

(7') $Au = f(x,u)$ in Ω, $u|_\Gamma = 0$, $u > 0$ in Ω ; $u \in C(\bar{\Omega})$, $Au \in L^\infty(\Omega)$

Remark II.3. : μ can be choosen in the following way : we can always assume that $\bar{U} = \bar{I}$ where I is open, connected, $I \subset \Omega$ and $\partial I = \partial\Omega \cup S$ where S is a smooth (n-1) dimensional manifold, closed, included in Ω . Then denoting by μ_1 the smallest positive eigenvalue of A with Dirichlet conditions on I, we choose $\mu \geq \mu_1$. ■

Remark II.4. : We could assume that in (8'), U is any open subdomain of Ω , then (7') would have to be replaced by

$$\begin{cases} Au = f(x,u) \text{ in } \Omega, \quad u > 0 \text{ in } \Omega; \quad u \in L^\infty(\Omega), Au \in L^\infty(\Omega). \\ u_{|\Gamma_o} = 0 \text{ (in a sense to be precised, see [20] for the precise definition} \end{cases}$$

and also for the characterization of Γ_o where Γ_o is included in Ω). ∎

<u>Proof of Theorem II.2.</u> : As in the proof of Theorem II.1. we may assume that $f(x,.) \in C^1$ uniformly for $x \in \bar{\Omega}$. Next we remark that any positive solution of (7') satisfy :

$$0 \le u \le \frac{M}{\lambda} \text{ because of (9) and (10) (see [20]).}$$

Thus changing f one can always assume that $f(x,t) = f(x, t \wedge \frac{M}{\lambda})$. Hence we may choose K large enough so that $f(x,t) + Kt$ is nondecreasing for $t \ge 0$.

Now to conclude, we introduce the iterative method (see [20] , for the resolution of these problems).

(7'-n) $\quad K u^n + Au^n = K u^{n-1} + f(x,u^{n-1})$ in Ω, $u^n_{|\partial\Omega} = 0$, $u^n \in C(\bar{\Omega})$, $Au^n \in L^\infty(\Omega)$.

If we assume that u^o is a weak subsolution i.e. if

$$K u^o + Au^o \le f(x,u^o) + K u^o \text{ in } \mathcal{D}'(\Omega), \quad u^o_{|\partial\Omega} = 0, \quad u^o \in W^{1,\infty}(\Omega)$$

then $u^n \nearrow u \ge u_o$ (remark that $\|u_n\|_{L^\infty} \le \frac{1}{\lambda+K} [K \|u_{n-1}\|_{L^\infty} + C]$). Whence, the theorem is proved if we find some u^o satisfying :

$$Au^o \le f(x,u^o) \text{ in } \mathcal{D}'(\Omega), \quad u^o_{|\partial\Omega} = 0, \quad u^o \in W^{1,\infty}(\Omega), \quad u^o \ne 0.$$

But as in the proof of Theorem II.1, we consider v^1 solution of :

$$Av^1 = \mu v^1 \text{ in } I, \quad v^1_{|\partial I} = 0, \quad v^1 > 0 \text{ in } I \text{ (see remark II.2), and}$$

notice that for ε small enough $A \varepsilon v^1 = \mu\varepsilon v^1 \le f(x, \varepsilon v^1)$ in I.

Finally by Corollary I.1, if we define $u^o = 0$ on $\Omega - \bar{I}$, $u^o = \varepsilon v^1$ on $\bar{I}$; u^o is the subsolution we needed to conclude the proof. ∎

II.3. Maximum solutions and variations of the domain

In this section, we will describe on an easy example, a fairly general type of result, namely that the maximum solution of a non-linear problem, when it exists, is strictly increasing with the domain.

More precisely, let $\Omega_1 \subset \Omega_2 \subset \mathbb{R}^n$ be two bounded regular domains. We assume that there exist maximum positive solutions u_i of the non-linear problem

$$\begin{cases} A\,u_i = f(x,u_i) & \text{in } \Omega_i\,, \\ u_i|_{\partial\Omega_i} = 0, & \end{cases}$$

(i = 1,2). Then, one has $u_1 < u_2$ on Ω_1 (provided $\Omega_1 \subset\subset \Omega_2$).

Here, we restrict ourselves to the following example :

Theorem II.3. Let $\Omega_1 \subset\subset \Omega_2 \subset \mathbb{R}^n$ be two regular bounded domains. Let $f \in C^1(\bar{\Omega}_2 \times \mathbb{R})$ satisfy :

(11) $\exists M > 0$ such that $\forall x \in \Omega_2$, $f(x,M) \le 0$.

If there exists at least one positive (non trivial) solution of

(7.1) $\quad -\Delta u = f(x,u)$ in Ω_1, $\quad u|_{\partial\Omega_1} = 0$, $\quad u > 0$ in Ω_1 , then, there exists a maximum positive solution $\bar{u}_i$ of

(7.i) $\quad -\Delta \bar{u}_i = f(x,\bar{u}_i)$ in Ω_i, $\quad \bar{u}_i|_{\partial\Omega_i} = 0$, $\quad M > \bar{u}_i > 0$ in Ω_i.

(i = 1,2). Furthermore, $\bar{u}_1 < \bar{u}_2|_{\Omega_1}$.

Proof of theorem II.3.: It is well known that (11) and (7.1) imply the existence of a positive solution $\bar{u}_1$ of (7.1) such that $\bar{u}_1 < M$. Now, if we extend $\bar{u}_1$ by 0 to $\bar{\Omega}_2$, we have a non negative non trivial weak subsolution $\tilde{u}_1$ of (7.2), and it is well known that this fact combined with (11) imply the existence of a maximum positive solution $\bar{u}_2$ of (7.2) such that $\tilde{u}_1 < \bar{u}_2 < M$ in Ω_2. ■

Remark II.5. : This observation can also be derived by constructing a monotone decreasing iteration scheme starting with M (which is a supersolution). One can then check that at each step, the iterates remain above $\tilde{u}_1$. ■

II.4. An application to some degenerate Hamilton-Jacobi-Bellman equation.

In this section, we present a result concerning the following non linear problem :

$$(12) \qquad \sup_{v\in V} \{A(v)\, u(x) - f(x,v)\} = 0 \quad \text{in } \Omega, \quad u\big|_{\partial\Omega} = 0;$$

where Ω is abounded regular domain in $\mathbb{R}^n$ and where V is a closed convex set in $\mathbb{R}^m$. This non linear Dirichlet problem is called : Hamilton-Jacobi-Bellman equation with Dirichlet conditions, see N.V. Krylov [12] , P.L. Lions [13] and P.L. Lions - J.L. Menaldi [15] for a general exposition.

We assume that $A(v) = -\sum_{i,j} a_{ij}(x,v) \dfrac{\partial^2}{\partial x_i \partial x_j} + \sum_i b_i(x,v) \dfrac{\partial}{\partial x_i} + c(x,v)$

and that $a_{ij}(x,v)$, $b_i(x,v)$, $c(x,v)$, $f(x,v)$ are functions defined on $\mathbb{R}^n \times V$ satisfying the following group of assumptions :

i/ $\left\{ \begin{array}{l} a_{ij}(x,v),\ b_i(x,v),\ c(x,v) \in C_b^2(\mathbb{R}^n) \text{ and remain in a bounded} \\ \text{subset of } C_b^2(\mathbb{R}^n) \text{ for } v \in V, \end{array} \right.$

ii/ $|\Phi(x,v) - \Phi(x,v')| \leq p(|v-v'|) \quad \forall \Phi = a_{ij}, b_i, c, f \quad \forall v,v' \in V,$
where p is a continuous function on $\mathbb{R}_+$ such that p is non decreasing and $p(0) = 0$,

iii/ $|f(x,v) - f(x',v)| \leq C|x-x'| \quad \forall x,x' \in \mathbb{R}^n \quad \forall v \in V,$

iv/ $c(x,v) \geq \lambda > 0 \quad \forall x \in \mathbb{R}^n \quad \forall v \in V,$

v/ $\forall v \in V \quad \forall x \in \mathbb{R}^n \quad \forall \xi \in R^n \quad \sum_{i,j} a_{ij}(x,v)\, \xi_i \xi_j \geq 0,$
$\exists v \in C_b(\mathbb{R}^n;V) \quad \exists \nu > 0 \quad \forall x \in \mathbb{R}^n \quad \forall \xi \in \mathbb{R}^n \quad \sum_{i,j} a_{ij}(x,v(x))\, \xi_i \xi_j \geq \nu |\xi|^2,$

vi/ $f(x,v) \geq 0 \quad \forall x \in \mathbb{R}^n \quad \forall v \in V.$

We shall denote this group of assumptions by assumption (A). We recall now a few results proved in P.L. Lions [13] and in P.L. Lions - J.L. Menaldi [15] : if $\lambda > \lambda_0$ (which depends explictly on the coefficients $a_{ij}(x,v), b_i(x,v)$) there exists a unique function $u(x)$ which satisfies

$$(12')\quad \begin{cases} u \in W^{1,\infty}(\mathbb{R}^n), \quad \forall v \in V \quad A(v)u \le f(v) \quad \text{in } \mathcal{D}'(\mathbb{R}^n), \\ \forall \tilde{u} \in W^{1,\infty}(\mathbb{R}^n) \text{ satisfying } \forall v \in V \quad A(v)\tilde{u} \le f(v) \quad \text{in } \mathcal{D}'(\mathbb{R}^n), \text{ then } \tilde{u} \le u. \end{cases}$$

Now if we introduce a function $q \in C_b^2(\mathbb{R}^n)$ such that :

$$q(x) = 0 \text{ on } \bar{\Omega}, \quad q(x) > 0 \text{ if } x \notin \bar{\Omega}$$

and if we consider u_ε solution of (12') where $A(v)$ is replaced by $A(v) + \frac{1}{\varepsilon} q$ (i.e. approximation by penalization of the domain) then (see [15] $u_\varepsilon(x) \downarrow u(x) \in W_0^{1,\infty}(\Omega)$ (when $\varepsilon \downarrow 0_+$). Furthermore $u(x)$ satisfies

$$\forall v \in V \quad A(v)u \le f(v) \quad \text{in } \mathcal{D}'(\Omega); \quad u \ge 0 \quad \text{in } \bar{\Omega}.$$

We now prove that u satisfies in Ω a similar characterization to (12') :

<u>Theorem II.4.</u> see [15] : Under assumption (A), if $u(x)$ is the function introduced above, then $u(x)$ is the unique maximum nonnegative sub-solution of (12)i.e.

$$\forall v \in V, \quad A(v)u \le f(v) \quad \text{in } \mathcal{D}'(\Omega); \quad u \in W^{1,\infty}(\Omega), \ u_{|\partial\Omega} = 0, \quad u \ge 0 \quad \text{in } \Omega$$

and if $\tilde{u} \in W_0^{1,\infty}(\Omega)$, $\tilde{u} \ge 0$ in Ω; $\forall v \in W$, $A(v)\tilde{u} \le f(v)$ in $\mathcal{D}'(\Omega)$, then $\tilde{u} \le u$.

<u>Proof</u> : In view of the results recalled above, we just have to prove that if $\tilde{u} \in W_0^{1,\infty}(\Omega)$, $\tilde{u} \ge 0$ on Ω and $\forall v \in W^{1,\infty}$, $A(v)\tilde{u} \le f(v)$ in $\mathcal{D}'(\Omega)$, then $\tilde{u} \le u$.
But this follows easily from corollary I.1; indeed if we extend $\tilde{u}$ by 0 (denoting still by $\tilde{u}$ the resulting function) we have $\tilde{u} \in W^{1,\infty}(\mathbb{R}^n)$ and by Corollary I.1. : $\forall v \in V \quad A(v)\tilde{u} \le \tilde{f}(v)$ in $\mathcal{D}'(\mathbb{R}^n)$; where $\tilde{f}(v) = 1_{\bar{\Omega}}(x)\, f(x,v)$.
Now remark that by assumption (A) $\tilde{f}(v) \le f(v)$ in $\mathbb{R}^n$; as $\frac{1}{\varepsilon} p(x)\tilde{u} = 0$ in $\mathbb{R}^n$, we have $\forall \varepsilon > 0 \quad \forall v \in W \quad (A(v) + \frac{1}{\varepsilon} p)\tilde{u} \le f(v)$ in $\mathcal{D}'(\mathbb{R}^n)$.
Then by (12') $\tilde{u} \le u_\varepsilon \quad \forall \varepsilon > 0$ and we conclude taking $\varepsilon \to 0$. ■

III. Combining order and topological arguments

In this section we follow and extend a method of P.H. Rabinowitz [17] which will give in very general situations the existence of two positive solutions of problems like :

(15) $\quad Au = f(x,u)$ in Ω ; $\quad u_{|\partial\Omega} = 0$; $\quad u > 0$ in Ω .

Where Ω is a bounded regular domain in $\mathbb{R}^n$, and A is a second order uniformly elliptic operator : we assume that $A = -\sum_{i,j} a_{ij}(x) \frac{\partial^2}{\partial x_i \partial x_j} +$

$$+ \sum_i b_i(x) \frac{\partial}{\partial x_i} + c(x).$$

where $a_{ij} = a_{ji} \in C(\bar{\Omega})$, $b_i \in L^\infty(\Omega)$, $c \in L^\infty(\Omega)$ and

(16) $\quad \exists \nu > 0 \quad \forall x \in \bar{\Omega} \quad \forall \xi \in R^n \qquad \sum_{i,j} a_{ij}(x) \xi_i \xi_j \geq \nu |\xi|^2,$

(17) $\qquad c(x) \geqslant 0.$

Let $f(x,u)$ be a measurable locally bounded function on $\bar{\Omega} \times \mathbb{R}$, locally Lipschitz continuous in u uniformly for x in $\bar{\Omega}$, such that $f(x,0) = 0$ and

(18) $\qquad \overline{\lim_{u \to 0_+}} \frac{f(x,u)}{u} < \mu_1, \quad$ uniformly in $x \in \bar{\Omega}$,

where μ_1 is the smallest positive eigenvalue of A with Dirichlet conditions on Ω .

Theorem III.1. : We assume (16) - (17) - (18) - Furthermore we assume the existence of $\underline{u}$, $\bar{u}$ such that $\underline{u}, \bar{u} \in W^{2,p}(\Omega)$ $(p > n)$ and

(15') $\qquad A\bar{u} \geq f(x,\bar{u})$ a.e. in Ω ; $\bar{u} \geq 0$ on $\partial\Omega$ (resp. $A\underline{u} \leq f(x,\underline{u})$, $\underline{u} = 0$ on $\partial\Omega$)

(19) $\qquad 0 \leq \underline{u} \leq \bar{u}$ in Ω

(20) $\qquad \underline{u}, \bar{u}$ are not solutions of (15).

Then there exist u_1, u_2 solutions of (15) in $W^{2,p}(\Omega)$ $(\forall p < +\infty)$ such that u_1 is the maximum solution of (15) such that $u_1 \leqslant \bar{u}$, and we have $u_1 \geqslant \underline{u}$. The second solution u_2 satisfies : $u_2 \leq u_1$ and $\exists x \in \Omega \quad u_2(x) < \underline{u}(x)$.

Remark III.1. This result is to be compared with [17] . The method we present below is just a direct adaptation of the argument of P.H. Rabinowitz in [17] . One can also find a proof, based on a critical point method, in [2.] However, the context in [2,17] is slightly different, the nonlinearity depending on a parameter. Furthermore, there, the operator A has a variational structure and it is assumed that there exists $\bar{z}$ such that $f(x,z) \leq 0$, $\forall x \in \bar{\Omega}$, $\forall z \geq \bar{z}$. This assumption allows one to have strong a priori bounds on all the solutions of the equation. This feature is absent in our hypotheses.

Proof : The proof will be divided in several steps :

1/ existence of u_1,
2/ restatement of the problem,
3/ existence of u_2,
4/ proof of some lemmas.

1. Existence of u_1 . The first solution is obtained by a standard iteration scheme. Let $m = \|\bar{u}\|_{C(\bar{\Omega})}$, so that $0 \le \bar{u} \le m$. Define $\tilde{f}$ by

$$\tilde{f}(x,t) = \begin{cases} f(x,t) & \text{if } t \in [0,m] \\ f(x,m) & \text{if } m \le t \\ 0 & \text{if } t \le 0 \end{cases}$$

Let $K < 0$ be a constant such that $\tilde{f}(x,t) + Kt$ is nondecreasing with respect to $t \in \mathbb{R}$. The iteration scheme is defined by $u^o = \bar{u}$ and

$$(15.n) \quad \begin{cases} Au^{n+1} + Ku^{n+1} = f(x,u^n) + Ku^n \quad \text{a.e. in } \Omega \\ \\ u^{n+1}{}_{|\partial\Omega} = 0, \quad u^{n+1} \in W^{2,p}(\Omega) \qquad (\forall\, p < +\infty) \end{cases} .$$

Then, the sequence (u^n) is easily seen to be nonincreasing, $0 \le \ldots \le u^{n+1} \le u^n \le \ldots\ u^o = \bar{u}$, and to converge to the maximum solution u_1 of (15) such that $u_1 \le \bar{u}$, and $u_1 \geqslant \underline{u}$.

2. Restatement of the problem : The existence of a second solution u_2 is derived by constructing a solution $u_2 \le \bar{u}$ (whence $u_2 \le u_1$) such that $\exists x \in \Omega$, $u_2(x) < \underline{u}(x)$ (i.e. $u \not\geq \underline{u}$), thus $u_2 \ne u_1$. This is obtained by a topological degree argument of P.H. Rabinowitz [17] . To that effect, define the operators F_t, $0 \le t \le 1$ as follows :

For $v \in C^1_o(\bar{\Omega}) = \{v \in C^1(\bar{\Omega}),\ v_{|\partial\Omega} = 0\}$, $u = F_t v$ is the solution of

$$\begin{cases} Au + tKu = tf(x,v) + tKv \quad \text{in } \Omega \\ u_{|\partial\Omega} = 0, \quad u \in W^{2,p}(\Omega), \quad \forall p < +\infty. \end{cases}$$

Using the L^p estimate (Agmon-Douglis-Nirenberg [1]), $F_t v$ is easily seen to be a compact operator from $(t,v) \in [0,1] \times C^1_o(\bar{\Omega})$ into $C^1_o(\bar{\Omega})$. Moreover F_t is order preserving (cf. [14]), i.e. if $v_1 \le v_2$, then $F_t v_1 \le F_t v_2$. The problem is now to find a fixed point of $F \equiv F_1$ "which is not above $\underline{u}$. We require the following two lemmas whose proofs are given in part 4/ of this section.

Lemma III.1. $\exists R > 0$ such that for $\forall v \in C^1_o(\bar{\Omega})$, $0 \le v \le \bar{u}$, $\forall t \in [0,1]$, one has $\|F_t v\|_{C^1(\bar{\Omega})} < R$.

Lemma III.2. $\exists \alpha > 0$ such that if $u \in C_0^1(\bar{\Omega})$ satisfies $u \geq 0$ and $u \leq F_t u$ for some $t \in [0,1]$, then, either $\|u\|_{C^0(\Omega)} > \alpha$ or $u \equiv 0$.

2. Existence of u_2 : Let us introduce the following open sets in $C_0^1(\bar{\Omega})$:

$$Q = \{u \in C_0^1(\bar{\Omega}),\ \|u\|_{C^1(\bar{\Omega})} < R,\ u(x) < \bar{u}(x) \text{ in } \Omega,\ \frac{\partial u}{\partial \nu}(y) > \frac{\partial \bar{u}}{\partial \nu}(y),$$

if $\bar{u}(y) = 0$, $y \in \partial\Omega\}$, where ν is the unit outward normal on $\partial\Omega$;

$$\mathcal{O} = \{u \in Q,\ \|u\|_{C^0(\bar{\Omega})} < \alpha\}$$

$$\tilde{Q} = Q - \bar{\mathcal{O}} = \{u \in Q,\ \|u\|_{C^0(\bar{\Omega})} > \alpha\}$$

$$\mathcal{G} = \{u \in Q,\ u > \underline{u} \text{ in } \Omega,\ \frac{\partial \underline{u}}{\partial \nu} < \frac{\partial u}{\partial \nu} \text{ on } \partial\Omega\}.$$

We are going to prove that $d(I-F,\mathcal{G},0) = +1$, $d(I-F,\mathcal{O},0) = +1$, $d(I-F,Q,0) = +1$ where d denotes the Leray-Schauder degree and we shall conclude remarking that in view of lemma II.2: $\mathcal{G} \subset \tilde{Q}$ and $d(I-F,\tilde{Q},0) = d(I-F,Q,0) - d(I-F,\mathcal{O},0) = 0$;

thus, $d(I-F,\tilde{Q}-\bar{\mathcal{G}},0) = -1$ and there exists a solution $u_2 \in \tilde{Q}$ which is not in $\bar{\mathcal{G}}$.

First, we check that the degrees are well defined : indeed if $u = F_t u$, $t \in [0,1]$ then by the strong maximum principle either $u \equiv 0$, or $u > 0$ in Ω and by lemma III.1. $\|u\|_{C^1(\bar{\Omega})} < R$, by lemma III.2 $\|u\|_{C^0(\bar{\Omega})} > \alpha$ and as $\underline{u}$, $\bar{u}$ are not solutions of (15) we finally have $u \notin \bar{Q}-Q$, $u \notin \bar{\mathcal{G}}-\mathcal{G}$, $u \notin \bar{\tilde{Q}} - \tilde{Q}$.

The fact that there are no solutions of $u = F_t u$ on ∂Q, $\partial\mathcal{O}$ and $\partial\tilde{Q}$ also shows that the degrees $d(I-F_t,Q,0)$, $d(I-F_t,\mathcal{O},0)$, $d(I-F\ ,\tilde{Q},0)$ are well defined and independent of $t \in [0,1]$ (homotopy invariance property of the topological degree).

Now, we calculate these degrees :

i/ $d(I-F,Q,0) = d(I-F_t, Q,0) = d(I-F_0,Q,0) = d(I,Q,0) = +1$

ii/ $d(I-F,\mathcal{O},0) = d(I-F_t, \mathcal{O},0) = d(I-F_0,\mathcal{O},0) = d(I,\mathcal{O},0) = +1$

iii/ let Φ be any element in $\mathcal{O}$ we define $G_t v = tFv + (1-t)\Phi$, as is convex and F maps $\bar{\mathcal{O}}$ into $\mathcal{O}$, G_t maps $\bar{\mathcal{O}}$ into $\mathcal{O}$ thus $d(I-F,\mathcal{O},0) = d(I-G_t,\mathcal{O},0) = d(I-G_0,\mathcal{O},0) = +1$.

4. Lemma III.1. is an immediate consequence of the Agmon-Douglis-Nirenberg estimate [1] .

Proof of lemma III.2. : By assumption (18), $\exists\alpha > 0$ $\exists k < \mu_1$ such that if $0 \leq u \leq \alpha$ $\forall x \in \bar{\Omega}$ $f(x,u) \leq ku$.

Now if $u \in C_0^1(\bar{\Omega})$, $u \geq 0$ and if for some $t \in [0,1]$ $u \leq F_t u = v$, we have

$$\begin{cases} Av + tKv = tf(x,u) + tKu & \text{in } \Omega \\ v|_{\partial\Omega} = 0 . \end{cases}$$

If we assume moreover that $\|u\|_{C^0(\bar{\Omega})} \leq \alpha$, then $Av + tKv \leq tKu + tku$

$$\leq tKv + kv$$

or $Av \leq kv$ in Ω, $v = 0$ on $\partial\Omega$. And this implies (see [14] for the case of continuous coefficients of the operator A) : $v \leq 0$ in Ω , hence $u \equiv 0$. ■

Remark III.2. : An application of theorem III.1. will be given in section IV. ■

IV. A local method for N.L.K.G. type equations.

The results contained in this section are developed in [6] . We shall just prove some results explaining how the techniques introduced above may be used to solve equation (1). Concerning the physical motivation of equation (1) we mention that it arises when looking for stationary wares in non linear classical field equations of the type Klein-Gordon or non linear Schrödinger equations (see [7], [19], [8] , for a more precise description). We first state the result we want to prove here.

Theorem IV.1. : Let $n \geq 2$ an let $1 < p < q < +\infty$, let λ, μ, m be positive constants. We assume that

(2) $\exists \zeta > 0$ such that $\lambda \dfrac{\zeta^{p+1}}{p+1} - \mu \dfrac{\zeta^{q+1}}{q+1} - m \dfrac{\zeta^2}{2} > 0$.

Then there exists a positive solution of (1) $u \in C^\infty(\mathbb{R}^n)$, radial, decreasing ($u(x) < u(y)$ if $|x| > |y|$), satisfying :

$$\forall \alpha \; \exists C_\alpha , \; \delta_\alpha > 0, \; |D^\alpha u(x)| \leq C_\alpha e^{-\delta_\alpha r} \quad (\text{where } r = |x|) :$$

(1) $\quad - \Delta u = \lambda u^p - \mu u^q - m u \quad \text{in } \mathbb{R}^n$.

Remark IV.1. In [5] , we prove that (2) is a necessary condition to have non-trivial solution of (1). More general results are proved in [4-7-8] , but the case considered here is a kind of model case. ■

In this section we consider a local approach for solving (1) i.e. first solving (1) but in a ball B_R (of radius R) with Dirichlet condition on ∂B_R, and then letting $R \to +\infty$. We thus want to solve :

(1-R) $\quad - \Delta u = \lambda u^p - \mu u^q - mu$ in B_R, $u|_{\partial B_R} = 0$, $u \geq 0$ in B_R.

We have the following result :

Theorem IV.2. : Under the assumptions of theorem IV.1., $\exists R_0 > 0$ such that

i/ if $R < R_0$ there exists no trivial solution of (1-R);

ii/ if $R \geq R_0$ there exists a maximum positive solution $\bar{u}_R$ of (1-R) which is radial decreasing. When $R \nearrow +\infty$, $\bar{u}_R \nearrow \beta$ uniformly on compact sets where β is the maximum positive solution of $\lambda\beta^p - \mu\beta^q - m\beta = 0$;

iii/ if $R > R_o$ there exists a positive radial decreasing solution u_R of (1.R) such that $\exists \xi_R \in B_{R_o}$ $u_R(\xi_R) < \bar{u}_{R_o}(\xi_R)$.

There exists $R_n \nearrow \infty$, $u_{R_n} \to u$ uniformly on compact sets, where u is a positive radial, non increasing solution of

$$(1) \qquad -\Delta u = \lambda u^p - \mu u^q - mu \quad \text{a.e. in } \mathbb{R}^n$$

such that $u(x) \searrow 0$ as $|x| \nearrow +\infty$.

Remark IV.2. : It is easy to deduce theorem IV.1. from theorem IV.2 and we shall not do it here (see [6]). ∎

The proof will be divided in several steps :

1/ if R is small enough there is no non trivial solution,

2/ if there exists a non trivial solution in $B_{R'}$, then $\forall R \geq R'$ there exists $\bar{u}_R$,

3/ existence of u_R,

4/ Convergence of u_R when $R \nearrow +\infty$,

5/ convergence of $\bar{u}_R$ when $R \nearrow +\infty$.

First we modify the function $g(u) = \lambda |u|^{p-1}u - \mu |u|^{q-1}u - mu$ in the following way :

$\tilde{g}(u) = g(u^+ \wedge \beta)$ i.e. $\tilde{g}(u) = 0$ if $u \leq 0$, $\tilde{g}(u) = g(u)$ if $0 \leq u \leq \beta$, $\tilde{g}(u) = 0$ if $u \geq \beta$ where $0 < \alpha < \beta$ are the two non negative solution of $g(t) = 0$ (they exist because of (2).).

1/ If u is a non trivial solution of

$$\begin{cases} -\Delta u = \tilde{g}(u) & \text{in } \Omega \\ u_{|\partial\Omega} = 0 \end{cases} \quad \text{then} \quad u > 0 \text{ in } \Omega$$

and $u < \beta$ in $\bar{\Omega}$,

for every domain Ω. Remark that the same holds for any positive solution of $-\Delta u = g(u)$ in Ω, $u = 0$ on $\partial\Omega$, so non trivial solutions for $\tilde{g}$ and positive solutions for g are the same. Now if $0 \le u \le \beta$ $g(u) = u\,h(u)$ where $h(u) \le c$ independent of Ω. Thus if R is small enough such that the smallest eigenvalue $\mu_1(B_R)$ of $-\Delta$ with Dirichlet conditions on B_R satisfies $\mu_1(B_R) > c$, there is no non trivial solution of (1.R).

2/ Next assume that for some R' there exists some non trivial solution of

$$\begin{cases} -\Delta u = \tilde{g}(u) & \text{in } B_{R'}, \\ u|_{\partial B_{R'}} = 0. \end{cases}$$

Then, by Corollary I.1, u, extended by 0 in $B_R - B_{R'}$, is a weak subsolution in B_R, i.e.

$$\begin{cases} -\Delta u \le g(u) & \text{in } B_R \quad \text{(weakly)} \\ u|_{\partial B_R} = 0, & \text{for every } R \ge R'. \end{cases}$$

This implies easily the existence of a maximum positive solution $\bar{u}_R$ of (1.R) for every domain B_R such that $R \ge R'$. Moreover $\beta > \bar{u}_R \ge u$.
Remark also that in view of theorem II.3. one has $\bar{u}_{R_1} \le \bar{u}_{R_2}$ if $R' \le R_1 \le R_2$.
Finally we notice that because of (2), the function

$$w = \begin{cases} \zeta \text{ on } B_{R-1} & (R > 1) \\ \zeta(R-1) & \text{if } |x| = r \in [R-1, R] \end{cases}$$

satisfies for R large enough :

$$\tilde{S}(w) = \int_{B_R} \{\tfrac{1}{2}|\nabla w|^2 - \tilde{G}(w)\}\,dx < 0 \quad (\tilde{G}(t) = \int_0^t g(s)ds).$$

Therefore, the minimization problem : find u such that

$$\tilde{S}(u) = \inf_{v \in H_0^1(B_R)} \tilde{S}(v)$$

gives a non trivial solution of (1.R) for R large enough.

In conclusion we have proved that

$$R_o = \inf \{R > 0, \exists \text{ positive solution of } (1.R)\} > 0, \text{ and}$$

i/ if $R < R_o$ there is no non trivial solution of $-\Delta u = \tilde{g}(u)$ in B_R, $u_{|\partial B_R} = 0$

ii/ if $R \geq R_o$ there exists a maximum positive solution $\bar{u}_R$ of (1.R) and $\bar{u}_R \nearrow$ as $R \nearrow$.

Actually there is one point which remains to be proved : the existence of a positive solution of $(1.R_o)$.

But if we consider $\bar{u}_{R_o} = \lim_{R \downarrow R_o} \downarrow \bar{u}_R$, $\bar{u}_{R_o}$ is a solution of $(1.R_o)$ which is non trivial since $\max_{B_R} \bar{u}_R > \alpha$ and this implies $\max_{B_{R_o}} \bar{u}_{R_o} \geq \alpha$.

3/ Existence of u_R for $R > R_o$: first we introduce

$$\begin{cases} v(x) = \bar{u}_{R_o}(x) & \text{if } |x| \leq R_o \\ \quad = 0 & \text{if } |x| > R_o \end{cases}$$

By Corollary I.1. v is a subsolution of(1.R) for all $R > R_o$. On the other hand β is a supersolution of (1.R) for all R. Furthermore, $v < \beta$ and neither v, nor β is a solution of (1.R); Thus, applying theorem III.1, (let us remark that (18) is satisfied because $\overline{\lim_{u \to 0_+}} \frac{g(u)}{u} = g'(0) = -m < \mu_1(B_R), \forall R$) we deduce the existence of u_R positive solution of (1.R) such that $\exists \xi_R \in B_R$ $u_R(\xi_R) < v(\xi_R)$ (whence $\xi_R \in B_{R_o}$).

Now the fact that u_R can be chosen radial decreasing is a technical detail : we have to reproduce the proof of. Th. III.1. in the context of radial functions and we need to work in the space $E = \{u \in C^2(\bar{B}_R),\ u_{|\partial B_R} = 0,\ u \text{ is radial}\}$ with the open set Q being replaced by

$$Q = \{u \in E,\ \|u\|_{C^2(\bar{B}_R)} < K,\ u < \beta \text{ in } \Omega,\quad u'(r) < 0 \text{ if } r \neq 0,$$
$$u''(0) < 0\}.$$

This introduces a few technical difficulties which are essentially overcome by the following lemma which we shall not prove here (see [6]) and which also shows that u_R is radial decreasing.

<u>Lemme IV.1.</u> : Let $f(x)$ be a Lipschitz continuous, radial, non increasing, non negative function of x in B_R, then the solution u of

$$-\Delta u = f(x) \text{ in } B_R,\qquad u_{|\partial B_R} = 0,$$

either is 0 (and $f \equiv 0$), or is in $C^2(\bar{B}_R)$, radial, decreasing ($u'(r) < 0$ if $r \neq 0$, $u''(0) < 0$), and positive in B_R.

4/ Convergence of u_R : as u_R takes its values in $[0,\beta]$, it is easy to extract a subsequence $R'_k \to +\infty$ such that $u_{R'_k} \to u$ uniformly on compact sets and in $W^{2,p}_{loc}(\mathbb{R}^n)$ ($\forall p < +\infty$). Thus u is non negative, radial, non increasing and is a solution of

$$(1)\qquad -\Delta u = \lambda u^p - \mu u^q - mu \quad \text{a.e. in } \mathbb{R}^n.$$

First we remark that as $u_R(0) = \max_{B_R} u_R$ we have $u(0) = \lim u_{R'_k}(0) \geq \alpha$, and this shows that u is positive.

Next we denote by $L = \lim_{|x|\uparrow+\infty} \downarrow u(x)$, as $0 \leq u \leq \beta$ we have $L \in [0,\beta]$.

Furthermore as

$$\exists \xi_{R'} \in B_{R_0}\quad u_{R'}(\xi_{R'}) < v(\xi_{R'})$$

we have $\min_{|\xi| \leq R_0} u(\xi) \leq \max_{\bar{B}_{R_0}} v < \beta$ and thus, $L < \beta$.

We want to prove $L = 0$, and we are going to prove first that $g(L) = 0$, implying either $L = 0$, or $L = \alpha$. Then we will eliminate the possibility that $L = \alpha$.

*/ We prove $g(L) = 0$: one can write (1) in the following form (remark that $u \in C^\infty(\mathbb{R}^n)$)

$$- u'' - \frac{n-1}{r} u' = g(u) \qquad \forall\, r \geq 0.$$

Multiply this equation by u', and integrate between 0 and R to find

$$(3) \qquad - \frac{1}{2} u'^2(R) - \int_0^R \frac{n-1}{r} u'^2(r)dr = G(u(R)) - G(u(0))$$

thus, $\int_0^{+\infty} \frac{n-1}{r} u'^2(s)ds < +\infty$ and $u'^2(R) \underset{R \to +\infty}{\longrightarrow} \theta \geq 0$ (by(3)). But then, $\theta = 0$ (for the previous integral to be finite).

Now this implies $- u''(R) \underset{R \to +\infty}{\longrightarrow} g(L)$ and as u' is bounded we see that $g(L) = 0$.

*/ From the preceding proof we see that $- G(L) + G(u(0)) = + \int_0^{+\infty} \frac{n-1}{r} u'^2(r)dr$

On the other hand, we have

$$G(u_{R'_k}(0)) = \int_0^{R'_k} \frac{n-1}{r} u'^2_{R'_k}(r)dr + \frac{1}{2}(u'_{R'_k}(R'_k))^2$$

thus, $G(u_{R'_k}(0)) \geq \int_0^K \frac{n-1}{r} u'^2_{R'_k}(r)dr$, $\forall R'_k \geq K \geq 0$.

By Fatou's lemma this inequality implies :

$$G(u(0)) \geq \int_0^K \frac{n-1}{r} u'^2(r)dr \qquad \forall K \geq 0,$$

and we deduce $G(L) \geq 0$.

As $L = 0$ or α and $G(\alpha) < 0$, we have necessarily $L = 0$; that is

$$\lim_{|x| \nearrow +\infty} |u(x)| = 0$$

5/ Convergence of $\bar{u}_R$: We know that $\bar{u}_R \nearrow \bar{u}$ which is a solution of (1), positive, radial, non increasing : $0 < \bar{u} \leq \beta$ and $\bar{u} \in C^\infty(\mathbb{R}^n)$. Let us denote by $L' = \lim_{|x| \nearrow +\infty} \downarrow \bar{u}(x)$, the same argument as above shows that $L' = 0$ or β. We assume that $L' = 0$ and we prove this implies a contradiction.

We introduce $\tilde{u}_R$ the solution (at least for R large enough, see part (2) of the same proof) of

$$\tilde{S}(\tilde{u}_R) = \inf_{v \in H_0^1(B_R)} \tilde{S}(v) , \qquad \tilde{u}_R \in H_0^1(B_R);$$

$\tilde{u}_R$ is a positive solution of (1.R) and of course $0 \le \tilde{u}_R \le \bar{u}_R \le \bar{u}$.

As we have assumed that $\bar{u}(x) \searrow 0$ as $|x| \nearrow +\infty$, it is easy to show (cf [19]) by an argument of ordinary differential equations that

$$\exists C \quad \exists\, \delta > 0 \;,\quad |\bar{u}(x)| \le C\, e^{-\delta r} \quad (\text{where} \quad r = |x|).$$

(this remains true even for non radial solutions of (1), which go to zero at infinity see [7]). Thus by Lebesgue theorem : if R'_k is such that $\tilde{u}_{R'_k} \xrightarrow[R'_k \to +\infty]{} \tilde{u}$ a.e. in $W^{2,p}_{loc}(\mathbb{R}^n)$ $(\forall p < +\infty)$

$$\lim_{R' \to +\infty} \int_{\mathbb{R}^n} \tilde{G}(\tilde{u}_{R'_k})\, dx \to \int_{\mathbb{R}^n} \tilde{G}(\tilde{u})\, dx.$$

Thus by Fatou's lemma :

$$\int_{\mathbb{R}^n} \frac{1}{2} |\nabla \tilde{u}|^2 - \tilde{G}(\tilde{u})\, dx \le \lim_{R'_k \to +\infty} \inf \; \tilde{S}(\tilde{u}_{R'_k}) \le 0 \quad (\text{see part 2})).$$

On the other hand $\tilde{u}$ is a positive solution of (1), radial, non increasing (indeed each of the $\tilde{u}_R$ because of the Schwarz symmetrisation process is known to be radial, non increasing, see [6]), $\tilde{u}(0) = \lim_{R' \to +\infty} \tilde{u}_{R'}(0) \ge \alpha$ and $\lim_{|x| \nearrow +\infty} \searrow \tilde{u}(x) \le \lim_{|x| \nearrow +\infty} \searrow \bar{u}(x) = 0$. Thus it is easy to show (cf [5] , [6]) that

$$\forall \alpha \quad \exists C_\alpha ,\; \delta_\alpha > 0 \quad |D^\alpha \tilde{u}(x)| \le C_\alpha\, e^{-\delta_\alpha r} \quad \text{where} \quad r = |x|$$

and this implies by the well known Pohozaev identity, recalling that $\tilde{u}$ is a solution of (1) (cf [16], [19] , [7]) :

$$\int_{\mathbb{R}^n} \frac{1}{2} |\nabla \tilde{u}|^2 - G(\tilde{u})\, dx > 0 \quad , \text{ which gives the contradiction.} \;\blacksquare$$

<u>Remark IV.3.</u> : The existence of such a "minimal radius" R_0 is from the physical view point an interesting phenomena, which is related to a form of the Heisenberg uncertainty principle (see [18] for the precise formulation).

BIBLIOGRAPHY

[1] S. Agmon, A. Douglis and L. Nirenberg : Estimates near the boundary for solutions of elliptic partial differential equations satisfying general boundary conditions; Comm. Pure and Applied Math. 12 (1959) p. 623-727 Part. I.

[2] A. Ambrosetti and P.H. Rabinowitz : Dual variational methods in critical point theory and applications; J. Funct. Anal. 14 (n°4) (1973) p. 349-381. Cf. also P.H. Rabinowitz : Variational methods for non-linear eigenvalue problems in Eigenvalues of non-linear problems, C.I.M.E. lecture notes, Ed. Cremonese, Roma 1974.

[3] D.G. Aronson and L.A. Peletier : Large time behaviour of solutions of the porous media equation. To appear.

[4] H. Berestycki et P.L. Lions : Existence d'ondes solitaires dans des problèmes non-linéaires du type Klein-Gordon. C.R.A.S. Paris - Ser. A. 287 (1978) p.503-506.

[5] H. Berestycki and P.L. Lions : Existence of a ground state in non linear equations of the type Klein-Gordon . To appear in Variational Inequalities Ed. Gianessi, Cottle, Lions, J. Wiley, New-York.

[6] H. Berestycki and P.L. Lions : A local approach to the existence of solitary waves for nonlinear Klein-Gordon equations.

[7] H. Berestycki and P.L. Lions : Existence of solitary waves in non linear Klein-Gordon equations. Part I : The ground state. Part II : Existence of infinitely many bound states. To appear.

[8] H. Berestycki and P.L. Lions : Existence d'ondes solitaires dans des problèmes non linéaires du type Klein-Gordon. C.R.A.S. Paris 288 (1979) p. 395-398.

[9] C.M. Brauner and B. Nicolaenko : See their contribution in this volume.

[10] H. Brézis and P.L. Lions : To appear in Bull. U.M.I.

[11] A.M. Il'in, A.S. Kalashnikov and O.A. Oleinik : Linear equations of the second order of parabolic type. Russian Math. Surveys 17 (1962) n°3, p. 1. - 143.

[12] N.V. Krylov : Control of a solution of a stochastic integral equation. th. Proba. Appl. Vol. XVII. n°1 (1972), p. 114-131.

[13] P.L. Lions : Contrôle de diffusions dans $\mathbb{R}^n$. To appear.

[14] P.L. Lions : Problèmes elliptiques du 2ème ordre non sous forme divergences. To appear in Proc. Roy. Sec. Edim.

[15] P.L. Lions and J.L. Menaldi : Control of stochastic equations and Bellman equations. To appear. See also : Problèmes de Bellman avec le contrôle dans les coefficients de plus haut degré. C.R.A.S. Paris série A, 287, (1978), p. 409-412.

[16] S.I. Pohozaev : Eigen functions of the equation $\Delta u + \lambda f(u) = 0$. sov. Math. Pokl. S (1965) p. 1408-1411.

[17] P.H. Rabinowitz : Pairs of positive solutions of non linear elliptic partial differential equations, Indiana Univ. Math. J. 23 (1973/74), p. p. 173-186.

[18] A.F. Rañada. and L. Vásquez : Kinks and the Heisenberg un certainty principle. To appear in Phys. Rev. D.

[19] W.A. Strauss : Existence of solitary wares on higher dimensions. Comm. Math. Phys. 55 (1977), p. 149-162.

[20] D.W. Stroock and S.R.S. Varadhan : On degenerate elliptic - parabolic operators of second order and their associated diffusions. Comm. Pure Appl. Maths. 25 (1972), p. 651-713.

H. Berestycki

Anal. Num. Université Paris VI
Tour 55-65, 5ème Etage
4, Place Jussieu
75230 PARIS-CEDEX 05
FRANCE

P. L. Lions

E.N.S. 45, rue d'Ulm
75230 PARIS-CEDEX 05

et Anal. Num. Université Paris VI
Tour 55-65, 5e étage
4, place Jussieu
75230 PARIS-CEDEX 05

MULTIPLE SOLUTIONS OF A BIFURCATION PROBLEM

C. BOLLEY
Institut National des Sciences Appliquées de Rennes
Laboratoire d'Analyse Numérique
B.P. 14 A 35031 - RENNES CEDEX

INTRODUCTION

Let us consider the Dirichlet's problem associated with the nonlinear elliptic equation $Au - \lambda u + P(u) = 0$, where P is some nonlinear function such that $P(0) = 0$. We investigate the nontrivial solutions of this problem, and more particularly the solutions which bifurcate from the trivial solution on a multiple eigenvalue. In a numerical analysis, we study an iterative scheme which converges to a stable solution or an instable solution according to the hypotheses. We give numerical examples in dimension one and two.

I - EXISTENCE AND NUMBER OF THE SOLUTIONS

Let Ω be a regular bounded and connected open set of $\mathbb{R}^m$ where $m \leq 3$. We consider the boundary value problem

$$(1.1.) \quad \begin{cases} Au - \lambda u + P(u) = 0 & \text{in } \Omega \\ u \in H^2(\Omega) \cap H^1_0(\Omega) \end{cases}$$

where $A = -\sum_{i,j=1}^{m} a_{ij} \frac{\partial^2}{\partial x_i \, \partial x_j}$ is an elliptic operator such that $a_{ij} = a_{ji} \in \mathbb{R}$ $(i,j=1,\dots,m)$; λ is a real parameter and P is a numerical function in $C^1(\mathbb{R})$ with $P(0) = P'(0) = 0$ and such that the function f defined by $f(x) = \frac{P(x)}{x}$ is a strictly increasing function on $\mathbb{R}^+$, strictly decreasing on $\mathbb{R}^-$ tending to infinity when x tends to $\pm\infty$.

We write (u,λ) each element of $H^2(\Omega) \cap H^1_0(\Omega) \times \mathbb{R}$ where u is a solution of (1.1) associated with the parameter λ.

We are interested with bifurcating solutions of (1.1) which come from the points $(0,\lambda)$. It's well known that bifurcation can appear on the point $(0,\lambda)$ only if λ belongs to the spectrum of the operator A for the Dirichlet's problem.

Let λ_0 be the first eigenvalue of A : it is a simple eigenvalue.

Proposition 1 : The point $(0,\lambda_0)$ is a bifurcating point for the problem (1.1). Two branches of bifurcation are associated with it : one of them corresponds to a positive solution, the other corresponds to a negative

solution. Moreover the positive solution is the unique non-negative non-trivial solution and the negative solution is the unique non - positive non-trivial solution.
These two solutions exist only when $\lambda > \lambda_o$ and are asymptotically stable. The trivial solution is asymptotically stable when $\lambda > \lambda_o$ and instable when $\lambda > \lambda_o$.

This proposition results essentially of the work of P.H. RABINOWITZ [4] and [5]. The uniqueness is shown by use of the methods of STAKGOLD-PAYNE [7]. See C. BOLLEY [1] for the stability of the solutions.

We can state precisely other solutions of the problem (1.1) in some particular cases.

Example 1 : $\Omega =]0,a[$ where $a > 0$; $A = -\Delta$

All the eigenvalues λ_i, $i \in \mathbb{N}$, of A are simple. Then it results of M.A.KRASNOSELSKII [3] or P.H. RABINOWITZ [4], that two branches of bifurcation appear from each point $(0,\lambda_i)$ $i \in \mathbb{N}$. The solutions which appear from $(0,\lambda_1)$ change sign once on $]0,a[$; the solutions which come from $(0,\lambda_2)$ change sign twice, etc...

Example 2 : $\Omega =]0,1[\times]0,1[$; $A = -\Delta$. Suppose f is even.

The eigenvalues of $-\Delta$ for the Dirichlet's problem, are the elements of the set $\mathcal{S} = \{\lambda(\ell,k) = \pi^2(\ell^2+k^2)$, ℓ et $k \in \mathbb{N}^*\}$

Proposition 2 : Let $\lambda_i \in \mathcal{S}$ be of multiplicity p ; then :

i) the point $(0,\lambda_i)$ is a bifurcating point and at least 2p branches of solutions of the problem (1.1) bifurcate from this point.

ii) moreover if λ_i can be written : $\lambda_i = \lambda(\ell,2\ell)$, $\ell \in \mathbb{N}^*$, at least 2p+4 branches bifurcate from the point $(0,\lambda_i)$.

For example $5\pi^2$ is a double eigenvalue of $-\Delta$, and the four following solutions

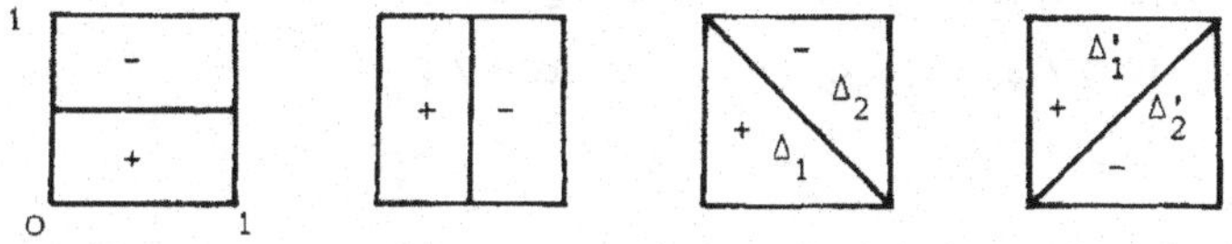

Figure 1

and the opposite ones, appear from the point $(0,5\pi^2)$.

Proof

i) if λ_i is an eigenvalue of multiplicity p, p pairs of integers (ℓ_j, k_j) $j=1,\dots,p$ exist such that :

$$\lambda_i = \lambda(\ell_j, k_j) \quad j=1,\dots,p$$

Let (ℓ_j, k_j) be one of these pairs. Ω is decomposed into $\ell_j \times k_j$ rectangles $D_{n,m}$ $(n=0,\dots,\ell_j-1\ ;\ m=0,\dots,k_j-1)$ of sides $\frac{1}{\ell_j}$ and $\frac{1}{k_j}$:

$$D_{n,m} = \left]\frac{n}{\ell_j}, \frac{n+1}{\ell_j}\right[\times \left]\frac{m}{k_j}, \frac{m+1}{k_j}\right[$$

Then λ_i is the first eigenvalue of $-\Delta$ on each rectangle $D_{n,m}$. It results from proposition 1 that one solution u_{nm} of (1.1) which is positive on D_{nm}, equal to zero on ∂D_{nm}, and the solution $-u_{nm}$, bifurcate when λ crosses the value λ_o. We show immediately, by means of symetry arguments, that the function u defined by :

for $(n,m) \in \{0,\dots,\ell_j-1\} \times \{0,\dots,k_j-1\}$, for $(x,y) \in \bar{D}_{nm}$:

$$u(x,y) = (-1)^{n+m} \, u_{nm}(x,y)$$

is a solution of (1.1) on Ω ; $-u$ is another solution.

If $(\ell_{j'}, k_{j'})$ is another pair of integers, Ω is decomposed in another way, so two different solutions are associated to that pair.

It results from proposition 1 that these solutions form 2p branches of bifurcation, when λ changes.

ii) a) Let us consider first the case where λ_i is equal to the second eigenvalue of $-\Delta$ in $\Omega =]0,1[\times]0,1[$, that is to say : $\lambda_i = 5\pi^2 = \lambda(1,2) = \lambda(2,1)$. λ_i is a double eigenvalue associated with the two following orthonormal eigenfunctions :

$$\text{for } (x,y) \in \Omega \quad \phi_1(x,y) = 2 \sin \pi x \ \sin 2\pi y$$
$$\phi_2(x,y) = 2 \sin 2\pi x \ \sin \pi y$$

$5\pi^2$ is also the first eigenvalue of $-\Delta$ in each following triangular domain :

$$\Delta_1 = \{(x,y) \in \Omega \ / \ \phi_1 + \phi_2 > 0\} = \{(x,y) \in \Omega \ / \ x+y < 1\}$$
$$\Delta_2 = \{(x,y) \in \Omega \ / \ \phi_1 + \phi_2 < 0\} = \{(x,y) \in \Omega \ / \ x+y > 1\}$$
$$\Delta_1' = \{(x,y) \in \Omega \ / \ \phi_1 - \phi_2 > 0\} = \{(x,y) \in \Omega \ / \ x-y < 0\}$$
$$\Delta_2' = \{(x,y) \in \Omega \ / \ \phi_1 - \phi_2 < 0\} = \{(x,y) \in \Omega \ / \ x-y > 0\}$$

(see figure 1).

It results from proposition 1 the existence of a solution $\tilde{u}_1$ of (1.1) on Ω which is positive on Δ_1, negative on Δ_2, and also of a solution $\tilde{u}_2$ on Ω which is positive on Δ_1', negative on Δ_2' ; $-\tilde{u}_1$ and $-\tilde{u}_2$ are two other solutions. These four solutions are added to the four constructed in i).

b) Now, if λ_i is written $\lambda_i = \lambda(\ell,2\ell)$ with $\ell > 1$, the square Ω is decomposed into ℓ^2 squares C_ℓ of side $\frac{1}{\ell}$. Then, λ_i is equal to the second eigenvalue of $-\Delta$ in each square C_ℓ : this brings us back to the preceding case a).

It results that four solutions of the form of $\tilde{u}_1$, $\tilde{u}_2$, $-\tilde{u}_1$ and $-\tilde{u}_2$ appear on each square C_ℓ, as soon as λ crosses the value λ_i ; by juxtaposition, these solutions will give some solutions in Ω only if the derivatives have the same values on the sides of the squares C_j : then, four solutions in Ω are added to the 2p solutions constructed in ii).

rollary 1 : Suppose Ω is a rectangle $\Omega_{ab} =]0,a[\times]0,b[$ a,b > 0 and λ_i belongs to

$$\mathcal{S}_{ab} = \left\{ \lambda'(\ell,k) = \pi^2\left(\frac{\ell^2}{a^2} + \frac{k^2}{b^2}\right),\ \ell, k \in \mathbb{N}^* \right\}$$

then

i) the proposition 2i) is still true.

ii) moreover if $\frac{a}{b}$ is a rational number of the form $\frac{a}{b} = \frac{p_1}{q_1}$ where p_1 and $q_1 \in \mathbb{N}^*$ and if λ_i can be written $\lambda_i = \lambda'(\ell_1 p_1, 2\ell_1 q_1)$, $\ell_1 \in \mathbb{N}^*$, at least 2q+4 branches of solutions bifurcate from the point $(0,\lambda_i)$.

For example : suppose $a = p_1 = 3$, $b = q_1 = 2$ and $\ell_1 = 1$.
n the four following solutions bifurcate from the point $(0,5\pi^2)$:

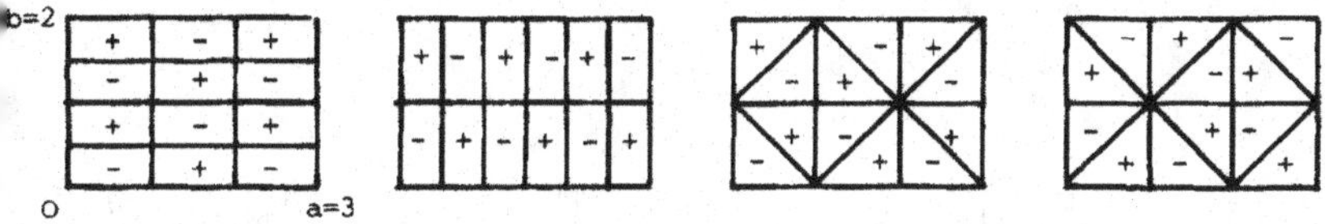

Figure 2

gether with the four opposite solutions.

oof

the proof is the same as that of the proposition 2 i).

the rectangle Ω_{ab} is decomposed into $\ell_1 p_1 \times \ell_1 q_1$ squares C of side $\frac{a}{\ell_1 p_1} = \frac{b}{\ell_1 q_1}$;
then, λ_i is equal to the second eigenvalue of $-\Delta$ in each square C;
we can end the demonstration like in ii) b of the proposition 2.

In some cases, we can state the exact number of bifurcating solutions. We need a liminary lemma. Ω is here every regular bounded open set and λ_i is a double eigenvalue of $-\Delta$ on Ω , with two associated orthonormal eigenfunctions ϕ_1 and ϕ_2.

Let (u_λ, λ_h) be a family of solutions of (1-1) such that : $||u_h||_{L^2(\Omega)}$ tends to zero and λ_h tends to λ_i when h tends to zero.

u_h can be written :

(1.2.) $$u_h = \varepsilon_h(\alpha_h \phi_1 + \beta_h \phi_2) + v_h$$

where v_h belongs to $H^2(\Omega) \cap H^1_o(\Omega) \cap M$ with $M = \left\{ v \in L^2(\Omega) / (v,\phi_i)_{L^2(\Omega)} = 0 \quad i=1,2 \right\}$ and $\alpha_h^2 + \beta_h^2 = 1$.

Even if it means having to take subsequences, we suppose that the sequences (α_h) and (β_h) converge to the values α and β when h tends to zero. Then :

Lemma 1 : Under the hypotheses stated above, α and β must satisfy :

(1.3.) $$\beta\,\big(P(\alpha\phi_1+\beta\phi_2),\phi_1\big)_{L^2(\Omega)} = \alpha\big(P(\alpha\phi_1+\beta\phi_2),\phi_2\big)_{L^2(\Omega)}$$

Proof

(u_h,λ_h) is a solution of (1.1.), so :

$$A\,u_h - \lambda_h u_h + P(u_h) = 0$$

and from (1.2.) :

(1.4.) $$\begin{cases} A\,v_h - \lambda_i v_h = (\lambda_h - \lambda_i)\,u_h - P(u_h) \\ \text{with } v_h \in H^2(\Omega) \cap H^1_o(\Omega) \cap M \end{cases}$$

Therefore :

$$||v_h||_{H^2(\Omega)\cap H^1_o(\Omega)} \leq ||(A-\lambda_i I)^{-1}||_{\mathcal{L}(L^2(\Omega)\cap M,\,H^2(\Omega)\cap H^1_o(\Omega)\cap M)} \cdot \big(|\lambda_h-\lambda_i|\cdot||u_h||_{L^2(\Omega)} + ||P(u_h)||_{L^2(\Omega)}\big)$$

$$\leq C\,\big(\,|\lambda_h-\lambda_i| + ||f(u_h)||_{L^2(\Omega)}\big)\cdot ||u_h||_{L^2(\Omega)}$$

and

(1.5.) $$\frac{||v_h||_{H^2(\Omega)\cap H^1_o(\Omega)}}{||u_h||_{L^2(\Omega)}} \quad \text{tends to zero when h tends to zero}$$

Moreover, the problem (1.4.) is uniquely solvable if and only if

$$((\lambda_h - \lambda_i)\,u_h - P(u_h)\,,\,\phi_j)_{L^2(\Omega)} = 0 \qquad \text{for } j = 1,2$$

or, which is equivalent :

$$\begin{cases} (\lambda_h - \lambda_i)\,\varepsilon_h\,\alpha_h = (P(u_h,\phi_1)_{L^2(\Omega)} \\ (\lambda_h - \lambda_i)\,\varepsilon_h\,\beta_h = (P(u_h),\phi_2)_{L^2(\Omega)} \end{cases}$$

Thus :

$$\beta_h(P(u_h),\phi_1)_{L^2(\Omega)} = \alpha_h(P(u_h),\phi_2)_{L^2(\Omega)}$$

and from (1.5.), we have

$$\beta(P(\alpha\phi_1 + \beta\phi_2),\phi_1)_{L^2(\Omega)} = \alpha(P(\alpha\phi_1+\beta\phi_2),\phi_2)_{L^2(\Omega)}$$

when h tends to zero.

Proposition 3 : Let $\Omega =]0,1[\times]0,1[$
Suppose λ_i is a double eigenvalue of $-\Delta$ such that $\lambda_i = \lambda(\ell,2\ell)$
$\ell \in \mathbb{N}^*$ and i) $P(x) = x^{2n+1}$ $n \in \mathbb{N}^*$

or ii) $P(x) = x|x|$

then exactly eight solutions bifurcate from the point $(0,\lambda_i)$.

Proof

The eigenvalue λ_i is associated with the two orthonormal eigenfunctions :

$$\begin{aligned} \phi_1(x,y) &= 2\sin \ell\pi x \quad \sin 2\ell\pi y \\ \phi_2(x,y) &= 2\sin 2\ell\pi x \quad \sin \ell\pi y \end{aligned} \qquad \text{for } (x,y)\in\Omega$$

From lemma 1, it is necessary to find the pairs (α,β) satisfying (1.3.) to determine the bifurcating solutions.

In case i) the equation (1.3.) is written :

$$\sum_{m=0}^{2n+1} \gamma^{m+1}\, C^m_{2n+1} \iint_\Omega \phi_1^{2n+2-m}\,\phi_2^m dx\,dy = \sum_{m=0}^{2n+1} \gamma^m\, C^m_{2n+1} \iint_\Omega \phi_1^{2n+1-m}\,\phi_2^{m+1}\,dx\,dy$$

when $\alpha \neq 0$ and with $\gamma = \frac{\beta}{\alpha}$

Let us calculate $a_m = \iint_\Omega \phi_1^{2n+2-m}\,\phi_2^m\,dx\,dy$ for $m = 0,\dots, 2n+2$

$$a_m = 2^{4n+4}\int_0^1 (\sin\ell\pi x)^{2n+2}\,(\cos\ell\pi x)^m dx.\int_0^1 (\sin\ell\pi y)^{2n+2}(\cos\ell\pi y)^{2n+2-m}\,dy$$

therefore, when m is odd, $a_m = 0$: that shows in particular that the pairs $(\alpha,\beta)=(1,0)$ and $(0,1)$ are solutions.

When m is even, $m = 2k$:

$$a_{2k} = 2^{4n+6} \int_0^{1/2} (\sin \pi x)^{2n+2} (\cos \pi x)^{2k} dx \int_0^{1/2} (\sin \pi y)^{2n+2} (\cos \pi y)^{2n+2-m} dy$$

$$= \frac{2^{4n+4}}{\pi^2} \quad \beta(n + \frac{3}{2}, k + \frac{1}{2}) \cdot \beta(n + \frac{3}{2}, n + \frac{3}{2} - k)$$

$$= \frac{2^{4n+4}}{\pi^2} \left[\Gamma(n + \frac{3}{2}) \right]^2 \frac{(2n+2-2k)! \quad (2k)!}{(n+1-k)! \ (2n+2-k)! \ k! \ (n+1+k)!}$$

where Γ and β are the eulerienne functions of first and second kind.

Therefore, equation (1.3.) is equivalent to

$$(1.6.) \quad \begin{cases} \sum_{k=0}^{n} \gamma^{2k+1} \quad c_k = 0 \\ \text{with} \quad c_k = \dfrac{2k-n}{k!(n-k)! \ (n+k+2)! \ (2n+2-k)!} \end{cases}$$

Now, we have :

$$c_k = - c_{n-k} \qquad k = 0,\ldots,n$$

$$\text{and} \quad c_k < 0 \qquad \text{when } k < \frac{n}{2}$$

Therefore :

$$(1.6.) \iff \sum_{k=0}^{\left[\frac{n-1}{2}\right]} c_k \ \gamma^{2k+1}(\gamma^{2n-4k} - 1) = 0$$

Or $$\sum_{k=0}^{\left[\frac{n-1}{2}\right]} c_k \ \gamma^{2k+1}(\gamma^{2n-4k-2} + \gamma^{2n-4k-4} + \ldots + 1) \ (\gamma^2 - 1) = 0 \qquad \text{with } c_k < 0.$$

The only values of γ satisfying that equation are 0,1 and -1 ; therefore, the pairs (α,β) solutions when h tend to zero are : (0,1), (0,-1), (1,0), (-1,0), $(\frac{1}{\sqrt{2}}, \frac{1}{\sqrt{2}})$, $(-\frac{1}{\sqrt{2}}, -\frac{1}{\sqrt{2}})$, $(\frac{1}{\sqrt{2}}, -\frac{1}{\sqrt{2}})$ and $(-\frac{1}{\sqrt{2}}, \frac{1}{\sqrt{2}})$ that is to say the eight solutions found in the proposition 2 ii).

In case ii), the equation (1.3.) is written :

$$\int_\Omega (\phi_1 + \gamma\phi_2) \ |\phi_1 + \gamma\phi_2| \ (\gamma\phi_1 - \phi_2) \ dx \ dy = 0$$

or (1.7.) $$\int_{\Omega_1} (\phi_1 + \gamma\phi_2)^2 \ (\gamma\phi_1 - \phi_2) dx \ dy - \int_{\Omega_2} (\phi_1 + \gamma\phi_2)^2 (\gamma\phi_1 - \phi_2) \ dx \ dy = 0$$

with

$$\Omega_1 = \{ (x,y) \in \Omega \ / \ \phi_1 + \gamma\phi_2 > 0 \}$$

$$\Omega_2 = \{ (x,y) \in \Omega \ / \ \phi_1 + \gamma\phi_2 < 0 \}$$

Suppose γ is included between 0 and 1 : the other solutions will be deduced by symetry or exchange of the variables. We have :

$$\int_{\Omega_1} \phi_1^3 \, dx\, dy = 8\int_0^1 (\sin\pi x)^3 \int_0^{\frac{1}{\pi}.\mathrm{Arc}\cos(-\gamma\cos x)} (\sin 2\pi y)^3 \, dy\, dx$$

$$= \frac{64}{9\pi^2} - \frac{64}{35\pi^2}\gamma^4 + \frac{128}{189\pi^2}\gamma^6$$

$$= -\int_{\Omega_2} \phi_1^3 \, dx\, dy = -8\int_0^1 (\sin\pi x)^3 \int_{\frac{1}{\pi}.\mathrm{Arc}\cos(-\gamma\cos\pi x)}^1 (\sin 2\pi y)^3 \, dy\, dx$$

$$\int_{\Omega_1} \phi_1^2\,\phi_2 dx\, dy = 8\int_0^1 (\sin\pi x)^2 \sin 2\pi x \int_0^{\frac{1}{\pi}.\mathrm{Arc}\cos(-\gamma\cos\pi x)} (\sin 2\pi y)^2 \sin\pi y \, dy\, dx$$

$$= \frac{256}{105\pi^2}\gamma^3 - \frac{256}{315\pi^2}\gamma^5 = -\int_{\Omega_2} \phi_1^2\,\phi_2 \, dx\, dy$$

$$\int_{\Omega_1} \phi_1\,\phi_2^2 \, dx\, dy = -\int_{\Omega_2} \phi_1\,\phi_2^2 \, dx\, dy = \frac{64}{15\pi^2} - \frac{128}{35\pi^2}\gamma^2 + \frac{64}{63\pi^2}\gamma^4$$

and at last :

$$\int_{\Omega_1} \phi_2^3 \, dx\, dy = -\int_{\Omega_2} \phi_2^3 \, dx\, dy = \frac{256}{35\pi^2}\gamma - \frac{256}{189\pi^2}\gamma^3$$

Hence the following equation :

$$\gamma(\gamma^2 - 1)\,(\gamma^4 - 6\gamma^2 + 21) = 0$$

The only solutions included between 0 and 1 are $\gamma = 0$ and $\gamma = 1$. Therefore, we obtain the eight pairs (α,β) of the case i) : the only solutions of (1.1.) are the eight solutions constructed in the proposition 2 ii).

<u>Corollary 2</u> : Suppose Ω is a rectangle $\Omega_{ab} =]0,a[\times]0,b[$ where $\frac{a}{b}$ is a rational number of the form $\frac{a}{b} = \frac{p_1}{q_1}$, p_1 and $q_1 \in \mathbb{N}^*$, and λ_i is a double eigenvalue of $-\Delta$ in Ω_{ab} such that $\lambda_i = \lambda'(\ell_1 p_1, 2\ell_1 q_1)$ $\ell_1 \in \mathbb{N}^*$.

if i) $P(x) = x^{2n+1}$ $n \in \mathbb{N}^*$

or ii) $P(x) = x|x|$

then exactly eight solutions bifurcate from the point $(0,\lambda_i)$.

<u>Proof</u>

Let $C = \frac{p_1}{a} = \frac{q_1}{b}$. The eigenvalue λ_i is associated with the two following orthonormal eigenfunctions :

$$\tilde{\phi}_1(x,y) = \frac{2}{ab} \sin \ell_1 C\pi x \quad \sin 2\ell_1 C\pi\, y$$

$$\tilde{\phi}_2(x,y) = \frac{2}{ab} \sin 2\ell_1 C\pi x \quad \sin \ell_1 C\pi\, y \qquad \text{for } (x,y) \in \Omega_{ab}$$

In case i), we find immediately that for $m = 0,\dots, 2n+2$:

$$\tilde{a}_m = \iint_{\Omega_{ab}} \tilde{\phi}_1^{2n+2-m}\ \tilde{\phi}_2 \, dx\, dy = \frac{1}{(ab)^{2n+1}}\ a_m$$

Therefore, the equation (1.3.) is, in this case, equivalent to the one obtained under the hypotheses of the proposition 3 i).

In case ii), let :

$$\bar{\Omega}_{ab} = \prod_{n=0}^{\ell_1 p_1 - 1} \prod_{m=0}^{\ell_1 q_1 - 1} C_{nm}$$

where

$$C_{nm} = \left[\frac{n}{C\ell_1}, \frac{n+1}{C\ell_1}\right] \times \left[\frac{m}{C\ell_1}, \frac{m+1}{C\ell_1}\right]$$

The equation (1.3.) is written here :

$$\sum_{n=0}^{\ell_1 p_1 - 1} \sum_{m=0}^{\ell_1 q_1 - 1} \left[\int_{C^1_{nm}} (\tilde{\phi}_1 + \gamma\tilde{\phi}_2)^2 (\gamma\tilde{\phi}_1 - \tilde{\phi}_2)\, dx\, dy - \int_{C^2_{nm}} (\tilde{\phi}_1 + \gamma\tilde{\phi}_2)^2\ (\gamma\tilde{\phi}_1 - \tilde{\phi}_2)\, dx\, dy\right] = 0$$

where $C^1_{nm} = \{ (x,y) \in C_{nm} \ /\ \tilde{\phi}_1 + \gamma\tilde{\phi}_2 > 0 \}$

$C^2_{nm} = \{ (x,y) \in C_{nm} \ /\ \tilde{\phi}_1 + \gamma\tilde{\phi}_2 < 0 \}$

We suppose that $\gamma = \frac{\beta}{\alpha}$ is included between 0 and 1. We have in the square C_{nm}:

. when n et m are even :

$$\tilde{\phi}_1 + \gamma\tilde{\phi}_2 > 0 \iff y < \frac{1}{\ell_1 C\pi} \text{ Arc cos } (-\gamma \cos \ell_1 C\pi x) + \frac{m}{\ell_1 C} \equiv g^1_{m,\gamma}(x)$$

. when n is even and m is odd :

$$\tilde{\phi}_1 + \gamma\tilde{\phi}_2 > 0 \iff y < \frac{m+1}{\ell_1 C} - \frac{1}{\ell_1 C\pi} \text{ Arc cos } (-\gamma \cos \ell_1 C\pi x) \equiv g^2_{m,\gamma}(x)$$

. when n is odd and m is even

$$\tilde{\phi}_1 + \gamma\tilde{\phi}_2 > 0 \iff y > g^1_{m,\gamma}(x)$$

. when n and m are odd :

$$\tilde{\phi}_1 + \gamma\tilde{\phi}_2 > 0 \iff y > g^2_{m,\gamma}(x)$$

The sign of $\tilde{\phi}_1 + \gamma\tilde{\phi}_2$ is given by :

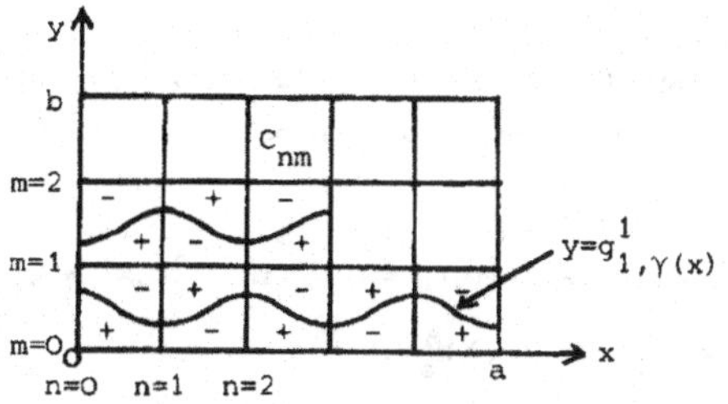

Figure 3

Let $I^i_{nm} = \int_{C^i_{nm}} (\tilde{\phi}_1 + \gamma\tilde{\phi}_2)^2 \, (\gamma\tilde{\phi}_1 - \tilde{\phi}_2)\, dx\, dy \qquad i = 1,2$

when m is even, $m = 2m'$, let :

$$x' = \ell_1 C\, x - n \quad \text{and} \quad y' = \ell_1 C\, y - 2m'$$

then, with the notations of the proof of the proposition 3, and for $i = 1,2$:

$$I^i_{2n',2m'} = \frac{1}{(ab)^3} \frac{1}{(\ell_1 C)^2} \int_{\Omega_1} (\phi_1 + \gamma\phi_2)^2 \, (\gamma\phi_1 - \phi_2) \; dx'\, dy' \quad \text{where } n = 2n'$$

$$I^i_{2n'+1,2m'} = \frac{-1}{(ab)^3} \frac{1}{(\ell_1 C)^2} \int_{\Omega_{3-i}} (\phi_1 + \gamma\phi_2)^2 \, (\gamma\phi_1 - \phi_2)\, dx'\, dy' \quad \text{where } n' = 2n+1$$

When m is odd, $m = 2m'+1$, let :

$$x' = \ell_1 C\, x - n \quad , \quad y' = 2m' + 2 - \ell_1 C\, y$$

then :

$$I^i_{2n',2m'+1} = - \frac{1}{(ab)^3} \frac{1}{(\ell_1 C)^2} \int_{\Omega_{3-i}} (\phi_1 + \gamma\phi_2)^2 \, (\gamma\phi_1 - \phi_2)\, dx'\, dy' \quad \text{where } n = 2n'$$

$$I^i_{2n'+1,2m'+1} = \frac{1}{(ab)^3} \frac{1}{(\ell_1 C)^2} \int_{\Omega_i} (\phi_1 + \gamma\phi_2)^2 \, (\gamma\phi_1 - \phi_2)\, dx'\, dy' \quad \text{where } n = 2n'+1$$

Therefore, the equation (1.3.) is equivalent to the one obtained in case ii) of the proposition 3.

II - NUMERICAL CALCULATION OF THE SOLUTIONS OF THE PROBLEM (1.1.)

In this section, we shall deal with the numerical calculation of the solutions of the problem (1.1.), and more particularly with the stable solutions.

Let us remember that an asymptotic stable solution is, from the definition, the limit in $L^\infty(\Omega)$ when t tends to infinity, of the solution u(t) of the evolution problem :

$$(2.1.)\quad \begin{cases} \frac{\partial}{\partial t} u(t) + Au(t) - \lambda u(t) + P(u(t)) = 0 & \text{in }]0,\infty[\times \Omega \\ u(0) = u_o \quad \text{given in } H^2(\Omega) \cap H_o^1(\Omega) \\ u(t) \in C^1([0,\infty[\,;\, L^2(\Omega)) \cap C^o([0,\infty[;\ H^2(\Omega) \cap H_o^1(\Omega)) \end{cases}$$

We discretise in time this evolution problem. The scheme is the following :

$$(2.2.)\quad \begin{cases} \text{given } u_o \in L^\infty(\Omega) \ , \\ u_{n+1} \in H^2(\Omega) \cap H_o^1(\Omega) \quad \text{is constructed such that :} \\ \dfrac{u_{n+1} - u_n}{\Delta t} + A\, u_{n+1} - \lambda u_{n+1} + f(u_n)\, u_{n+1} = 0 \quad \text{in } \Omega \end{cases}$$

which is equivalent to :

$$(2.2)' \qquad A\, u_{n+1} + \left(\frac{1}{\Delta t} - \lambda + f(u_n) \right) u_{n+1} = \frac{u_n}{\Delta t} \quad \text{in } \Omega$$

We show immediately that if u_o belongs to $L^\infty(\Omega)$ and if $\Delta t > 0$ verifies $\Delta t(\lambda - \lambda_o) < 1$, then there exists a unique sequence (u_n) belonging to $H^2(\Omega) \cap H_o^1(\Omega)$, solution of the equations (2.2.) ; moreover this sequence belongs to $W^{2,p}(\Omega) \cap \overset{\circ}{W}{}^{1,p}(\Omega)$ for $p \in]1,\infty[$ and $n \geq 1$.

That scheme is studied with more details in C. BOLLEY [1].

The first result of convergence is the following :

<u>Proposition 4</u> :

a) When $\lambda \leq \lambda_o$: given $u_o \in L^\infty(\Omega)$, there exists $\Delta t_o > 0$ such that for all $\Delta t > 0$ with $0 < \Delta t \leq \Delta t_o$, then the sequence (u_n) converges to zero in $W^{2,p}(\Omega)$ for all $p \in]1,+\infty[$.

b) When $\lambda > \lambda_o$: given $u_o \in L^\infty(\Omega)$, bounded from below by a lower solution ϕ of (1.1.) which is non negative, non trivial, then there exists $\Delta t_1 > 0$ such that for all Δt with $0 < \Delta t \leq \Delta t_1$, the sequence (u_n) converges to the positive solution of (1.1.) in $W^{2,p}(\Omega)$, for all $p \in]1,+\infty[$.

In each case, the limit of the sequence is the unique non negative solution which is asymptotically stable.

Proof

In case a) there exists a lower-solution ϕ which bounds u_o from below on $\bar{\Omega}$: we can choose a sufficiently small negative constante ; and in case b) one such lower-solution which is non-negative, non-trivial exists by hypotheses. In both cases, there exists an upper-solution ψ which bounds u_o from above : we can choose a sufficiently large positive constant.

By means of the maximum principle, we show that for all sufficiently small Δt, and for all $n \in \mathbb{N}$:

$$\phi \leqslant u_n \leqslant \psi \quad \text{in } \Omega$$

Therefore, the sequence (u_n) is bounded in $L^\infty(\Omega)$ and from the equation (2.2.), it is bounded in $W^{2,p}(\Omega) \cap \overset{\circ}{W}{}^{1,p}(\Omega)$ for $p \in]1,+\infty[$. We can extract from it, a subsequence $(u_{n_k})_k$ which weakly converges in $W^{2,p}(\Omega) \cap \overset{\circ}{W}{}^{1,p}(\Omega)$ to an element u. By taking the limit in the equation (2.2.) for the sequence (u_{n_k}), we show that u satisfies to:

$$\begin{cases} Au - \lambda u + P(u) = 0 & \text{in } \Omega \\ u \in H^2(\Omega) \cap H^1_o(\Omega) \end{cases}$$

Therefore :

when $\lambda \leqslant \lambda_o$: u is the trivial solution of (1.1.)

when $\lambda > \lambda_o$: u being bounded from below like all the sequence (u_{n_k}) by the function ϕ , u is the positive solution of (1.1.).

Then we easily show that in both cases, all the whole sequence converges in $W^{2,p}(\Omega)$.

When the problem (1.1.) satisfies a supplementary hypothesis about the number of its solutions, we have :

Proposition 5 : Let $u_o \in L^\infty(\Omega)$, ϕ a lower-solution of (1.1.) and ψ an upper-solution of (1.1.) such that $\phi \leqslant u_o \leqslant \psi$ on Ω. If the problem (1.1.) has at most a finite number of solutions between ϕ and ψ , then there exists $\Delta t_2 > 0$ such that for all Δt with $0 < \Delta t \leqslant \Delta t_2$, the sequence (u_n) converges in $W^{2,p}(\Omega)$ for $p \in]1,+\infty[$, to a solution of (1.1.) which is between ϕ and ψ.

The proof of the proposition 5 is based on the following lemma.

Lemma 2 : Let (y_n) be a sequence belonging to a metric space E and such that :

i) from each subsequence of the sequence (y_n), we can select a converging subsequence.

ii) the sequence (y_n) has a finite number of points of closure.

iii) the sequence $(d(y_{n+1}, y_n))$ tends to zero in E when n tends to infinity.

Then the sequence (y_n) converges in E.

The proof of the lemma is immediate.

We have another result of convergence to stable solutions :

Proposition 6 : Let u be a solution of problem (1.1.) such that the eigenvalues of the operator $A - \lambda I + P'(u) I$, which is defined from $H^2(\Omega) \cap H^1_o(\Omega)$ to $L^2(\Omega)$, be strictly positive. If $\Delta t > 0$ satisfies $\Delta t(\lambda - \lambda_o) < 1$ and $\Delta t(P'(u) - f(u)) \leqslant 1$ on $\bar{\Omega}$ then u is an attractive point in $H^2(\Omega) \cap H^1_o(\Omega)$ for the iterative scheme defined by the equations (2.2.).

Remark : It results from the theorem of Lyapunov (see D.H. SATTINGER [6]) that under the hypotheses of the proposition 6, u is a stable solution of the problem (1.1.)

Proof of the proposition 6

Consider the operator T defined from $H^2(\Omega) \cap H^1_o(\Omega)$ into itself, as soon as $\Delta t(\lambda - \lambda_o) < 1$, by :

$$\forall\, v \in H^2(\Omega) \cap H^1_o(\Omega) \qquad T(v) = \frac{1}{\Delta t}\left(\frac{1}{\Delta t} I + A - \lambda I + f(v) I\right)^{-1} v$$

The equation (2.2.) is written :

$$u_{n+1} = T(u_n) \quad .$$

U is a fixed point of T if and only if u is a solution of (1.1.).

We show that T has a Frechet derivative $T'(u) \in \mathcal{L}(H^2(\Omega) \cap H^1_o(\Omega))$ at each fixed point of T ; T'(u) is defined by :

$$T'(u) = \left(A - \lambda I + \frac{1}{\Delta t} I + f(u)\, I\right)^{-1} \left(\frac{1}{\Delta t} I - P'(u)\, I + f(u)\, I\right) .$$

Under the hypotheses of the proposition 6, the spectral radius $\rho(T'(u))$ of T'(u) is stricly smaller than 1 : so it results from the theorem of Ostrowski (see KITCHEN [2]) that u is an attractive point in $H^2(\Omega) \cap H^1_o(\Omega)$ for the iterative scheme defined by the equation (2.2.), that is to say that there exists a neighbourhood S of u in $H^2(\Omega) \cap H^1_o(\Omega)$ such that if u_o belongs to S, the sequence (u_n) converges to u in $H^2(\Omega) \cap H^1_o(\Omega)$.

Under other hypotheses on Δt, we can show that the sequence (u_n) converges to an instable solution of (1.1.) More precisely :

<u>Proposition 7</u> : If λ is not an eigenvalue of A et if Δt satisfies

$$\Delta t > \frac{2}{\lambda-\lambda_i} \quad (i \in \mathbb{N}) \quad \text{and}$$

$$\Delta t > \|(A - \lambda I)^{-1}\|_{\mathcal{L}(L^2(\Omega), H^2(\Omega) \cap H^1_o(\Omega))}$$

where λ_i, $i \in \mathbb{N}$, are the eigenvalues of A, then the trivial solution is an attractive point in $H^2(\Omega) \cap H^1_o(\Omega)$ for the iterative scheme defined by the equations (2.2.).

Here we are no more under the hypothesis $\Delta t(\lambda - \lambda_o) < 1$ which gives the existence and uniqueness of the sequence (u_n). However we show that the operator T of the proposition 6, and its Frechet derivative T'(O), are still defined. We have :

$$T'(o) = (\Delta t\, A - \lambda\Delta t\, I + I)^{-1}$$

it's a compact operator belonging to $\mathcal{L}(H^2(\Omega) \cap H^1_o(\Omega))$. Let μ_i, $i \in \mathbb{N}$, be its eigenvalues, then :

$$\rho(T'(O)) = \sup_{i \in \mathbb{N}} |\mu_i| = \sup_{i \in \mathbb{N}} \left| \frac{1}{1 - \Delta t(\lambda - \lambda_i)} \right| < 1$$

We use, now, the theorem of Ostrowski.

III - <u>NUMERICAL RESULTS</u>

a) First we treat the following problem :

$$(3.1.) \quad \begin{cases} - u'' - \lambda u + u|u| = 0 & \text{in } \Omega =]0,1[\\ u \in H^2(0,1) \cap H^1_o(0,1). \end{cases}$$

The scheme given by the equations (2.2.) is completed by a discretization in space : a finite difference method is used.

First, the initial data is $u_o(x) = 20\, x(1-x)$ for $x \in [0,1]$. Given some values of λ, we make λ change.

<u>When $\lambda = 0$</u> $(\lambda < \lambda_o = \pi^2)$, the method converges to the zero solution. The convergence is all the faster as Δt is larger.

When $\Delta t = 10$, u_{20} is equivalent to 10^{-40}.

When $\lambda > \lambda_o$: we have the following table :

λ	Δt	Behaviour of the sequence u_n
10	0,3	very slow convergence to the positive solution
	5	convergence with 4 iterations to the positive solution
	7	convergence with 16 iterations to the positive solution
	9	unstabilities
	15	convergence to zero
15	0,1	very slow convergence to the positive solution
	0,2	convergence with 4 iterations to the positive solution
	0,3	convergence with 16 iterations to the positive solution
	0,4	unstabilities
	0,5	convergence to zero

Figures 4 and 5 show the variations of the positive solution with λ .

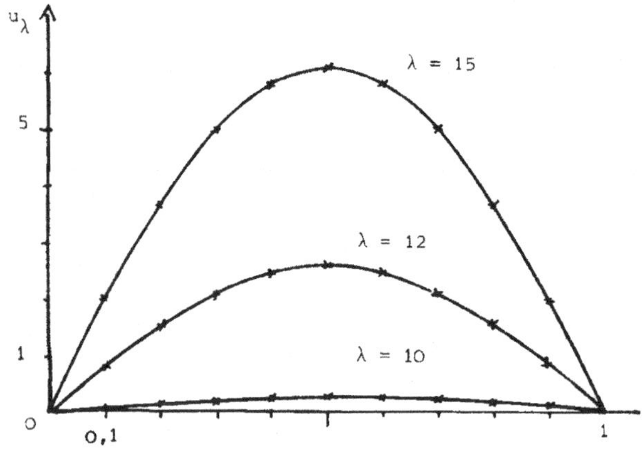

Figure 4

The positive solution for some values of λ .

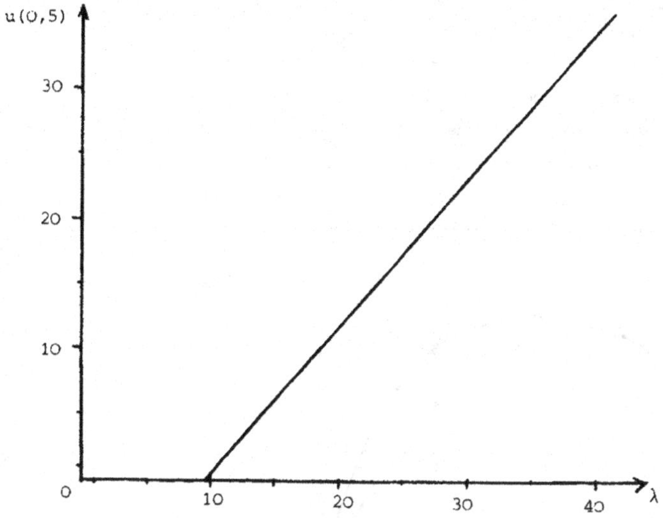

Figure 5

Variations of the positive solution at the point 0,5 with λ

b) Research of solutions which change sign in Ω

We treat the problem (3.1.) for λ in a neighbourhood of the second, then the third eigenvalue of $-\Delta$, that is to say $4\pi^2$ and $9\pi^2$.

At the point $(0,4\pi^2)$, there appears a solution $\hat{u}$ which is negative in $]0,\frac{1}{2}[$, positive in $]\frac{1}{2},1[$ and zero on $\frac{1}{2}$. $-\hat{u}$ is another solution of (3.1.) ; these are instable solutions.

When $\lambda = 40$ and when u_o changes of sign on the point 0,6, we have :

Δt	Behaviour of the sequence u_n
0,05	Convergence to the positive solution
0,5	Convergence with 5 iterations to the solution $\hat{u}$
1,5	Divergence
2,5	Convergence to zero

At the point $(0,9\pi^2)$, there appears a solution $\tilde{u}$ which is negative in $]0,\frac{1}{3}[$ and $]\frac{2}{3},1[$, positive in $]\frac{1}{3},\frac{2}{3}[$ and zero at the points $\frac{1}{3}$ and $\frac{2}{3}$.

From an u_o which changes sign once on Ω, at the point 0,6, we have, if $\lambda = 90$.

Δt	Behaviour of the sequence u_n
0,01	Convergence to the positive solution
0,03	Convergence to the solution $\hat{u}$
0,1	Convergence to the solution $\tilde{u}$
0,5	Convergence to zero

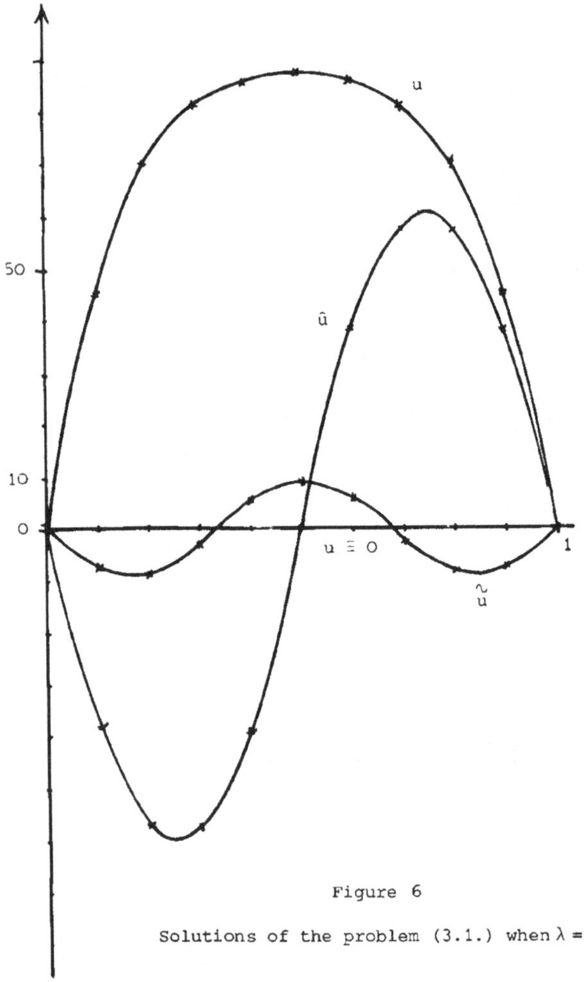

Figure 6

Solutions of the problem (3.1.) when $\lambda = 90$

c) Case where the operator A has multiple eigenvalues

We consider the problem (3.1.) in a square :

$$(3.2.)\quad \begin{cases} -\Delta u - \lambda u + u|u| = 0 & \text{in} \quad \Omega =]0,1[\times]0,1[\\ u \in H^2(\Omega) \cap H_0^1(\Omega) \end{cases}$$

From propositions 2 and 3 there appears a positive solution and a negative solution from the point $(0,2\pi^2)$, then solutions represented by

from the point $(0,5\pi^2)$, the solutions 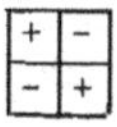 and 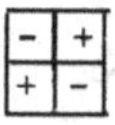 from the point $(0,8\pi^2)$, and so on ...

Like in the previous examples, if u_o is positive in Ω and if $\lambda > \lambda_o = 2\pi^2$, the sequence (u_n) converges to the positive solution of (3.2.) when $\Delta t < \frac{1}{\lambda - \lambda_o}$, and to the zero solution when Δt is larger.

Now if u_o is negative in the square $\Omega' =]0,3;0,7[\times]0,3;0,7[$, and positive in $\Omega \setminus \bar{\Omega}'$, we have the following results :

When $\lambda = 80$

Δt	Behaviour of the sequence u_n
0,01	Convergence to the positive solution
0,04	Convergence to the solution 
0,2	Convergence to the solution 
0,3	Convergence to the solution 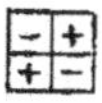
0,8	Convergence to zero

When $\lambda = 180$

Δt	Behaviour of the sequence u_n
0,005	Convergence to the positive solution
0,04	Convergence to the solution 
0,1	Convergence to the solution 
0,2	Convergence to the solution 
0,5	Convergence to zero

BIBLIOGRAPHY

[1] C. BOLLEY *"Solutions numériques de problèmes de bifurcation"*. To appear.

[2] J. KITCHEN *"Concerning the Convergence of Iterates to Fixed Points"*. Studia Math 27, pp. 247-249. 1966.

[3] M.A. KRASNOLESKII *"Topological Methods in the Theory of Nonlinear Integral Equations"*. Macmilan, New York, 1965.

[4] P.H. RABINOWITZ *"Some Global Results for Nonlinear Eigenvalue Problems"*. J. of Funct. Anal. 7, pp. 487-513, 1971.

[5] P.H. RABINOWITZ *"On Bifurcation from Infinity"*. J of Diff. Equ. 14, pp. 462-475, 1973.

[6] D. SATTINGER *"Monotone Methods in Nonlinear Elliptic and Parabolic Boundary Value Problems"*. Indiana Univ. Math. J. Vol. 21, N° 11, 1972.

[7] I. STAKGOLD, L.E. PAYNE *"Nonlinear Problems in Nuclear Reactor Analysis"*. Springer Verlag, Lecture Notes N° 322.

ON NONLINEAR EIGENVALUE PROBLEMS WHICH EXTEND INTO FREE BOUNDARIES PROBLEMS

C.M. BRAUNER(*) and B. NICOLAENKO(**)

==========

Summary

We consider a class of nonlinear elliptic eigenvalue problems, in an arbitrary bounded regular domain in $\mathbb{R}^n$, with multiple bending points (infinite in some cases). We associate with them a family of perturbed problems ; the study of the corresponding singular perturbation enables us to extend the limiting elliptic problem into a free boundary problem. The latter also admits an infinite number of free boundary solutions in some cases of hyperspherical geometries.

0. INTRODUCTION

The study of isothermal reaction models in chemical catalyst kinetics leads to the following Langmuir - Hinshelwood model where absorption is taken into effect

(0.1) $$\Delta u = \lambda u^m \left(\frac{\varepsilon+1}{\varepsilon+u}\right)^{m+k}, \quad u/_{\partial\Omega} = 1, \text{ where } u \geqslant 0;\ \lambda > 0,\ k > 0,\ m \geqslant 1,$$

and ε small > 0 correspond to absorption parameters. Ω is a bounded regular domain in $\mathbb{R}^n$.

Similarly, in enzyme kinetics, one is led to the consideration of

(0.1)' $$\Delta u = \lambda \frac{u^m}{\varepsilon + u^{m+k}}, \quad u/_{\partial\Omega} = 1$$

(*) Laboratoire de Mathématiques-Informatique-Systèmes, Ecole Centrale de Lyon, 69130 Ecully, France.

(**) Département de Mathématiques, Université de Paris-Sud, 91405 Orsay, France, and University of California, Math. Anal. group T7, Los Alamos Scientific Laboratory, 87545 Los Alamos, New Mexico USA.

and similar conditions on the parameters.

Typically, Ω is a catalyst pellet or an enzyme membrane ([3] [4] [7]), and u represents the concentration of a reactant or a substrate diffusing across the boundary $\partial\Omega$. The non-homogenous boundary conditions reflect given non trivial concentrations at the boundary.

The nonlinear eigenvalue problems (0.1) (0.1)' are related to non-convex, non-monotone nonlinearities, and are not of classical types investigated in CRANDALL-RABINOWITZ [17] [18], KEENER-H.B. KELLER [19].

Formally, the limit problem associated to (0.1) (0.1)' when ε is set equal to zero is

$$(0.2) \qquad \Delta u = \frac{\lambda}{u^k} \quad , \quad u/_{\partial\Omega} = 1 \quad , \quad k > 0 \; ,$$

which belongs to the class of generalized Emden-Fowler equations (ARIS [3]). For an arbitrary domain, problem (0.2) has first been investigated in [10](1). The first section of this paper will present a survey of these results. In particular, we demonstrate the existence of a critical λ_* , $0 < \lambda_* < +\infty$, such that (0.2) does not admit any classical solution $\in C^{2,\alpha}(\bar{\Omega})$ for every $\lambda > \lambda_*$. Indeed by reformulating (0.2) into a problem with Dirichlet conditions, we recover a strictly monotone and stricly convex nonlinearity, which is not however a continuous application of $\mathbb{R}^+$ into $\mathbb{R}^+$.

Classicaly there exists a maximal branch solution of (0.2), $A = (\lambda, u)$, $(0, 1) \in A$, which can be continued into a simply connected component in the sense of LERAY-SCHAUDER [21] and RABINOWITZ [27]. This component is unbounded in $C^{2,\alpha}(\bar{\Omega})$, because of the existence of λ_* . More precisely, along this component, min $(u(\lambda))$ goes to zero, and the Laplacian becomes unbounded. As seen in [10], this can happen at λ_* . Only under further conditions on n and k, can λ_* be a regular bending point in $C^{2,\alpha}(\bar{\Omega})$.

Problems (0.1) and (0.2) are fundamentaly different ; indeed one can easily verify that, for fixed ε, (0.1) admits a regular solution for every λ. We demonstrate the stronger result that the perturbed problem possesses a simply connected branch which extends a maximal branch, locally unique in a neighbourhood of $\lambda = 0$, $u = 1$. This component is defined for every finite λ, and its projection along $\mathbb{R}^+$ is unbounded ([21] [27]). Thus we can expect a singular limit of the component for $\lambda > \lambda_*$ as $\varepsilon \to 0$.

---

(1) See also [8], [9].

This paper is devoted to a thorough investigation of this unusual singular perturbation problem.

In the first part, we investigate the limit problem (0.2). We recall our previous general results for an arbitrary domain concerning the maximal branch A and its behaviour in the neighbourhood of $\lambda = \lambda_*$. We present detailed global results for spherical geometries (see also [11]) and show that, at least in the latter case, the simply connected component extending A globally converges to an <u>endpoint</u> (λ_c, u_c) while unbounded in $C^{2,\alpha}(\bar{\Omega})$. We give a general definition of an endpoint which covers previously known cases (LERAY [20], CALLEGARI-H.B. KELLER-REISS [15], STEWARTSON [29]).

In the second part, we present general results on the perturbed problem (0.1). In a third part, we give a priori estimates on the convergence of branches as $\varepsilon \to 0$. The fundamental difficulty of the perturbation problem is that different branches do not have the same asymptotic behaviour as $\varepsilon \to 0$: For $\lambda < \lambda_*$, some are regular yet for every $\lambda > \lambda_*$ all branches are singular. Classical a priori estimates are too coarse to distinguish between the two cases. In fact there is a dichotomy between $k < 1$ and $k \geqslant 1$.

In a fourth part we demonstrate singular convergence of all solutions of (0.1) to a null limit, for $\lambda > \lambda_*$, large enough, but only for $k \geqslant 1$ (There is a classical boundary layer situation). Nevertheless, because of the above mentionned indeterminacy, we first must prove the result for spherical domains ; then we use a general lemma of comparison ([S]) to establish the result for more general domains.

In a fifth part, for $0 < k < 1$, $\lambda > \lambda_*$, large enough, we demonstrate that branches converge to solutions of a free boundary value problem as $\varepsilon \to 0$. The limit is null is some set, of positive measure, strictly contained in Ω. Boundary conditions on $\partial\Omega$ are conserved and the limit equation (0.2) is only satisfied locally in a neighbourhood of $\partial\Omega$.

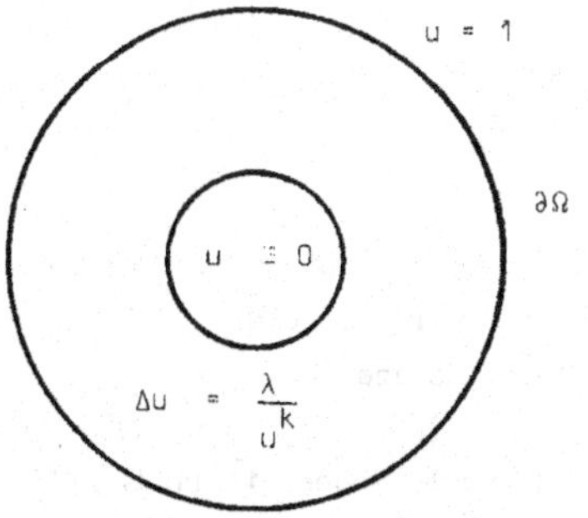

Figure 1

The free boundary value problem is not a classical one as the Laplacian is unbounded at the interface. Only in spherical geometry can we demonstrate classical transmission conditions at the interface (continuity of the function and its gradient). In general, the problem cannot be characterized by standard variational inequations. Methods rely on mixture of monotonicity and variational techniques. Again to demonstrate the existence of a nontrivial subset where the limit $u \equiv 0$, we must establish the result first for spherical domains and then use a comparison lemma ([6]). Chemically, the set $u \equiv 0$ corresponds to a frozen reaction zone where absorption prevents diffusion. The reaction takes place only in some neighbourhood of $\partial\Omega$ and the catalyst pellet is inefficient. Similar phenomena occur in the enzymatic membranes context for some enzyme choices.

Interesting mathematical problems are opened by this specific unusual free boundary value problem which can indeed possesses a infinite number of solutions for critical values of λ.

Finally, we consider the case $k = 0$ which is of special interest.

1. THE LIMIT PROBLEM

1.1 General case

Let Ω be a regular bounded domain in $\mathbb{R}^n$. We call "limit problem" problem (0.2), which is the formal limit of (0.1) (0.1)' when $\varepsilon \to 0$, i.e.

$$(1.1) \qquad \Delta u = \frac{\lambda}{u^k} \;,\quad u/_{\partial\Omega} = 1 \;,\; \lambda > 0 \;,\; k > 0$$

We look for positive solutions of (1.1), which verify $u \leqslant 1$ from the maximum principle.

Remark 1.1 : This problem can be renormalized as

$$(1.1)' \qquad \begin{cases} -\Delta v = \dfrac{\lambda}{(1-v)^k} \;, & v/_{\partial\Omega} = 0 \\ u = 1 - v \;, & 0 \leqslant v \leqslant 1 \end{cases}$$

The application $z \to \dfrac{1}{(1-z)^k}$ is stricly monotone, convex on $[0,1[$, but unbounded. Usual methods for nonlinear eigenvalue problems (CRANDALL-RABINOWITZ [17] [18], AMANN [2], KEENER-KELLER [19] must be adapted.

We recall results presented in [10] and to appear in [12]. Let $\mathcal{O}$ be the open set of $W^{2,r}(\Omega)$, $r > \frac{n}{2}$, defined by

(1.2) $\quad \mathcal{O} = \{ u \in W^{2,r}(\Omega) \;,\; \exists \eta > 0 \;,\; u(x) > \eta \}$

We obtain the existence of a maximal branch $A = (\lambda, u(\lambda)) \in \mathbb{R} \times \mathcal{O}$, for $\lambda \in [0,\lambda_*[$, $u(0) = 1$. This branch cannot be continued beyond $\lambda_* \leqslant \mu_0$, where μ_0 is the smallest eigenvalue of $-\Delta$. Every $u(\lambda)$ on the branch is in fact in $C^\infty(\bar{\Omega})$. Furthermore, the application $\lambda \to u(\lambda)$ is decreasing ; the smallest eigenvalue of the Frechet derivative of (1.1), taken at every element of A, is > 0. In fact one has the following global result :

Lemma 1.1 : For $\lambda > \lambda_*$, there exists no solution of (1.1) in the open set $\mathcal{O}$. In $\lambda = \lambda_*$, any solution is unique.

Extending a nonlinear variationnal method of CRANDALL-RABINOWITZ [18], we obtain the following theorems valid for every regular domain, $\forall\, n \geqslant 1$.

Theorem 1.1 : $\forall\, k > 0$, there exists a unique limit $u_* = u(\lambda_*)$ such that $u(\lambda) \searrow u_*$, $\lambda \nearrow \lambda_*$, in the weak topology of $W^{2,5}(\Omega)$. In fact one has strong convergence in $W^{1,5}(\Omega)$.

Theorem 1.2 : In $L^5(\Omega)$, one has

(1.3) $$\Delta u_* = \frac{\lambda}{u_*^{\,k}} \;,\quad u_*/_{\partial\Omega} = 1 .$$

Similar results have been reestablished by MIGNOT-PUEL [25] in a general framework.

Only under some conditions is (λ_*, u_*) a regular bending point of problem (1.1) in $C^j(\bar{\Omega})$, $\forall\, j \geqslant 1$, in the sense of CRANDALL-RABINOWITZ [17] [18], KEENER-KELLER [19].

We set

(1.4) $$\begin{cases} \alpha = \alpha(k) = (k+1)\left(1 + \sqrt{\dfrac{k}{k+1}}\right) \\ \mu(k) = \dfrac{2\alpha - 1 + k}{k+1} \end{cases}$$

($\mu(k)$ is a monotone increasing function from 1 to 5 - as k goes from 0 to $+\infty$).

We set also

(I.5) $$\begin{aligned} N_2(k) &= [\mu(k)] \text{ if } \mu(k) \notin \mathbb{N}, \\ &= \mu(k)-1 \text{ if } \mu(k) \in \mathbb{N}. \end{aligned}$$

Theorem 1.3 : If $n \leqslant N_2(k)$, $u(\lambda) \to u_*$ in $C^\infty(\bar{\Omega})$ and $u_* \in \mathcal{O}$. Moreover, (λ_*, u_*) is a regular bending point.

For example, if $k = 1$, $\alpha = \mu = 2 + \sqrt{2}$ and $N_2 = 3$.

Theorem 1.4 : If $n \leqslant N_2(k)$, the solutions of (1.1) near (λ_*, u_*) take the form of an arc $C = (\lambda(s), \tilde{u}(s))$, $|s| < \delta$, for some δ, such that

(i) The application $s \to (\lambda(s), \tilde{u}(s))$ is analytic of $]-\delta, +\delta[$ into $\mathbb{R} \times C^j(\bar{\Omega})$, $\forall j \geqslant 1$;

(ii) $\lambda(0) = \lambda_*$, $\lambda'(0) = 0$, $\lambda''(0) < 0$;

(iii) $\tilde{u}(0) = u_*$, $\tilde{u}'(0) = \psi$, where ψ is the eigenfunction > 0 of

$$(1.6) \qquad -\Delta\psi = \frac{\lambda_* k}{(1-u_*)^{k+1}} \psi \ , \quad \psi/_{\partial\Omega} = 0 \ .$$

From general theorems (RABINOWITZ [27], LERAY-SCHAUDER [21]) on connected components of solutions of nonlinear problems such as (1.1), we obtain, for $\forall n$, $\forall k > 0$:

Corollary 1.1 : There exists a connected component $(\lambda, u(\lambda))$ extending the maximal branch A, such that $0 \leqslant \lambda \leqslant \lambda_*$ and $u(\lambda)$ is unbounded in $C^{2,\alpha}(\bar{\Omega})$; specifically min $u(\lambda) \to 0$.

Indeed the above connected component does end at (λ_*, u_*) whenever the latter is not a regular bending point. Then the component reduces to A and in fact we have an endpoint in the following sense :

Definition 1.1 : An endpoint $(\lambda_{critical}, u_{critical}) = (\lambda_c, u_c)$ is a limit point of a connected component of positive regular solutions of a nonlinear eigenvalue problem iff

(i) Frechet - differentiability is lost at (λ_c, u_c) ;

(ii) u_c is not in $C^2(\bar{\Omega})$;

(iii) u_c belongs to some F - space.

For instance, in the case mentionned above, $\lambda_* = \lambda_c$ and u_c belongs to $W^{2,5}(\Omega)$; the Laplacian is unbounded though in $L^5(\Omega)$ at worst (uniformly in $n \geqslant 1$, $k > 0$). Concrete examples are found in spherical geometries.

This notion of endpoint is quite general and does occur also in problems of fluid mechanics (equation of Falkner-Skan [13]). J. LERAY has already observed in his Thesis ([20]) such endpoint phenomena in a similar problem with $k = -2m$ $(m \geq 1)$. See also CALLEGARI-KELLER-REISS [15] for a presentation of membrane buckling.

1.2 Global results in spherical geometries

Singular endpoints and regular bending points both occur in spherical geometries when $\bar{\Omega} = \left\{ x \in \mathbb{R}^n,\ 0 \leq r = \left(\sum_{i=1}^{n} x_i^2\right)^{1/2} \leq 1 \right\}$, for spherically symetric solutions.

For $k > 0$, $n \geq 1$, we introduce the notations

$$(1.7) \qquad \beta = \beta(k) = \frac{2}{k+1}, \quad \lambda_c = \lambda_c(n,k) = \beta(\beta + n - 2) \text{ and one has the}$$

Lemma 1.2 : For $\lambda = \lambda_c$, problem (1.1) admits the singular solution $u_c = r^\beta$; $u_c \notin C^2(\bar{\Omega})$ and $u_c \in H^p(\Omega)$ iff $n > 2p - \frac{4}{k+1}$.

Now we set

$$(1.8) \qquad n_{\pm}(k) = \frac{6k+2}{k+1} \pm \sqrt{\frac{k}{k+1}}$$

Theorem 1.5 :

(i) If $n_-(k) < n < n_+(k)$, then problem (1.1) admits an infinite number of regular bending points, $\lambda_1, \lambda_2, \ldots$, with $\lambda_i \to \lambda_c$ and $u(\lambda_i) \to u_c$ in $H^p(\Omega)$.

(ii) If $1 < n \leq n_-(k)$ or $n \geq n_+(k)$, then (1.1) admits no bending point and in fact $\lambda_c = \lambda_*$. (1)

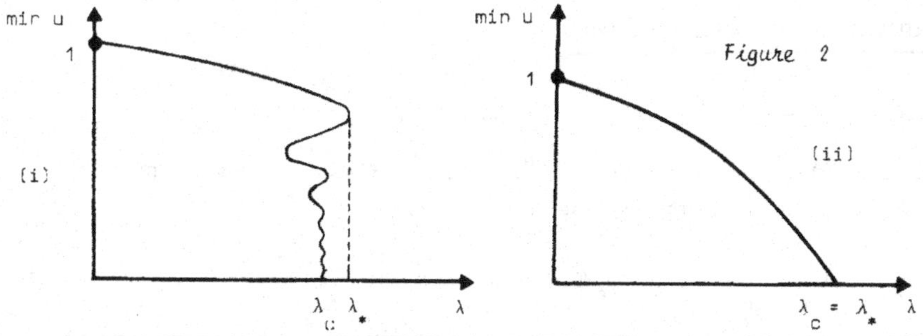

Figure 2

(1) see [8] [9] for the case $n = 1$.

Corollary 1.2 : In spherical geometries, there always exists a unique endpoint (λ_c, u_c) in the sense of definition 1.1 such that $(\lambda, u(\lambda)) \to (\lambda_c, u_c)$ at least in $\mathbb{R}^+ \times W^{2,1}(\Omega)$, for $n \geq 2$.

The proof of these results will be found in Appendix 1. The technique consists in reducing exactly (1.1) to an initial value problem for an ordinary differential equation (O.D.E.).

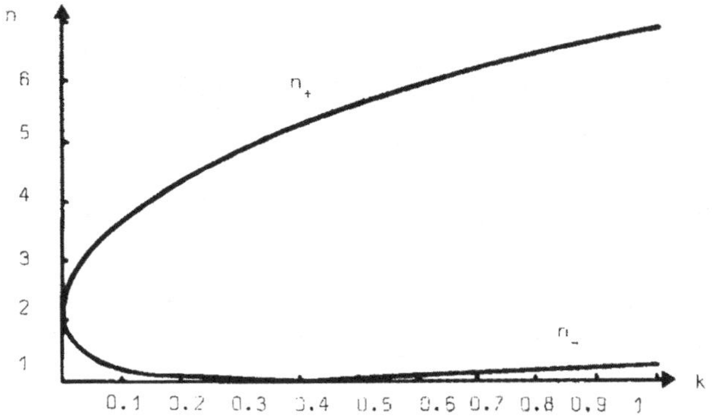

Figure 3

Remark 1.2 : λ_1 (first bending point) is of course λ_* and we show that $\lambda_1 = \lambda_* \geq \lambda_i$, $i = 2, 3, \ldots$

Remark 1.3 : The existence of an endpoint in arbitrary domains, whenever λ_* is a regular bending point, is an important open question. The only available a priori global estimate is $0 \leq u \leq 1$.

2. THE PERTURBED PROBLEM ($\varepsilon > 0$ fixed)

2.1 General results

Lets Ω be a regular bounded domain in $\mathbb{R}^n$. For $\varepsilon > 0$ fixed, we consider problem (0.1)' and we limit ourselves to the case $m = 1$ [1] :

$$(2.1) \qquad \Delta u_\varepsilon = \lambda \frac{u_\varepsilon}{\varepsilon + u_\varepsilon^{1+k}} \quad , \quad u_\varepsilon/_{\partial\Omega} = 1 ,$$

[1] The case $m > 1$ will be detailed in a forthcoming paper. Entirely similar results extend to $\Delta u_\varepsilon = \lambda u_\varepsilon(\varepsilon + 1 / \varepsilon + u_\varepsilon)^{1+k}$, $u_\varepsilon/_{\partial\Omega} = 1$.

where λ and k are positive parameters.

Theorem 2.1 : Let $\varepsilon > 0$, $k > 0$ fixed. Then

(i) $\forall \lambda > 0$, problem (2.1) has at least one solution $u_\varepsilon = u_\varepsilon(\lambda)$ in $C^\infty(\bar{\Omega})$, with $0 < u_\varepsilon \leqslant 1$;

(ii) $\forall \lambda > 0$, there exist a maximal solution $\bar{u}_\varepsilon(\lambda)$ and a minimal solution $\underline{u}_\varepsilon(\lambda)$;

(iii) The mappings $\lambda \to \bar{u}_\varepsilon(\lambda)$ and $\lambda \to \underline{u}_\varepsilon(\lambda)$ are monotone decreasing on $\mathbb{R}_+$.

Proof : We use a classical method of monotone sequences (see for example [1] [28]). For $\lambda > 0$ fixed, we consider the mapping $\psi \to T\psi$ defined by

$$(2.2) \qquad \Delta(T\psi) = \lambda \frac{T\psi}{\varepsilon + \psi^{1+k}} \quad , \quad (T\psi)/_{\partial\Omega} = 1$$

We verify that T is a monotone increasing mapping. Starting with the lower solution $u_0 = 0$, one constructs by recurrence a monotone increasing sequence u_n by setting $u_n = Tu_{n-1}$. This sequence is bounded by 1, hence convergent, and its limit is the minimal solution $\underline{u}_\varepsilon(\lambda)$.In a similar fashion, one constructs a monotone decreasing sequence $u^{(n)}$, starting with the upper-solution $u^{(o)} = 1$. Its limit is the maximal solution $\bar{u}_\varepsilon(\lambda)$. The assertion (iii) is demonstrated by recurrence.

Finally, note that $u_\varepsilon(\lambda) \geqslant \phi_\varepsilon(\lambda)$ where $\phi_\varepsilon(\lambda)$ is the solution of

$$(2.3) \qquad \Delta\phi_\varepsilon = \frac{\lambda \phi_\varepsilon}{\varepsilon} \quad , \quad \phi_\varepsilon/_{\partial\Omega} = 1 \ ,$$

then $u_\varepsilon > 0$ in $\bar{\Omega}$ for λ finite.

■

2.2 A local uniqueness result

In this section we will prove that problem (2.1) has a unique solution for λ small or large enough.

Theorem 2.2 : For $\varepsilon > 0$, $k > 0$ fixed, problem (2.1) has a unique solution in a neighbourhood of $\lambda = +\infty$ and $\lambda = 0^+$.

Proof :

(i) First we notice by elementary reasoning (LIONS [23]) that $u_\varepsilon(\lambda) \to 0$ a.e., as $\lambda \to +\infty$, with a boundary layer in a vicinity of $\partial\Omega$. Let u_1 and u_2 be two different solutions of (2.1), and $w = u_1 - u_2$. The function w verifies

$$(2.4) \qquad \Delta w = \frac{\lambda u_1}{\varepsilon + u_1^{1+k}} - \frac{\lambda u_2}{\varepsilon + u_2^{1+k}} = \lambda w \frac{\varepsilon - k\theta^{k+1}}{(\varepsilon + \theta^{k+1})^2}$$

where $\theta \in]u_1, u_2[$ and $w/_{\partial\Omega} = 0$.

The smallest eigenvalue of (2.4) is given by

$$(2.5) \qquad \nu(\lambda) = \inf_{\psi \in D} \left[\int_\Omega (\nabla\psi)^2 \, dx + \lambda \int_\Omega \frac{(\varepsilon - k\theta^{k+1})}{(\varepsilon + \theta^{k+1})^2} \psi^2 \, dx \right]$$

where $D = \left\{ \psi \in H_0^1(\Omega) , \int_\Omega \psi^2 \, dx = 1 \right\}$.

Because $\dfrac{\varepsilon - k\theta^{k+1}}{(\varepsilon + \theta^{k+1})^2} \psi^2 \to \dfrac{\psi^2}{\varepsilon}$ as $\lambda \to +\infty$,

and $\left| \dfrac{\varepsilon - k\theta^{k+1}}{(\varepsilon + \theta^{k+1})^2} \psi^2 \right| \leq Cst . \psi^2$, we may use Lebesgue's theorem and see that

$$\int_\Omega \frac{(\varepsilon - k\theta^{k+1})}{(\varepsilon + \theta^{k+1})^2} \psi^2 \, dx \to \frac{1}{\varepsilon} .$$

Hence we are sure that, for λ sufficiently large, $\nu(\lambda) \geq \mu_0$ where μ_0 is the smallest eigenvalue of $-\Delta$. Then the spectrum is > 0 and $w = 0$.

(ii) We use a similar reasoning when λ is small. We note that

$$\nu(\lambda) \geq \inf_{\psi \in D} \left[\int_\Omega (D\psi)^2 \, dx + \lambda \int_\Omega \frac{\varepsilon - k}{\varepsilon^2} \psi^2 \, dx \right]$$

$$\geq \mu_0 + \lambda \frac{\varepsilon - k}{\varepsilon^2}$$

the latter being > 0 is λ small enough.

■

Therefore, in a neighbourhood of $(\lambda = 0, u = 1)$ in $\mathbb{R}^+ \times C^\infty(\bar{\Omega})$, there exists a unique solution $u_\varepsilon(\lambda) \equiv \bar{u}_\varepsilon(\lambda)$ (hence identical to the maximal one), which can be extended into a simply connected component of positive solutions of (2.1) as in CRANDALL-RABINOWITZ [17], LERAY-SCHAUDER [21].

Theorem 2.2 : Problem (2.1) possesses a simply connected component $\mathcal{B}_\varepsilon = (\lambda, u_\varepsilon)$, $(0, 1) \in \mathcal{B}_\varepsilon$; this component has an unbounded projection on $\mathbb{R}^1_+$.

Proof : Indeed for every λ, $0 \leq \lambda < +\infty$, $\phi_\varepsilon(\lambda) \leq u_\varepsilon(\lambda) \leq 1$. For finite λ, the component stays bounded in $C^j(\bar{\Omega})$, $\forall j \geq 1$. Since we have demonstrated the uniqueness for large λ, fixed ε, we are insured that the component corresponds to this local unique solution. ■

Remark 2.1 : We have a generalized end point in the sense that $\lambda_c = +\infty$, $u_c = 0$ in $L^p(\Omega)$ strong.

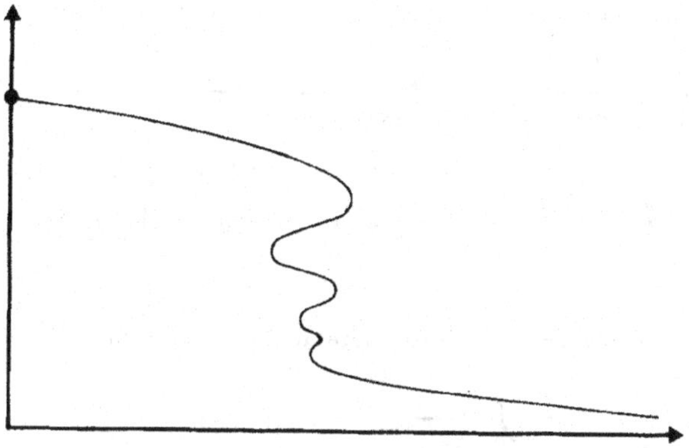

Figure 4

Remark 2.2 : If B_ε has a first bending point $\lambda_*(\varepsilon)$, then the preceding demonstration also yields the existence of a second bending point $\lambda_{**}(\varepsilon)$ (using local uniqueness for λ fixed large enough). Of course, in some cases, there are no bending points.

2.3 A variational formulation

We present a variational charaterization of some solutions which indeed will yield this simply connected component B_ε whenever the latter is locally unique in λ.

We set

$$(2.6) \qquad \phi_\varepsilon(z) = \int_0^z \frac{y}{\varepsilon + y^{1+K}} \, dy$$

and we introduce the functional

$$(2.7) \qquad J_\varepsilon(v) = \frac{1}{2} \int_\Omega (\nabla v)^2 \, dx + \lambda \int_\Omega \phi_\varepsilon(v) \, dx$$

for $v \in K$, where

$$(2.8) \qquad K = \{ v \in H^1(\Omega) \, , \, 0 \leq v \leq 1 \text{ a.e.}, \, v/_{\partial\Omega} = 1 \}$$

Theorem 2.3 : Let $\varepsilon > 0$, $\lambda > 0$, $k > 0$ fixed. Then there exists an element $u \in K$

such that $J_\varepsilon(u) = \inf_{v\in K} J_\varepsilon(v)$.

Proof : The crucial point is to prove that the mapping $v \to \phi_\varepsilon(v)$ is continuous from $L^2(\Omega)$ into $L^1(\Omega)$. Let $v_n \to v$ in $L^2(\Omega)$ and a.e. One has for almost every x

$$\phi_\varepsilon(v_n)\ (x) = \int_0^{v(x)} \frac{y}{\varepsilon + y^{1+k}}\, dy + \int_{v(x)}^{v_n(x)} \frac{y}{\varepsilon + y^{1+k}}\, dy$$

then $\phi_\varepsilon(v_n) \to \phi_\varepsilon(v)$ a.e., the second integral being bounded by $\frac{1}{2\varepsilon}\,(v_n(x) - v(x))^2$.
Moreover, $\phi_\varepsilon(v_n) \leqslant \int_0^1 \frac{y}{\varepsilon}\, dx = \frac{1}{2\varepsilon}$, and we conclude by Lebesgue's theorem. ∎

Corollary 2.1 : u verifies the following variational inequation

$$(2.9)\quad \begin{cases} \int_\Omega \nabla u.\ \nabla(v-u)\ dx + \lambda \int_\Omega \frac{u}{\varepsilon + u^{1+k}}\,(v - u)\ dx \geqslant 0 \\ \forall\, v \in K \end{cases}$$

Indeed the functional J_ε is Gâteaux-differentiable.

Remark 2.3 : It is easy to reformulate (2.9) in term of functions vanishing at the boundary :

Setting $\hat{v} = 1 - v$, $\hat{u} = 1 - u$, $\hat{K} = \{\hat{v} \in H^1_0(\Omega),\ 0 \leqslant \hat{v} \leqslant 1 \text{ a.e.}\}$, one has $\hat{u} \in \hat{K}$ and

$$(2.10)\quad \begin{cases} \int_\Omega \nabla\hat{u}.\nabla(\hat{v} - \hat{u})\ dx \geqslant \lambda \int_\Omega \frac{u}{\varepsilon + u^{1+k}}\ (\hat{v} - \hat{u})\ dx \\ \forall\, \hat{v} \in \hat{K} \end{cases}$$

As $u/(\varepsilon + u^{1+k})$ belongs to $L^\infty(\Omega)$, it is well known (see for instance [??]) that $\hat{u} \in W^{2,p}(\Omega)$, $\forall p \geqslant 1$, and so is u.

Corollary 2.2 : $u \in K \cap W^{2,p}(\Omega)$, $\forall p \geqslant 1$, and verifies the variational inequation

$$(2.11)\quad \begin{cases} \int_\Omega -\Delta u.\ (v - u)\ dx + \lambda \int_\Omega \frac{u}{\varepsilon + u^{1+k}}\ (v - u)\ dx \geqslant 0 \\ \forall\, v \in K \end{cases}$$

Theorem 2.4 : Let $\varepsilon > 0$, $\lambda > 0$, $k > 0$ fixed, u given by theorem 2.3. Then u is solution of problem (2.1).

Proof : See for instance PUEL [26].

Corollary 2.3 : $\forall\ \varepsilon > 0$, $\lambda > 0$, $k > 0$ fixed, problem (2.1) has at least one solution which realizes the infimum of J_ε in K . We will note it $u^J_\varepsilon(\lambda)$. One has

(2.12) $$\underline{u}_\varepsilon(\lambda) \leqslant u^J_\varepsilon(\lambda) \leqslant \bar{u}_\varepsilon(\lambda)$$

Whenever we have no uniqueness at fixed λ, we do not know in general wether $u^J_\varepsilon(\lambda)$ is identical to either $\underline{u}_\varepsilon$ or $\bar{u}_\varepsilon$.

3. FIRST CONVERGENCE RESULTS FOR BRANCHES OF THE PERTURBED PROBLEM

It is clear that the branch $\mathcal{B}_\varepsilon$ does singularly converge for $\lambda > \lambda_*$, as $\varepsilon \to 0$, since the limit problem has no regular solution in this case. In this chapter, we shall give general convergence results, and also uniform convergence results wich will be valid only in some domain in λ.

3.1 First a priori estimates

In this section we give a result valid for every branch, $\forall\ k > 0$.

Theorem 3.1 : Let $\lambda > 0$, $k > 0$ fixed. Then, as $\varepsilon \to 0$, the sequence $u_\varepsilon(\lambda)$ is bounded in $H^1_{loc}(\Omega)$. Furthermore, $\Delta u_\varepsilon(\lambda)$ is bounded in $L^1_{loc}(\Omega)$.

Proof : Let $\phi \in \mathcal{D}(\Omega)$ fixed. We denote $v_\varepsilon = u_\varepsilon \cdot \phi$, which verifies

(3.1) $$\begin{cases} -\Delta v_\varepsilon + \lambda \dfrac{v_\varepsilon}{\varepsilon + u_\varepsilon^{1+k}} = -2\,\nabla u_\varepsilon \cdot \nabla\phi - u_\varepsilon \cdot \Delta\phi \\ v_\varepsilon/_{\partial\Omega} = 0 \end{cases}$$

After multiplying by v_ε and integrating over Ω, it follows

$$|\nabla v_\varepsilon|^2_{L^2(\Omega)} + \lambda \int_\Omega \frac{v_\varepsilon^2}{\varepsilon + u_\varepsilon^{1+k}}\,dx = -2\sum_{i=1}^{n}\int_\Omega \frac{\partial u_\varepsilon}{\partial x_i}\,\frac{\partial\phi}{\partial x_i}\,u_\varepsilon\,\phi\,dx$$
$$- \int_\Omega u_\varepsilon^2\ \Delta\phi \cdot \phi\,dx$$

Indeed the first term of the second member is equal to

$$-\sum_{i=1}^{n}\int_\Omega \frac{\partial}{\partial x_i}\,u_\varepsilon^2 \cdot \frac{\partial\phi}{\partial x_i}\,\phi\,dx = \sum_{i=1}^{n}\int_\Omega u_\varepsilon^2\,\frac{\partial}{\partial x_i}\left(\frac{\partial\phi}{\partial x_i}\phi\right)\,dx$$

Since $0 \leqslant u_\varepsilon \leqslant 1$, we conclude that the sequence v_ε is bounded in $H^1_0(\Omega)$. That is true for every $\phi \in \mathcal{D}(\Omega)$, hence the result.

Moreover,

$$|\Delta u_\varepsilon \cdot \phi|_{L^1(\Omega)} = \int_\Omega \Delta u_\varepsilon \cdot \phi \, dx = - \sum_{i=1}^{n} \int_\Omega \frac{\partial u_\varepsilon}{\partial x_i} \frac{\partial \phi}{\partial x_i} \, dx \leqslant Cst(\phi) .$$

$\blacksquare$

Corollary 3.1 As $\varepsilon \to 0$, a subsequence of $u_\varepsilon(\lambda)$ converges in $L^p(\Omega)$ strong, $\forall p \geqslant 1$, and a.e.

Proof : Let K a compact $\subset$ K. There exists $\phi \in \mathcal{D}(\Omega)$ such that $0 \leqslant \phi \leqslant 1$, $\phi/_K = 1$. As $u_\varepsilon . \phi$ is bounded in $H_0^1(\Omega)$, one can extract a subsequence converging in $L^2(\Omega)$ and a.e. Then u_ε converges a.e. in K. Now let (K_n) a countable overlapping of Ω. By mean of a diagonal process, one can extract a subsequence converging a.e. in $K_n, \forall n$, hence in Ω, and in $L^p(\Omega)$ strong by Lebesgue's Theorem.

$\blacksquare$

Remark 3.1 : Since the above estimate is not valid near the boundary, we can expect a boundary layer phenomenon, at least for some values of parameters λ and k.

For the maximal and minimal solutions, $\bar{u}_\varepsilon$ and $\underline{u}_\varepsilon$, we have additionally monotone convergence :

Lemma 3.1 : Let $\varepsilon_1 < \varepsilon_2$, then $\bar{u}_{\varepsilon_1} \leqslant \bar{u}_{\varepsilon_2}$ and $\underline{u}_{\varepsilon 1} \leqslant \underline{u}_{\varepsilon 2}$.

Proof : By mean of the recurrence process described in theorem 2.1

$\blacksquare$

Corollary 3.2 : As $\varepsilon \to 0$, the sequences $\bar{u}_\varepsilon$ and $\underline{u}_\varepsilon$ are decreasing.

Remark 3.2 : The latter result may be false for an intermediate solution (see [9]).

3.2 Uniform convergence cases

Theorem 3.2 : Let $k > 0$ fixed, Ω a regular bounded domain in $\mathbb{R}^n$. Then there exists a perturbed maximal branch A_ε, defined for ε small enough and $0 \leqslant \lambda \leqslant \lambda_*$, $(0.1) \in A_\varepsilon$, such that A_ε converge to A in $[0,\lambda_*[\times C^j(\Omega)$, $\forall j \geqslant 1$.

Proof : A_ε is unique in some local neighbourhood of (0.1). On $[0,\lambda_*[$, the maximal solution $u(\lambda)$ of (1.1) on A is a subsolution of (2.1). Thus $\Delta \bar{u}_\varepsilon(\lambda)$ is bounded both from above and below. $\bar{u}_\varepsilon(\lambda)$ is bounded in $W^{2,p}(\Omega)$, $\forall p \geqslant 1$, from which we obtain convergence in $C^j(\bar{\Omega})$, $\forall j \geqslant 0$. From Corollary 3.2, the limit must be the maximal solution $u(\lambda) \in A$ of (1.1).

$\blacksquare$

Remark 3.3 : For ε small enough, on $[0,\lambda_*[$, A_ε is identical to B_ε (see §2.1). Thus we have a partial result of uniform convergence of B_ε.

Now we will take the hypothesis where <u>the formal limit problem (1.1) has a regular bending point at $\lambda = \lambda_*$</u>. A sufficient condition is given in theorem 1.3.

Theorem 3.3 : Let $k > 0$ fixed, and suppose that λ_* exists. Then if ε is sufficiently small, there exists for $\bar{u}_\varepsilon(\lambda)$ a regular bending point $\lambda_*(\varepsilon)$, and an arc C_ε which converges locally in $\mathbb{R}^1 \times C^j(\bar{\Omega})$, $\forall\ j \geqslant 1$, when $\varepsilon \to 0$, to the limit arc C defined at theorem 1.4.

<u>Principle of the demonstration</u> : We adapt an argument from [17]. Consider (2.1) as a regular perturbation of (1.1) :

$$(2.1) \qquad G(\lambda, u, \varepsilon) = 0$$

$$(1.1) \qquad G(\lambda, u, 0) = 0$$

From Theorem (3.2), we conclude that, $\forall\, \eta_1$, $0 < \eta_1 < \lambda_*$, there exists a neighbordood V_1 of $\{\lambda, u(\lambda) : 0 \leqslant \lambda \leqslant \lambda_* - \eta_1\}$ and $\exists\ \varepsilon_1$ such that (2.1) has a solution $(\lambda, u(\lambda, \varepsilon))$ in $V_1 \times [0, \varepsilon_1]$. That is, the arc A_ε is in a V_1 - neighborhood of the limit arc A.

Next, from Theorem (1.4), solutions of (1.1) in a neighborhood of (λ_*, u_*) take the form of an arc $C = (\lambda(s), \tilde{u}(s))$. By using the implicit function theorem locally (smooth dependence on ε), in a manner similar to the demonstration of the local structure of a bending point [17], one obtains functions $\lambda(s, \varepsilon)$, $\tilde{u}(s, \varepsilon)$ defined on some square $S = \{(s, \varepsilon) : |s| < \eta_2, |\varepsilon| < \eta_2\}$, such that $G(\lambda, u, \varepsilon) = 0$. There is also an open neighborhood V_2 of (λ_*, u_*) such that if $G(\lambda, u, \varepsilon) = 0$, $(\lambda, u) \in V_2$ and $|\varepsilon| \leqslant \eta_2$, then $(\lambda, u) = (\lambda(s, \varepsilon), \tilde{u}(s, \varepsilon)$ for exactly one $(s, \varepsilon) \in S$. Equivalently, if $|\varepsilon| < \eta_2$, there exists an arc C_ε parametrized by s and ε, in a V_2-neighborhood of the limit arc C.

Thus we can assert that $\frac{\partial}{\partial s}\lambda(s, \varepsilon)$ has an unique zero $s(\varepsilon)$, $|s(\varepsilon)| < \eta_2$ for $|\varepsilon| < \eta_2$, corresponding to a regularly perturbed bending point of C_ε. Finally, one easily pieces together the curves A_ε and C_ε. ■

Corollary 3.3 : In spherical geometries, for $n_- < n < n_+$, we have the stronger result : in a neighborhood of every regular bending point λ_i of (1.1), $i = 1, \ldots, \infty$, there exists, for ε sufficiently small, a perturbed bending point of (2.1), with corresponding local convergence in $\mathbb{R}^1 \times C^j(\bar{\Omega})$, $\forall\ j \geqslant 0$, of the perturbed arc.

As $\varepsilon \to 0$, (2.1) possesses an increasingly large, countable and unbounded sequence of bending points.

Proof : The corollary is most easily verified by remarking that, in spherical geometries, the limit problem (1.1) and the perturbed problem (2.1) are both second- order nonlinear Sturm-Liouville problems, once expressed in radial coordinates. Moreover for every local branch such that min u $\geqslant$ cst, the Sturm-Liouville problem is regular and all the eigenvalues of the Frechet - derivatives are simple. The proof of Theorem 3.2 does extend immediately.

∎

Remark 3.4 : The above convergence results are, of course, not uniform in ε with respect to all bending points. In some sense, we have a singular perturbation even for such arcs, since, in the limit, there is an infinite number of regular bending points ; whereas, for $\varepsilon \neq 0$ their number is strictly finite.

Remark 3.5 : The above uniform convergence results are independent on k and only obtained for $\lambda < \lambda_*$. One notices that the demonstration methods either suppose the existence a priori of a limit bending point, or utilize a monotony method based on the maximal limit branch A.

3.3 The fundamental indeterminacy

We introduce the notation

(3.2) $$\varepsilon' = \varepsilon^{\frac{1}{k+1}}$$

Problem (2.1) may be rewritten in the following form

$$\Delta u_\varepsilon = \lambda \frac{u_\varepsilon}{\varepsilon'^{k+1} + u_\varepsilon^{k+1}} = \frac{\lambda}{\varepsilon'^{k}} \; \frac{\frac{u_\varepsilon}{\varepsilon'}}{1 + \frac{u_\varepsilon}{\varepsilon'}^{k+1}}$$

hence

(3.3) $$\varepsilon'^{k} \Delta u_\varepsilon = \frac{\frac{u_\varepsilon}{\varepsilon'}}{1 + \left(\frac{u_\varepsilon}{\varepsilon'}\right)^{k+1}} = F\left(\frac{u_\varepsilon}{\varepsilon'}\right)$$

As the sequence Δu_ε is bounded in $L^1_{loc}(\Omega)$ (theorem 3.1), and using the positivity of Δu_ε, it follows that $\varepsilon'^{k} \Delta u_\varepsilon \to 0$ a.e. (at least a subsequence), then

(3.4) $$F\left(\frac{u_\varepsilon}{\varepsilon'}\right) \to 0 \qquad \text{a.e.}$$

Unfortunately, we can only infer from (3.4)

(3.5) $\frac{u_\varepsilon}{\varepsilon'} \to 0$ or $+\infty$ a.e.

what we call the fundamental indeterminacy.

Lemma 3.2 : $\forall\, u_\varepsilon$ solution of (2.1), $u_\varepsilon/_{\varepsilon}^{\frac{1}{k+1}} \to 0$ or $\to +\infty$ a.e.

Indeed if $u_\varepsilon/\varepsilon' \to 0$ a.e., we would have $u_\varepsilon \to 0$ a.e. and a classical singular perturbation (boundary layer) situation, as $u_\varepsilon/_{\partial\Omega} = 1$. However, it is perfectly possible that u_ε has a strictly positive limit in some subdomain (where $u_\varepsilon/\varepsilon' \nearrow +\infty$).

Nevertheless, we shall give a partial result :

Theorem 3.4 : Let $\Omega_0 = \{x,\ u_\varepsilon(x) \to 0\}$, and $\Omega_\infty = \left\{x \in \Omega_0,\ \frac{u_\varepsilon(x)}{\varepsilon'} \to +\infty\right\}$.
Then mes $\Omega_\infty = 0$.

Proof : If $x \in \Omega_\infty$, $F\left(\frac{u_\varepsilon(x)}{\varepsilon'}\right) \sim \frac{1}{\left(\frac{u_\varepsilon(x)}{\varepsilon'}\right)^k}$ then $\Delta u_\varepsilon(x) \sim \frac{1}{u_\varepsilon^k(x)} \to +\infty$,
hence mes $\Omega_\infty = 0$ because the estimate on Δu_ε. ■

Remark 3.6 : By definition $\Omega_\infty \subseteq \Omega_0$.

Corollary 3.4 : For every $u_\varepsilon(\lambda)$ such that $\lambda > \lambda_*$, we have $\Omega_0 \neq \emptyset$.

In the following sections we shall prove the fundamental result that mes $\Omega_0 > 0$ at least for λ large enough.

4. CASE $k > 1$.

From now on, we shall point out a dichotomy between the cases $k \geq 1$ and $0 < k < 1$. In the case $k \geq 1$ boundary layer phenomenona will appear, whereas free boundaries will take place in the case $0 < k < 1$ (§5). We will first consider spherical geometries in order to use solutions in balls as supersolutions in the general case.

4.1 Spherical case

Theorem 4.1 : Let $\lambda > \lambda_*$, $k \geq 1$ fixed, and $\bar{u}_\varepsilon(\lambda)$ be the maximal solution of the perturbed problem (2.1). Then, as $\varepsilon \to 0$, $\bar{u}_\varepsilon(\lambda) \searrow 0$ in $H^1_{loc}(\Omega)$ weak and $L^p(\Omega)$ strong ($p \geq 1$), with a boundary layer in a neighbourhood of $\partial\Omega$.

Proof : After theorem 3.1 and corollaries 3.1 and 3.2, $\bar{u}_\varepsilon(\lambda) \searrow \chi$ in $H^1_{loc}(\Omega)$ weak and $L^p(\Omega)$ strong. In one (radial) dimension, $H^1_{loc}(\Omega)$ weak implies C^0_{loc} strong convergence of $\bar{u}_\varepsilon(\lambda)$ to $\chi(r)$ on the open set $0 < r < 1$. As $\lambda > \lambda_*$, $\chi(r)$ is equal to zero at some point in $0 < r < 1$. If not, problem (2.1) would have a regular solution for $\lambda > \lambda_*$ (cf. Cor. 3.4). Moreover, the only solution such that $u(0) = 0$, $u > 0$ elsewhere, is $u_c(\lambda_c)$.

Now we suppose that there exists r_0, $0 < r_0 < 1$ such that $\chi(r_0) > 0$. As the function $\chi(r) \in C^0(0,1)$, there exists an interval $I = [r_0 - \eta, r_0 + \eta]$ such that $\chi > 0$ on I and $\bar{u}_\varepsilon \searrow \chi$ in $C^\infty(I)$. As $\frac{du_\varepsilon}{dr} \geq 0$, one has $\frac{d\chi}{dr} \geq 0$ on I and $\forall\, r \geq r_0$, $r \in I$:

$$r^{n-1}\frac{\partial\chi(r)}{\partial r} - r_0^{n-1}\frac{\partial\chi(r_0)}{\partial r} = \lambda \int_{r_0}^{r} \rho^{n-1}\frac{d\rho}{\chi(\rho)} ;$$

hence $\frac{\partial\chi}{\partial r}$ is monotone, strictly increasing, strictly positive $\forall\, r > r_0$, $r \in I$; from which we infer that $\chi > 0$ on $[r_0, 1]$.

Let $r_c = \sup\{r \,;\, \chi(r) = 0\}$. By the above reasonning, we have $\chi \equiv 0$ on $[0, r_c]$, and $\chi > 0$ on $]r_c, 1]$. Indeed, supposing $\exists\, r_0 < r_c$, $\chi(r_0) > 0$, this would imply $\chi(r_c) > 0$. Let r_i be a decreasing sequence such that $r_i \searrow r_c$. In radial coordinates, one has

$$\text{(4.1)} \quad \frac{d\chi}{dr}(1) - \lambda \int_{r_i}^{1} \frac{\rho^{n-1}}{\chi^k}\, d\rho = r_i^{n-1}\frac{d\chi}{dr}(r_i)$$

The sequence $r_i^{n-1}\frac{d\chi}{dr}(r_i)$ is monotone decreasing, <u>positive</u>, then $\frac{d\chi}{dr}(r_c^+)$ and $\int_{r_c}^{1} \frac{\rho^{n-1}}{\chi^k}\, d\rho$ exist (bounded limits). For $\eta > 0$, we set $K = \sup\left\{\frac{d\chi}{dr}(r),\ r \in [r_c, r_c + \eta]\right\} > 0$. We have $\chi(\rho) \leq K(\rho - r_c)$ on $[r_c, r_c + \eta]$ and

$$+\infty > \int_{r_c}^{r_c+\eta} \frac{\rho^{n-1}}{\chi^k}\, d\rho \geq \frac{1}{K}\int_{r_c}^{r_c+\eta} \frac{\rho^{n-1}}{(\rho - r_c)^k}\, d\rho .$$ But this is obviously false if $k \geq 1$. Hence $\{r \,;\, r < 1 \text{ and } \chi(r) > 0\} = \emptyset$. ■

Corollary 4.1 : The result of theorem 4.1 holds for every solution of problem (2.1).

Corollary 4.2 : Let $\lambda > \lambda_*$ fixed. Then for ε small enough, problem (2.1) has an unique solution.

Proof : Let u_1 and u_2 two different solutions. With the notations of theorem 3.4, $\Omega_0 = \Omega$ and $\frac{u_i}{\varepsilon'} \to 0$ a.e. $(i=1,2)$, $\varepsilon' = \varepsilon^{1/k+1}$.

Let $w = u_1 - u_2$. As in the proof of theorem 2.2,

(4.2) $\Delta w = \lambda w \dfrac{\varepsilon - k\,\theta^{k+1}}{(\varepsilon + \theta^{k+1})^2}$, $\theta \in\ ]u_1, u_2[$ and the smallest eigenvalue of (4.2) is given by

(4.3) $$\nu(\lambda) = \inf_{\psi \in D} \left[\int_\Omega (\nabla\psi)^2 + \lambda \int_\Omega \frac{\varepsilon - k\theta^{k+1}}{(\varepsilon + \theta^{k+1})^2}\, \psi^2\, dx \right]$$

But we observe that

$$\lambda \int_\Omega \frac{\varepsilon - k\,\theta^{k+1}}{(\varepsilon + \theta^{k+1})^2}\, \psi^2\, dx = \frac{\lambda}{\varepsilon} \int_\Omega \frac{1 - k\,\frac{\theta}{\varepsilon}^{k+1}}{(1 + \frac{\theta}{\varepsilon}^{k+1})^2} \psi^2\, dx = \frac{\lambda}{\varepsilon} \int_\Omega \frac{1 - k(\frac{\theta}{\varepsilon'})^{k+1}}{(1 + (\frac{\theta}{\varepsilon'})^{k+1})^2} \psi^2\, dx$$

As $\frac{\theta}{\varepsilon'} \to 0$, from Lebesgue's Theorem, we insure that $\nu(\lambda) > \mu_0$ hence the uniqueness. ∎

4.2 A comparison lemma

To demonstrate similar results for arbitrary, sufficiently smooth domains, we will use a comparison lemma which allows for piecewise C^2 super - and subsolutions.

Lemma 4.1 : Let B_R be any ball of radius R , $B_R \subset \Omega$. Let $u_\varepsilon(\lambda, R)$ a solution of (2.1) in B_R. Then the function equal to $u_\varepsilon(\lambda, R)$ in B_R, and identical to 1 in $\bar\Omega/\bar B_R$, is a supersolution for problem (2.1) considered in the domain Ω.

Proof : Use radial coordinates in Ω, with origin at the center of B_R. Let ζ be the radial variable. Then, spherically symetric solutions $u_\varepsilon(\lambda, R)$ of (2.1) in B_R correspond to a nonlinear Sturm - Liouville problem. $u_\varepsilon(\lambda, R)$ can be uniquely extended in Ω/B_R by solving an initial value problem for $\zeta \geqslant R$:

(4.4) $$\frac{d^2 u_\varepsilon}{d\zeta^2} + \frac{n-1}{\zeta} \frac{d u_\varepsilon}{d\zeta} = \lambda \frac{u_\varepsilon}{\varepsilon + u_\varepsilon^{1+k}}$$

(4.5) $$u_\varepsilon\big|_{\zeta=R} = 1$$

(4.6) $$\frac{\partial u_\varepsilon}{\partial\zeta}\Big|_{R^-} = \frac{\partial u_\varepsilon}{\partial\zeta}\Big|_{R^+}$$

Let $\tilde u_\varepsilon$ be this unique extension of $u_\varepsilon(\lambda, R)$ in $\bar\Omega$. From the maximum principle, $\tilde u_\varepsilon > 1$ in $\bar\Omega/\bar B_R$, hence $\tilde u_{\varepsilon/\partial\Omega} > 1$. This insures that $\tilde u_\varepsilon$ is indeed a strict supersolution for (2.1) in Ω. Thus, $\exists\ u_\varepsilon(\lambda, \Omega)$, solution of (2.1) in Ω, such that $u_\varepsilon(\lambda, \Omega) < \tilde u_\varepsilon$. Precisely, $u_\varepsilon(\lambda, \Omega) < u_\varepsilon(\lambda, R)$ in $\bar B_R$, and $u_\varepsilon(\lambda, \Omega) \leqslant 1$ in $\bar\Omega/\bar B_R$. ∎

This lemma is a special case of a more general lemma demonstrated in [6]. Similar observations had already been made by Il'in - Kalashnikov - Oleinik [30] :

Lemma 4.2 : (H. Berestycki, P.L. Lions) [6] :

Let Ω_1 be a subdomain of Ω such that $\partial\Omega_1$ is regular and $\bar{\Omega}_1 \subset \Omega$. Let $\Omega_2 = \Omega/\bar{\Omega}_1$, with $\partial\Omega_2 = \partial\Omega \cup \partial\Omega_1$, and ν the unit normal outward to Ω_1.

Let $u_i \in H^2(\Omega_i)$, $f_i \in L^1(\Omega_i)$, $(i = 1, 2)$. Then the following hypothesis :

$\Delta u_i \leqslant f_i$ a.e. in Ω_i, $u_1 = u_2$ on $\partial\Omega_i$, $\frac{\partial u_1}{\partial \nu} \geqslant \frac{\partial u_2}{\partial \nu}$ on $\partial\Omega_1$, implies :

$\Delta u \leqslant f$ im $\mathcal{D}'(\Omega)$, where $u \in H^1(\Omega)$ is defined by $u = u_i$ on Ω_i and $f = f_i$ on Ω_i.

To recover lemma 4.1, set $\Omega_1 = B_R$, $u_1 \equiv u_\varepsilon(\lambda, R)$ in B_R, $u_2 = 1$. Berestycki and Lions have used this comparison lemma to construct stationary wave solutions of the nonlinear Klein - Gordon equation. By covering Ω with an <u>infinite</u> number of balls B_R of appropriate radius R, we shall be able to extend convergence results in B_R to similar ones in Ω.

Remark 4.1 : Let $\lambda_i(B_1)$ be the sequence of bending points for the limit problem (1.1). Then :

(4.7) $\lambda_i(B_R) = \lambda_i(B_1)/R^2$, $(\lambda_* \equiv \lambda_1)$, similarly

(4.8) $\lambda_c(B_R) = \lambda_c(B_1)/R^2$, whenever the latter exists.

4.3. Boundary Layer Results for $k \geqslant 1$, in an arbitrary domain

In view of remark 4.1, to generalize Theorem 4.1, we must consider a class of domains Ω with somewhat more regularity :

Definition 4.1 : Let Ω be a simply connected, bounded open set. The boundary $\partial\Omega$ is a "spherically regular" surface iff :

(a) $\partial\Omega$ and Ω are C^3 ;

(b) at every point γ of $\partial\Omega$, there exists some ball $B_{r,\gamma}$, of radius r, tangent to $\partial\Omega$ at γ, and $\bar{B}_{r,\gamma} \subseteq \bar{\Omega}$;

(c) $\inf_{\gamma \in \partial\Omega} r = R_{min} > 0$.

The class of "spherically regular" surfaces contains, for instance :

1) hyperellipsoids ; 2) envelopes of families of hyperspheres depending upon

(n-1) parameters, of class C^3 in $\mathbb{R}^n$ [31]. We conjecture that every surface of class C^3, without any imbilical point, and uniformly locally elliptic [31], is "spherically regular".

However, it is easily seen that "spherically regular" surfaces include very general non-convex C^3 surfaces :

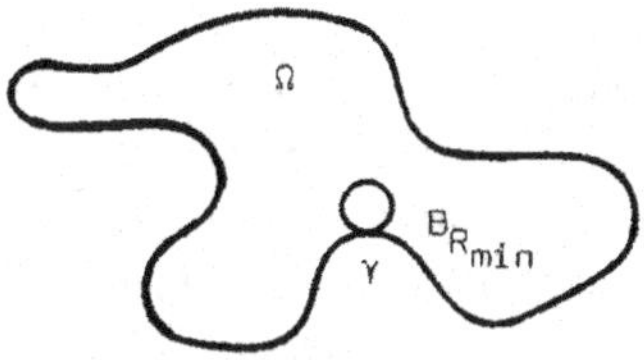

Figure 5

Counterexamples include surfaces which are only piecewise C^3 with cusps (e.g., hypercubes).

Theorem 4.2 : Let $k \geq 1$, Ω with $\partial\Omega$ "spherically regular". If $\lambda > \lambda_*(B_1)/R^2_{min}$, then all solutions $u_\varepsilon(\lambda) \to 0$ in $L^p(\Omega)$ strong (in fact a.e.) and $H^1_{loc}(\Omega)$ weak, as $\varepsilon \to 0$. Moreover, for fixed λ satisfying the above condition, problem (2.1) has an unique solution for ε small enough.

Proof : Cover Ω with an infinite number of balls $B_{R_{min}}$, either strictly $\subset \Omega$; or $\subseteq \Omega$ and tangent to $\partial\Omega$ at some point. The existence of such a covering is insured by the spherical regularity of $\partial\Omega$. For any such $B_{R_{min}}$, let :

$w(\lambda, \varepsilon) = \bar{u}_\varepsilon(\lambda, R_{min})$ in $B_{R_{min}}$, the latter being the maximal solution of (2.1) in $B_{R_{min}}$, and :

$w(\lambda, \varepsilon) \equiv 1$ in $\bar{\Omega}/B_{R_{min}}$.

Let $u_\varepsilon(\lambda, \Omega)$ be any solution of (2.1) in Ω. Then, from lemma 4.1 :

$0 \leq u_\varepsilon(\lambda, \Omega) < w(\lambda, \varepsilon)$ in Ω.

From Theorem 4.1, $\bar{u}_\varepsilon(\lambda, R_{min}) \searrow 0$ a.e. and in $L^p(B_{R_{min}})$ strong, provided that $\lambda > \lambda_*(B_{R_{min}}) = \lambda_*(B_1)/R^2_{min}$ (from Remark 4.1). Thus, for fixed λ satisfying the above condition, $u_\varepsilon(\lambda, \Omega) \to 0$ a.e. in every $B_{R_{min}}$, hence a.e. in Ω. One concludes with Lebesgue's Theorem. Uniqueness for ε small enough follows exactly as in Corollary 4.2.

■

Remark 4.2 : Using the above uniqueness, we can in fact improve theorem 4.2, replacing $\lambda_*(B_1)$ by $\lambda_c(B_1)$, as defined in (1.7). Whenever $\lambda_*(\varepsilon)$ is a genuine bending point, the demonstration is based on the existence and properties of a second bending point

$\lambda_{**}(\varepsilon)$, corresponding to this unique (for ε small enough) singular branch issued from $\lambda = \infty$. $\lambda_{**}(\varepsilon)$ again corresponds to the fundamental (simple) eigenvalue (of the Frechet derivative of (2.1)) crossing the origin of the spectral real axis.

5. CASE $0 < k < 1$

Again, in this case, we shall first demonstrate the existence of free boundary limits in spherical geometries. We use sharper estimates derived from first integrals of the second order O.D.E., corresponding to radial coordinates formulation. For arbitrary domains, we will establish the convergence of variational solutions defined in Section 2.3. We shall conclude by using the comparison lemmas 4.1 and 4.2, for very general domains of class C^3.

5.1 Spherical case

Let us rewrite the perturbed problem (2.1) in spherical coordinates

$$\frac{1}{r^{n-1}}\frac{d}{dr}\left(r^{n-1}\frac{du_\varepsilon}{dr}\right) = \frac{d^2u_\varepsilon}{dr^2} + \frac{n-1}{r}\frac{du_\varepsilon}{dr} = \lambda\frac{u_\varepsilon}{\varepsilon + u_\varepsilon^{1+k}} \tag{5.1}$$

$$\frac{du_\varepsilon}{dr}(0) = 0 \ , \quad u_\varepsilon(1) = 1 \tag{5.2}$$

Lemma 5.1 : $0 \leqslant \dfrac{du_\varepsilon}{dr}(r) \leqslant \sqrt{\dfrac{2\lambda}{1-k}}\, u_\varepsilon^{\frac{1-k}{2}}(r)$

Proof : We multiply (5.1) by $\dfrac{du_\varepsilon}{dr}$

$$0 \leqslant \frac{d^2u_\varepsilon}{dr^2}\frac{du_\varepsilon}{dr} + \frac{n-1}{r}\left(\frac{du_\varepsilon}{dr}\right)^2 = \frac{\lambda u_\varepsilon}{\varepsilon + u_\varepsilon^{1+k}}\frac{du_\varepsilon}{dr} \leqslant \frac{\lambda}{1-k}\frac{d}{dr}\left(u_\varepsilon^{1-k}\right)$$

hence

$$\frac{1}{2}\left(\frac{du_\varepsilon}{dr}(r)\right)^2 + \int_0^r \frac{n-1}{\rho}\left(\frac{du_\varepsilon}{dr}\right)^2 d\rho \leqslant \frac{\lambda}{1-k}\left(u_\varepsilon^{1-k}(r) - u_\varepsilon^{1-k}(0)\right) \tag{5.3}$$

$$\leqslant \frac{\lambda}{1-k}\, u_\varepsilon^{1-k}(r) \qquad \blacksquare$$

Corollary 5.1 : Let $\lambda > 0$, $0 < k < 1$ fixed. Then we can extract a subsequence also noted u_ε such that $u_\varepsilon \to u_0$ in $W^{1,\infty}(0,1)$ weak $*$, in $L^p(0,1)$ strong ($p \geqslant 1$) and in $C^\mu([0,1])$ ($0 < \mu < 1$).

Proof : Apply Sobolev imbeddings.

Remark 5.1 : If we specially consider $\overline{u}_\varepsilon$ and $\underline{u}_\varepsilon$, corollary 5.1 holds for the whole sequence.

Lemma 5.2 : $u_\varepsilon \to u_0$ in $W^{1,p}(0,1)$ strong ($p \geq 1$).

Proof : using the positivity of the Laplacian, $\left\| \frac{d}{dr} (r^{n-1} \frac{du_\varepsilon}{dr}) \right\|_{L^1(0,1)} =$
$\int_0^1 \frac{d}{dr} (r^{n-1} \frac{du_\varepsilon}{dr})\, d\rho = \frac{du_\varepsilon}{dr}(1) \leq cst$, hence $r^{n-1} \frac{du_\varepsilon}{dr}$ is bounded in $W^{1,1}(0,1)$; then $r^{n-1} \frac{du_\varepsilon}{dr} \to r^{n-1} \frac{du_0}{dr}$ in $L^1(0,1)$, since $W^{1,1}(0,1) \hookrightarrow L^1(0,1)$ is a compact injection. It follows that $\frac{du_\varepsilon}{dr} \to \frac{du_0}{dr}$ a.e., and we conclude by Lebesgue's theorem. ∎

As $u_0(r)$ belongs to $C^\mu[0,1]$, we can apply techniques used in the proof of Theorem 4.1, to cemonstrate the existence of a critical radius r_c :

Lemma 5.3 : There exists an unique critical radius $r_c(\lambda)$, $0 \leq r_c < 1$, such that :

$$(5.4)\quad \begin{cases} u_0 \equiv 0 \text{ on } [0,r_c], \\ u_0 > 0 \text{ on }]r_c, 1], \text{ and} \\ \int_{r_c}^1 \rho^{n-1} \frac{d\rho}{u_0^k} < \infty, \text{ and } \frac{du_0}{dr}(r_c^+) < \infty. \end{cases}$$

Proof : Exactly as in Theorem 4.1, except that the conclusion of the proof of Theorem 4.1 does not carry anymore if $0 < k < 1$. There is no more contradiction between $\int_{r_c}^1 \rho^{n-1} \frac{d\rho}{u_0^k} < \infty$ and $\frac{du_0}{dr}(r_c^+) < \infty$ simultaneously. ∎

Three cases can be distinguished :

(i) $\min u_0 > 0$, then $r_c = 0$ and u_0 is a solution of the limit problem ;

(ii) $\min u_0 = 0$, and $r_c = 0$, then $\lambda = \lambda_c$ and $u_0 = u_c$;

(iii) if $r_c > 0$, one has to consider a free boundary value problem (free B.V.P.), with transmission conditions to be determined.

We shall only consider the latter case, (ii) being regarded as a particular case.

Lemma 5.4 : $u_\varepsilon \to u_0$ in $C^\infty(]r_c, 1])$.

Proof : One has to prove the C^∞ convergence on every compact $\subset]r_c, 1]$, say $[r_1,1]$. If ε is small enough, $u_\varepsilon(r) \geq u_0(r) - \eta \geq u_0(r_1) - \eta$ (take for instance

$\eta = \frac{u_0(r_1)}{2}$). Then $\Delta u_\varepsilon(r) \geq \frac{\lambda}{(u_0(r_1)-\eta)^k}$, $\forall\, r \geq r_1$ and the convergence is regular on $[r_1, 1]$. ■

Lemma 5.5 : $u_\varepsilon \to u_0$ in $C^1[0,1]$.

Proof : On $[o,r_c]$, the result is a consequence of lemma 5.1 and corollary 5.1. The case $[r_c, r_c+\gamma]$ is trickier.

We start with $u_\varepsilon \to u_0$ in $C^\alpha(\bar\Omega)$, $0 < \alpha < 1$ (corollary 5.1), and we shall establish that $(r^{n-1}\frac{du_\varepsilon}{dr})^2$ is relatively compact in $C^0(\bar\Omega)$, which yields the desired result (except if $r_c = 0$, where a direct reasoning must be used).

We have :

$$\frac{d}{dr}\left(r^{n-1}\frac{du_\varepsilon}{dr}\right) = \lambda\, r^{n-1}\frac{u_\varepsilon}{\varepsilon + u^{1+k}} \;,\quad 0 \leq \frac{d}{dr}\left(r^{n-1}\frac{du_\varepsilon}{dr}\right) r^{n-1}\frac{du_\varepsilon}{dr} = \lambda\,(r^{n-1})^2\,\frac{u_\varepsilon}{\varepsilon + u_\varepsilon^{1+k}}\,\frac{du_\varepsilon}{dr}\;;$$

thus :

$$0 \leq \frac{1}{2}\frac{d}{dr}\left(r^{n-1}\frac{du_\varepsilon}{dr}\right)^2 \leq \frac{\lambda}{1-k}\frac{d}{dr}\left(u_\varepsilon^{1-k}\right).$$

But from lemma 5.3, the sequences $r^{n-1}\frac{du_\varepsilon}{dr}$ and u_ε^{1-k} are <u>monotone increasing</u>, hence :

$$\left|\left(r_1^{n-1}\frac{du_\varepsilon}{dr}(r_1)\right)^2 - \left(r_2^{n-1}\frac{du_\varepsilon}{dr}(r_2)\right)^2\right| \leq \frac{2\lambda}{1-k}\left|u_\varepsilon^{1-k}(r_1) - u_\varepsilon^{1-k}(r_2)\right|\;;$$

but :

$$\left|u_\varepsilon^{1-k}(r_1) - u_\varepsilon^{1-k}(r_2)\right| \leq \sup_{\substack{x,y\in[0,1]\\ x\neq y}}\frac{|x^{1-k} - y^{1-k}|}{|x-y|^{1-k}}\cdot\left|u_\varepsilon(r_1) - u_\varepsilon(r_2)\right|^{1-k}$$

$$\leq M\left|u_\varepsilon(r_1) - u_\varepsilon(r_2)\right|^{1-k},$$

where M is the Hölder norm of $x \to x^{1-k}$, which is Hölderian of order 1-k ; and :

$$\left|u_\varepsilon(r_1) - u_\varepsilon(r_2)\right|^{1-k} = \left(\frac{|u_\varepsilon(r_1) - u_\varepsilon(r_2)|}{|r_1 - r_2|^\alpha}\times|r_1 - r_2|^\alpha\right)^{1-k}$$

$$< \|u_\varepsilon\|_{C^\alpha}^{1-k}\times|r_1 - r_2|^{\alpha(1-k)}\;;$$

finally :

$$\left|\left(r_1^{n-1}\frac{du_\varepsilon}{dr}(r_1)\right)^2 - \left(r_2^{n-1}\frac{du_\varepsilon}{dr}(r_2)\right)^2\right| \leq \frac{2\lambda M}{1-k}\|u_\varepsilon\|_{C^\alpha}^{1-k}\times|r_1 - r_2|^{\alpha(1-k)}.$$

The sequence $(r^{n-1}\frac{du_\varepsilon}{dr})^2$ being bounded and equicontinuous, we conclude with Arzela-Ascoli's theorem. ■

We put our results in the following form :

Theorem 5.1 : $u_o \in C^1(\bar\Omega)$ is a solution of the free boundary value problem

(5.6) $u_o \equiv 0$ in $[0, r_c]$

(5.7) $\Delta u_o = \frac{\lambda}{u_o^k}$ in $]r_c, 1]$, $u_o(1) = 1$

(5.8) $u_o(r_c^-) = u_o(r_c^+)$

(5.9) $\nabla u_o(r_c^-) = \nabla u_o(r_c^+)$.

Because of $\int_{r_c}^1 \rho^{n-1} \frac{d\rho}{u_o^k} < \infty$ (lemma 5.3), and the transmission condition (5.9), we have $\Delta u_o \in L^1(\Omega)$, Ω being the unit ball.

Corollary 5.1 : For $\lambda > \lambda_*$, $r_c(\lambda)$ is strictly > 0 .

Proof : Any limit $u_o(\lambda)$ satisfies (5.6) ... (5.9) ; yet if $\lambda > \lambda_*$, it cannot satisfy the B.V.P. (1.1) globally in Ω . ∎

Lemma 5.6 : $\Delta u_\varepsilon \to \Delta u_o$ a.e. in Ω.

Proof : We have $\Delta u_\varepsilon \to \Delta u_o$ in $H^{-1}(\Omega)$. Let $\phi \in \mathcal{D}(\Omega)$; as $\varepsilon \to 0$,
$\lim \int_\Omega \Delta u_\varepsilon \phi \, dx = < \Delta u_o , \phi > \equiv - \int_\Omega \nabla u_o \nabla\phi \, dx = + \int_\Omega \Delta u_o \phi \, dx$;
to obtain the last equality, a distributional derivative jump formula must be rederived, using $u_o \in C^1(\bar\Omega)$; the only difficulty lies in the unboundedness of Δu_o, and the estimate $\int_{r_c}^1 \rho^{n-1} \frac{d\rho}{u_o^k} < \infty$ is crucial. Let ϕ be a test function equal to 1 in the ball B_c of radius r_c ; then, from the above result : $\int_{B_c} \Delta u_\varepsilon \, dx \to 0$.

As $\Delta u_\varepsilon \geq 0$, we have $\Delta u_\varepsilon \to 0$ in $L^1(B_c)$, and a subsequence converges a.e. in B_c. In $\bar\Omega / \bar B_c$, the result follows from lemma 5.4. ∎

Next, we need the following convergence result :

Proposition 5.1 : Let a sequence of functions $\{f_j\}_{j=1,\infty}$, such that :

i) $f_j \in L^1(\Omega)$; ii) $f_j \geq 0$; iii) $f_j \to f$ a.e. ;

iv) $\int_\Omega f_j \, dx \to \int_\Omega f \, dx$,

then $f_j \to f$ in $L^1(\Omega)$. [1]

Theorem 5.2 : $\Delta u_\varepsilon \to \Delta u_0$ in $L^1(\Omega)$.

Proof : It remains to prove the result in some interior domain containing the free boundary, say E. We have $\Delta u_\varepsilon \to \Delta u_0$ a.e. in E, $\Delta u_0 \in L^1(E)$; an choosing a test function equal to 1 in E, $\int_E \Delta u_\varepsilon \, dx \to \int_E \Delta u_0 \, dx$ (cf. lemma 5.6). The result follows from proposition 5.1.
■

Corollary 5.2 : $u_\varepsilon \to u_0$ in $W^{2,1}(\Omega)$.
To prove this corollary, one has only to express the partial derivatives.

The free boundary value problem (5.6) ... (5.9) may be studied directly by reducing it to an O.D.E. and resolving it by an exact shooting method (see Appendix 2). Then, with the notations of theorem 1.5, we have the following result describing the global structure of all solutions to the free B.V.P. :

Theorem 5.3 :

i) If $n_-(k) < n < n_+(k)$, then the free boundary value problem (5.6) ... (5.9) admits in the plan (r_c, λ) an infinite number of geometrical bending points $\hat\lambda_1, \hat\lambda_2, \dots$, with $\hat\lambda_i \to \lambda_c$ and $u_0(\hat\lambda_i) \to u_c$ in $H^p(\Omega)$. For $\lambda = \lambda_c$, the free boundary value problem possesses an infinite number of solutions (see Fig. 6)

ii) If $1 < n \leqslant n_-(k)$ or $n \geqslant n_+(k)$, there are no bending points.

iii) As $\lambda \to +\infty$, $u_0(\lambda) \to 0$ a.e. with a boundary layer in a neighbourhood of $\partial\Omega$.

($n_-(k)$, $n_+(k)$ have been defined in (1.8)).

Remark 5.2 : In fact, we have demonstrated that the limit elliptic B.V.P. (1.1) extends into a free B.V.P. beyond the end-point λ_c. The end-point λ_c (cf. definition 1.1) corresponds to a natural junction between the two problems, since the regularity of the free B.V.P. is at best $C^1(\bar\Omega) \cap W^{2,1}(\Omega)$. Moreover, $\forall n \geqslant 1$, $\forall \lambda > 0$, there exists at least one solution of the limit B.V.P. and/or one solution of the free B.V.P. In this sense, the problem of extending (1.1) beyond λ_* has been solved.

---

[1] We have not been able to explicitly locate this theorem in the litterature. We wish to thank H. BREZIS for pointing us a simple proof [14] : Let $g_j = \min(f_j, f)$; by Lebesgue's theorem, $g_j \to f$ in $L^1(\Omega)$, hence $(f - f_j)^+ \to 0$ in $L^1(\Omega)$. Finally note that $\int_\Omega |f - f_j| \, dx = 2 \int_\Omega (f - f_j)^+ \, dx - \int_\Omega (f - f_j) \, dx$.

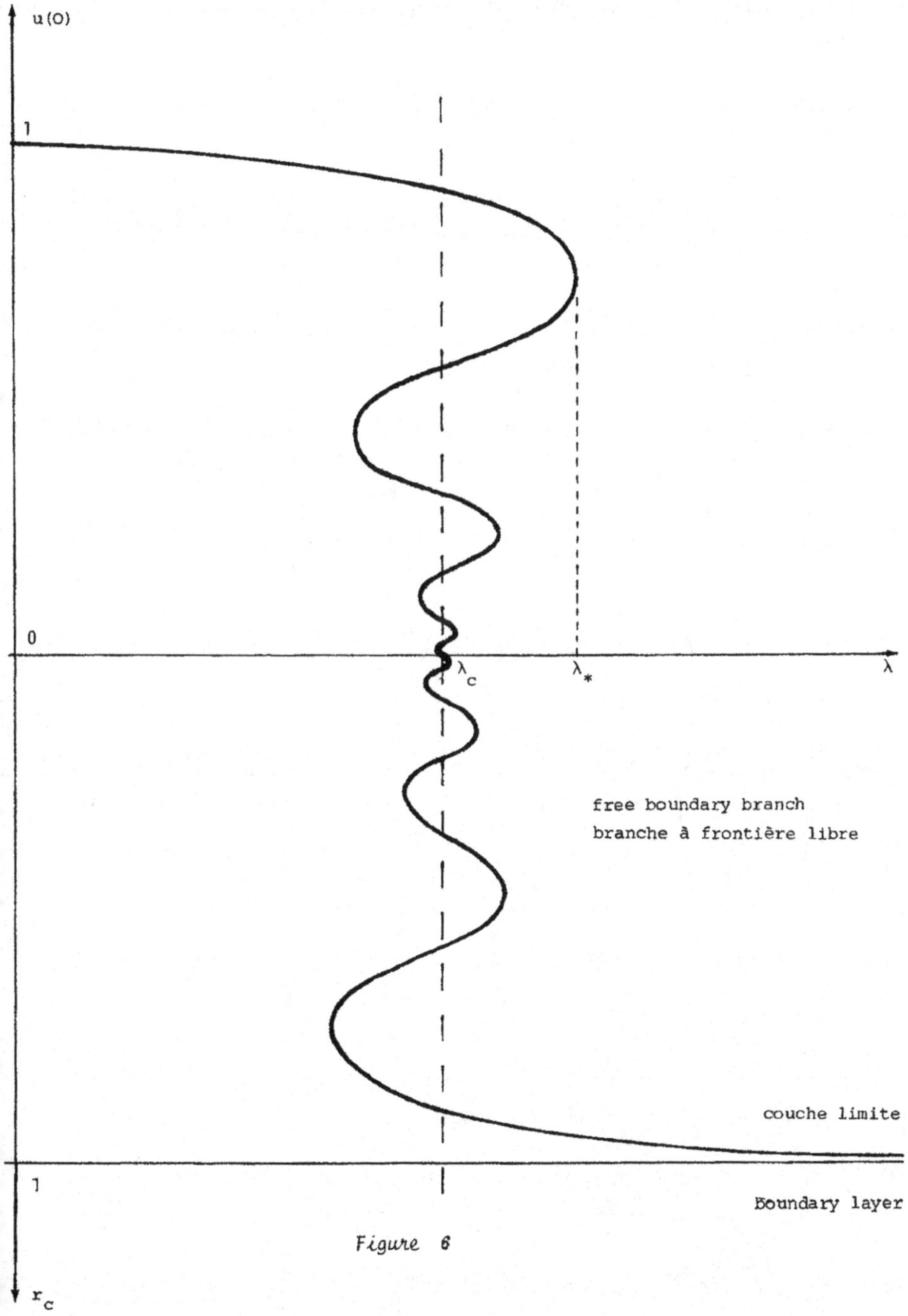
u(O)
1
0
λ_c
λ_*
λ
free boundary branch
branche à frontière libre
couche limite
1
Boundary layer
r_c

Figure 6

Moreover, in a neighborhood of λ_c, if one problem admits an arbitrarily large number of solutions, so does the second one.

Remark 5.3 : As $k \to 1^-$, $r_c \to 1$, and the free boundary degenerates into a boundary layer.

Remark 5.4 : Which of the multiple solutions of the free B.V.P. are actually limits of (2.1) as $\varepsilon \to 0$, is an important open question, as we have unicity only for λ large enough.

5.2 Convergence of the variational solutions

In this section, Ω is a regular bounded domain of $\mathbb{R}^n$. We consider the variational formulation introduced in §2.3 and prove the convergence in $H^1(\Omega)$ strong of solutions u_ε^J obtained by that way. Unfortunately, these results cannot be extended to other solutions.

For $0 < k < 1$, we introduce the functional

$$(5.10) \qquad J_0(v) = \frac{1}{2}\int_\Omega (\nabla v)^2\, dx + \frac{\lambda}{1-k}\int_\Omega v^{1-k}\, dx$$

for $v \in K = \left\{ v \in H^1(\Omega)\ ,\ 0 \leqslant v \leqslant 1\ ,\ v/_{\partial\Omega} = 1 \right\}$

Theorem 5.4 : Let $\lambda > 0$, $0 < k < 1$ fixed. Then, as $\varepsilon \to 0$, one can extract a subsequence also noted u_ε^J, converging in $H^1(\Omega)$ strong to an element u_0 which realizes the infimum of J_0 in K.

Proof : Since $J_\varepsilon(u_\varepsilon^J) \leqslant J_\varepsilon(1)$, $\frac{1}{2}\int_\Omega (\nabla u_\varepsilon^J)^2 \leqslant \lambda \int_\Omega \phi_\varepsilon(1) \leqslant \frac{\lambda \text{ mes } \Omega}{1-k}$

Then the sequence u_ε^J is bounded in $H^1(\Omega)$ and one can extract a subsequence $u_\varepsilon^J \to u_0$ in $H^1(\Omega)$ weak.

Let us show that

$$(5.11) \qquad \phi_\varepsilon(u_\varepsilon) \to \phi_0(u_0) \quad \text{in } L^1(\Omega) \text{ strong ;}$$

one has

$$\phi_\varepsilon(u_\varepsilon)(x) = \int_0^{u_\varepsilon(x)} \frac{y}{\varepsilon + y^{1+k}}\, dy \to \int_0^{u_0(x)} \frac{dy}{y^k} \quad \text{for a.e. } x$$

As $\phi_\varepsilon(u_\varepsilon) \leqslant \frac{1}{1-k}$, we can apply Lebesgue's theorem.

Now we can pass to the limit in the relation

(5.12) $\quad J_\varepsilon(u_\varepsilon) \leqslant J_\varepsilon(v) \quad \forall\, v \in K$

hence

(5.13) $\quad J_0(u_0) \leqslant J_0(v) \quad \forall\, v \in K$

and u_0 realizes the infimum of J_0.

It remains to prove the strong convergence : one verifies that $j_\varepsilon = \inf J_\varepsilon \to j_0 = \inf J_0$, then

(5.14)
$$\frac{1}{2}\int_\Omega (\nabla(u_\varepsilon - u_0))^2\, dx = j_\varepsilon - \int_\Omega \phi_\varepsilon(u_\varepsilon)\, dx + \frac{1}{2}\int_\Omega (\nabla u_0)^2\, dx - \int_\Omega \nabla u_\varepsilon\, \nabla u_0 dx$$
$$\to j_0 - \frac{1}{2}\int_\Omega (\nabla u_0)^2\, dx - \int_\Omega \phi_0(u_0)\, dx = 0 \quad ■$$

Remark 5.5 : In the above reasoning, the fact that $k \in\,]0,1[$ is crucial.

Now it is fundamental to give a nice caracterization of u_0 . Note that J_0 is not differentiable. Moreover, it is the sum of a convex plus a concave functional!

Theorem 5.5 : The function u_0 belongs to $H^1(\Omega)$ and verifies

(5.15)
$$\begin{cases} 0 \leqslant u_0 \leqslant 1 \quad \text{a.e.} \\ u_0/_{\partial\Omega} = 1 \\ -\, u_0^{1+k}\, \Delta u_0 + \lambda\, u_0 = 0 \end{cases}$$

Proof : As u_ε^J verifies (2.1), one has

(5.16) $\quad -\, [\varepsilon + (u_\varepsilon^J)^{1+k}]\ \Delta u_\varepsilon^J + \lambda\, u_\varepsilon^J = 0$

We see that $(u_\varepsilon^J)^{1+k} \to u_0^{1+k}$ at worst in $H^1(\Omega)$ weak, that one can pass to in limit in (5.16) at least in the distributions sense. ■

5.3 Existence of a nontrivial free boundary for arbitrary domains

The $H^1(\Omega)$ convergence of $u_\varepsilon^J(\lambda)$ to $u_0(\lambda)$ implies strong convergence of the traces on $\partial\Omega$ in $H^{1/2}(\partial\Omega)$; hence $u_0(\lambda)/_{\partial\Omega} = 1$, and there cannot be any boundary layer in a neighborhood of $\partial\Omega$.

Moreover, in (5.15), two situations can arise :

i) at $\forall$ x $\in \Omega$ where $u_o > 0$, we can divide both sides (at least formally) by u_o^{1+k}, and u_o satisfies the limit equation.

ii) or $u_o(x) = 0$, in which case (1.1) cannot be satisfied.

For λ large enough, we will in fact show that meas $\Omega_o = \{ x \in \Omega ; u_o(\lambda) = 0 \}$ is strictly positive, and that Ω_o contains a non trivial closed ball. Thus the free boundary $\partial\Omega_o$ is non trivial.

Following notations of Sections 4.2 and 4.3, let $B_{R_{max}}$ be the open ball of maximal **radius** R_{max} such that $B_{R_{max}} \subseteq \Omega$. Let $\bar{u}_\varepsilon(\lambda, R_{max})$ be the maximal solution of (2.1) in $B_{R_{max}}$. Then from the renormalization Remark 4.1 (Equs. (4.3)...(4.4)), and from Theorem 5.3, $\bar{u}_\varepsilon(\lambda, R_{max}) \equiv 0$ within some closed ball concentric to $B_{R_{max}}$.

Theorem 5.6 : Let Ω be any domain of class C^3 , $0 < k < 1$, and $\lambda > \lambda_*(B_1)/R^2_{max}$; let $u_o(\lambda)$ the variational solution given by Theorems 5.4 and 5.5. Then $\Omega_o = \{x \in \Omega ; u_o(x) = 0\}$ contains a closed ball concentric to $B_{R_{max}}$, of radius > 0 .

Proof : Use the comparison lemmas 4.1 and 4.2. Let $w(\lambda,\varepsilon) = \bar{u}_\varepsilon(\lambda, R_{max})$ in $B_{R_{max}}$, and

$$w(\lambda,\varepsilon) = 1 \quad \text{in } \bar{\Omega}/B_{R_{max}} ;$$

then

$$(5.17) \qquad 0 \leqslant u^J_\varepsilon(\lambda) < w(\lambda,\varepsilon) \quad \text{in } \Omega .$$

From theorems 5.1 and 5.3, $\bar{u}_\varepsilon(\lambda, R_{max}) \to 0$ in some closed ball concentric to $B_{R_{max}}$, of radius strictly > 0 , since $\lambda > \lambda_*(B_1)/R^2_{max}$ (cf. Corollary 5.1) moreover , in this concentric ball, the convergence takes place in C^1. Thus $\bar{u}^J_\varepsilon(\lambda) \to 0$ a.e. (in fact in H^1 strong) and $u_o(\lambda) \equiv 0$, with in this concentric ball. ■

Remark 5.6. : We can specify the size of this concentric ball. In fact, from the renormalization remark 4.1 (Equs. (4.3), (4.4)), its radius is $R_{max} \times \bar{r}_c(\lambda R^2_{max})$; $\bar{r}_c(\tilde{\lambda})$ is the free boundary radius for the maximal solution of the free B.V.P. (5.6) ...(5.9) in the unit ball B_1 , corresponding to the parameter value $\tilde{\lambda}$.

Remark 5.7 : In the above theorem, Ω is quite general, and $\partial\Omega$ need not be "spherically regular"!

Remark 5.8 : $\Omega_0 \subset \Omega$ strictly, and $\partial\Omega_0 \cap \partial\Omega = \emptyset$, since $u_\varepsilon^J(\lambda) \to u_0(\lambda)$ in $H^{1/2}(\partial\Omega)$ strong.

For domains with a "spherically regular" $\partial\Omega$, we can even generalize (iii) of Theorem 5.3 :

Theorem 5.7 : Let $0 < k < 1$, Ω with a "spherically regular" $\partial\Omega$. Then, as $\lambda \to +\infty$, $u_0(\lambda) \to 0$ a.e. (in fact in $L^p(\Omega)$, $\forall\ p \geq 1$).

Proof : As in Theorem 4.2, cover Ω with an infinite number of balls $B_{R_{min}}$, either $\subset \Omega$, or $\subseteq \Omega$ and tangent to $\partial\Omega$ at some point. For each of these balls, use lemmas 4.1 or 4.2, and (iii) of Theorem 5.3. Convergence of $u_0(\lambda)$ a.e. to zero as $\lambda \to +\infty$ follows, and one concludes with Lebesgue's theorem.

∎

Remark 5.9, Open Problems :

1) What is the optimal regularity of the free boundary $\partial\Omega_0$, for Ω smooth (of class C^3)? The global $H^1(\Omega)$ convergence estimate is too weak to yield C^0 and even less C^1 regularity as in hyperspheres. How are the classical transmission conditions replaced?

2) What is the adequate generalization of a Variational Inequality formulation, corresponding to the free B.V.P. implicitly defined in theorem 5.5? Recall that the functional J_0 in Theorem 5.4, is neither differentiable, nor convex.

3) Is there an end-point λ_c for (1.1) in all sufficiently regular Ω, and does it correspond to a matching with the free B.V.P., as in the case of hyperspheres (Theorem 5.3)?

Work on the above questions is in progress.

6. CASE $k = 0$

This specific example is most interesting and enlightening, since the free boundary value problem is characterized by a variational inequality (V.I.). Moreover, this case has a definite pedagogical interest (see Remark 6.4). Let Ω be a regular bounded open set in $\mathbb{R}^n$. When $k = 0$, the limit problem (1.1) simply becomes :

(6.1) $$\Delta u = \lambda \ , \ u/_{\partial\Omega} = 1 \ .$$

We immediately see that, for λ large enough, $u(\lambda)$ is locally < 0 (i.e. < 0 on a set of measure > 0). More precisely, let ξ be the solution of

(6.2) $$\Delta\xi = 1 \ , \ \xi/_{\partial\Omega} = 0$$

Clearly, $\xi < 0$, $\forall x \in \Omega$, and $u(\lambda) = \lambda\xi + 1$. Let $M = |\xi|_{L^\infty(\Omega)}$. The minimum of $u(\lambda)$ is $1 - \lambda M$; thus $u(\lambda)$ is locally < 0 as soon as $\lambda > \lambda_* = 1/_M$.

Example : Let Ω be the unit ball in $\mathbb{R}^n$, $u(\lambda) = 1 - \frac{\lambda}{2n}(1 - r^2)$, $\xi = -\frac{(1-r^2)}{2n}$
$M = \frac{1}{2n}$ and $\lambda_* = 2n$.

We see that λ_* is a very special case of an end-point for positive solutions of (6.1). There are two possibilities to extend the problem beyond λ_* :

(i) Extension by locally negative solutions, which, in general, shall be physically meaningless.

(ii) Extension by solutions to a free boundary value problem, obtained through the singular limit of a singular perturbation problem.

The perturbed problem becomes, as $k = 0$,

(6.3) $$\Delta u_\varepsilon = \lambda \frac{u_\varepsilon}{\varepsilon + u_\varepsilon} \ , \ u_\varepsilon/_{\partial\Omega} = 1 \ .$$

For fixed $\varepsilon > 0$, this problem admits an unique solution $u_\varepsilon > 0$, $\forall \lambda > 0$. One always has $u_\varepsilon(\lambda) \geqslant u(\lambda)$, where $u(\lambda)$ is solution of (6.1).

The crucial point is that the nonlinear term $\frac{u_\varepsilon}{\varepsilon + u_\varepsilon}$ is bounded by 1 (of course this result is false if $k > 0$). One immediately deduces the estimate $0 < \Delta u_\varepsilon < \lambda$, which yields that the sequence u_ε is bounded in the space $W^{2,p}(\Omega)$ $\forall p \geqslant 1$.

Theorem 6.1 : There exists an unique $u_0(\lambda) \in W^{2,p}(\Omega)$, $\forall p \geqslant 1$, such that $u_\varepsilon(\lambda) \searrow u_0(\lambda)$ in $W^{2,p}(\Omega)$ weak. We have $u_0/_{\partial\Omega} = 1$.

Corollary 6.1 : Let λ verify $0 < \lambda < \lambda_*$. Then $u_0(\lambda) = u(\lambda)$ and $u_\varepsilon \searrow u$ in $C^\infty(\bar{\Omega})$.

Proof : If $\lambda < \lambda_*$, $u(\lambda) \geq 1 - \lambda M > 0$; hence $u_\varepsilon \geq - \lambda M + 1$ and $\frac{u_\varepsilon}{\varepsilon + u_\varepsilon} \to 1$.
One concludes with Lebesgue's Theorem. ■

Remark 6.1 : The above demonstration still applies when $\lambda = \lambda_*$, for the set of points where u reaches its minimum, has null measure ; hence $\frac{u_\varepsilon}{\varepsilon + u_\varepsilon} \to 1$ a.e.

Corollary 6.2 : Let $\lambda > \lambda_*$. Then $u_0(\lambda)$ is null on a set of measure > 0 .

Proof : Suppose that $u_0 > 0$ a.e. in Ω , then $\frac{u_\varepsilon}{\varepsilon + u_\varepsilon} \to 1$ a.e. , and $u_0(\lambda) \equiv u(\lambda)$.
But $u < 0$ locally on a set of measure > 0, and we reach a contradiction. ■

More precisely, we shall demonstrate that u_0 is solution to a variational inequality (V.I.).

Theorem 6.2 : Let $K = \left\{ v \in H^1(\Omega) \text{ , } v \geq 0 \text{ , } v/_{\partial\Omega} = 1 \right\}$. Then $u_0(\lambda) \in K$, and is the solution of the V.I. :

(6.4) $\quad (\nabla u_0, \nabla(v-u_0)) + (\lambda, v-u_0) \geq 0 \quad \forall\, v \in K$.

Proof : Let $v \in K$. Multiply (6.3) by $(v-u_\varepsilon)$:

$$(-\Delta u_\varepsilon \text{ , } v-u_\varepsilon) + \lambda \left(\frac{u_\varepsilon}{\varepsilon + u_\varepsilon} \text{ , } v-u_\varepsilon \right) = 0 \text{ ;}$$

hence

$$(-\Delta u_\varepsilon \text{ , } v-u_\varepsilon) + (\lambda, v-u_\varepsilon) + \lambda \left(\frac{\varepsilon}{\varepsilon + u_\varepsilon} \text{ , } u_\varepsilon \right) = \lambda \left(\frac{\varepsilon}{\varepsilon + u_\varepsilon} \text{ , } v \right) \geq 0 \text{ ;}$$

and

$$\left(\frac{\varepsilon}{\varepsilon + u_\varepsilon} \text{ , } u_\varepsilon \right) = \varepsilon \int_\Omega \frac{u_\varepsilon}{\varepsilon + u_\varepsilon} \, dx \to 0 \text{ .}$$

In the limit, we have indeed :

$$(-\Delta u_0 \text{ , } v-u_0) + (\lambda, v-u_0) \geq 0 \text{ . } \quad \forall\, v \in K \text{ .}$$
■

Theorem 6.3 (LIONS [24]) : we have the following error estimate :

(6.5) $\quad ||u_0 - u_\varepsilon||_{H^1(\Omega)} \leq \text{Cst} \sqrt{\varepsilon}$

Proof : Multiply (6.3) by $u_0 - u_\varepsilon$, and take $v = u_\varepsilon$ into (6.4) :

$$(\nabla u_\varepsilon, \nabla(u_0-u_\varepsilon)) = (-\lambda, u_0-u_\varepsilon) + \lambda \left(\frac{\varepsilon}{\varepsilon + u_\varepsilon} \text{ , } u_0 - u_\varepsilon \right)$$

$$(\nabla u_0, \nabla(u_\varepsilon - u_0)) \geq (-\lambda, u_\varepsilon - u_0)$$

Hence

$$-|\nabla(u_\varepsilon - u_0)|^2 \geq \lambda \left(\frac{\varepsilon}{\varepsilon + u_\varepsilon}, u_0 - u_\varepsilon\right)$$

and

$$|\nabla(u_\varepsilon - u_0)|^2 \leq -\lambda \left(\frac{\varepsilon}{\varepsilon + u_\varepsilon}, u_0\right) + \lambda\left(\frac{\varepsilon}{\varepsilon + u_\varepsilon}, u_\varepsilon\right) \leq \lambda\, \varepsilon \text{ meas } \Omega .$$ ■

Remark 6.2 : The convergence of u to u_0 in $W^{2,p}(\Omega)$ weak, $\forall p \geq 1$, implies of course strong convergence in $C^{1,\alpha}(\bar{\Omega})$, $\forall \alpha$, $0 \leq \alpha < 1$. For $\lambda > \lambda^*$, meas $\Omega_0 = \{x \in \Omega ; u_0(x) = 0\} > 0$, which implies a regular free boundary. Transmission conditions are satisfied (continuity of u_0 and its gradient across the free boundary).

Sumarizing :

Corollary 6.3 : $\forall \lambda > \lambda_*$, the set $\Omega_0 = \{x \in \Omega ; u_0(x) = 0\}$ is bounded by a free boundary of class $C^{1,\alpha}$, $0 \leq \alpha < 1$. As $\lambda \to +\infty$, $\Omega_0 \to \Omega$.

Remark 6.3 : From this section's results, one could obtain more information on the free boundary limit $u_0(\lambda, k)$ of the perturbed problem considered in the avove §

$$(6.6) \qquad \Delta u_\varepsilon = \lambda \frac{u_\varepsilon}{\varepsilon + u_\varepsilon^{1+k}}, \quad u_\varepsilon/_{\partial\Omega} = 1, \; k > 0$$

(and also

$$(6.6)' \qquad \Delta u_\varepsilon = \lambda u_\varepsilon \left(\frac{\varepsilon + 1}{\varepsilon + u}\right)^{1+k}) ;$$

indeed, as $u_\varepsilon(\lambda, 0) \geq u_\varepsilon(\lambda, k)$, this insures u_0 that a free boundary exists for $\lambda > \lambda_*$ (λ_* defined for (6.1)). However, a practical computation of λ_* is not trivial, except in the case of balls.

Remark 6.4 : Conversely can equation (6.3) be considered as an approximation of the V.I. (6.4). This may lead to new demonstrations of classical results on V.I. We refer to BENSOUSSAN - LIONS [5] for a development of this point of view in a general framework.

APPENDIX 1

(Proof of Theorem 1.5)

========

The basic idea is to transform problem(2.1) into an ordinary differential equation (O.D.E.).

Let us introduce the following initial value problem (I.V.P.)

(A.1) $$\frac{d^2w}{d\rho^2} + \frac{n-1}{\rho}\frac{dw}{d\rho} = \frac{1}{w^k}, \quad 0 \leq \rho < +\infty$$

(A.2) $$w(0) = 1, \quad \frac{dw}{d\rho}(0) = 0$$

Lemma A.1 : The I.V.P. (A.1) (A.2) possesses a unique regular solution on $\mathbb{R}^+$.

Proof : We just have to prove the existence of a unique solution on some interval $[0, \rho_0]$. The continuation of the solution to $\mathbb{R}^+$ will follow from classical Theorems on O.D.E.

We know that problem (2.1) admits a unique solution $u(\lambda, r)$ in a neighbourhood of $\lambda = 0$, $u = 1$. Let us set $a = u(\lambda, 0)$, $\rho_0 = \sqrt{\lambda}/a$, and define

(A.3) $$w(\rho) = \frac{1}{a} u\left(\frac{\sqrt{\lambda}}{a}\rho\right) \text{ for } \rho \in [0, \rho_0]$$

Then $w(\rho)$ is a solution of (A.1) (A.2) on $[0, \rho_0]$. Reciprocally, if $w(\rho)$ is a solution of (A.1) (A.2) on $[0, \rho_0]$, we set

(A.4) $$u(r) = \frac{w(\rho_0 r)}{w(\rho_0)}, \quad \lambda = \frac{\rho_0^2}{w(\rho_0)^{k+1}}$$

and $u(r)$ is solution of (2.1). Hence every solution of (2.1) is obtained via the I.V.P. (A.1) (A.2). In particular the uniqueness on $[0, \rho_0]$ is guaranteed. ■

Lemma A.2 : As $\rho \to +\infty$,

(A.5) $$w(\rho) \sim \left(\frac{1}{\lambda_c}\right)^{\frac{1}{k+1}} \rho^{\frac{2}{k+1}}$$

Proof : We set $M = \left(\frac{1}{\lambda_c}\right)^{\frac{1}{k+1}}$, $\beta = \frac{2}{k+1}$

and consider the function defined by

(A.6) $$v(\rho) = \frac{1}{M} w(\rho)\, \rho^{-\beta}$$

Setting $\rho = e^s$, the function $v = v(s)$ verifies the equation

(A.7) $$v''(s) + (2\beta + n-2)\, v'(s) + \lambda_c \Big(v(s) - \frac{1}{v^k(s)}\Big) = 0$$

and we demonstrate that $v(s) \to 1$ as $s \to +\infty$. This is obtained by considering the integrodifferential equation, for any s_0 :

$$\frac{1}{2}\,(v'(s))^2 + (2\beta+n-2)\int_{s_0}^{s} (v'(\tau))^2\, d\tau + \lambda_c \Big(\frac{1}{2}\, v^2(s) - \frac{1}{1-k}\, v^{1-k}(s)\Big) = \text{constant}.$$

■

Now we consider the function

(A.8) $$\phi(s) = v(s) - 1$$

and we define (see Figure 3)

(A.9) $$n_\pm(k) = \frac{6k+2}{k+1} \pm \sqrt{\frac{k}{k+1}}$$

<u>*Lemma A.3*</u> : As $s \to +\infty$,

(i) if $n_- < n < n_+$, ϕ admits a infinite number of oscillations ;

(ii) if $1 < n \leqslant n_-$ or $n \geqslant n_+$, $\phi \to 0$ and keeps a constant sign.

<u>*Proof*</u> : ϕ verifies the system

(A.10) $$\begin{pmatrix}\phi\\ \phi'\end{pmatrix}' = \begin{pmatrix}0 & 1\\ -(k+1)\lambda_c & 2-n-2\alpha\end{pmatrix}\begin{pmatrix}\phi\\ \phi'\end{pmatrix} + \begin{pmatrix}0\\ \lambda_c R(\phi)\end{pmatrix}$$

where $R(\phi) = \dfrac{1}{(1+\phi)^k} - 1 + k\,\phi$

The linear approximation of (A.10) is

(A.11) $$\begin{pmatrix}\phi\\ \phi'\end{pmatrix}' = B\begin{pmatrix}\phi\\ \phi'\end{pmatrix}$$

where B is the above matrix.

The eigenvalues of B are the solutions of

(A.12) $\mu^2 + \mu(2\alpha + n - 2) + (k+1)\lambda_c = 0$

If $1 < n < n_-$, equation (A.12) admits 2 different real roots ; if $n = n_-$ or $n = n_+$, one double real root ; if $n_- < n < n_+$, two conjugate complex roots.

Now we consider a neighbourhood of the critical point $\phi = \phi' = 0$: If $1 < n \leq n_-$ and $n \geq n_+$, (0, 0) is a nodal point ; if $n_- < n < n_+$, (0, 0) is a spiral point. ■

APPENDIX 2

(Proof of Theorem 5.3)

=======

As in the above demonstration, we shall transform the free boundary value problem (5.6) ... (5.9) into a O.D.E.

Let us consider, for $0 < k < 1$, the I.V.P.

(A.13) $$\frac{d^2w}{d\rho^2} + \frac{n-1}{\rho}\frac{dw}{d\rho} = \frac{1}{w^k} , \quad 0 \leq \rho < +\infty,$$

(A.14) $$w(1) = 0 , \quad \frac{dw}{d\rho}(1) = 0$$

This I.V.P. has a regular solution on $]1, +\infty[$. The existence in the small is proved by considering the nonlinear integrodifferential equation

(A.15) $$\frac{1}{2} w'^2(\rho) + \int_1^\rho \frac{n-1}{\tau} w'^2(\tau)\, d\tau = \frac{1}{1-k} w^{1-k}(\rho)$$

To pass from the solution $w(\rho)$ on $[0, \rho_o]$ to the solution of the free boundary value problem, we define

(A.16) $$\begin{cases} r_c = \dfrac{1}{\rho_o} , \quad u \equiv 0 \quad \text{on} \quad [0, r_c] \\[2mm] u(r) = \dfrac{w(\rho_o r)}{w(\rho_o)} \quad \text{on} \quad]r_c, 1] \\[2mm] \lambda = \dfrac{\rho_o^2}{w(\rho_o)^{k+1}} \end{cases}$$

The asymptotic behaviour of the solution of (A.13) (A.4) is also given by lemmas A.2 and A.3 (only some constants are modified), hence the theorem. ■

REFERENCES

[1] H. AMMANN, On the existence of positive solutions of nonlinear elliptic boundary value problems, Ind. Univ. Math. J., 21, p. 125-146 (1971).

[2] H. AMANN, Fixed point equations and nonlinear eigenvalue problems in ordered Banach spaces, SIAM Review 18, p. 620-709 (1976).

[3] R. ARIS, The mathematical theory of diffusion and reaction, clarendon Press, Oxford (1975).

[4] H.T. BANKS, Modeling and control in the biomedical sciences, Lect. Notes in Biomathematics 6, Springer-Verlag (1975).

[5] A. BENSOUSSAN and J.L. LIONS, Application des inéquations variationelles en contrôle stochastique, Dunod (volume 2, to appear).

[6] H. BERESTYCKI and P.L. LIONS, This volume.

[7] C.M. BRAUNER, Perturbations singulières dans des systèmes non linéaires et applications à la biochimie, Thèse, Université Paris-Sud (1975).

[8] C.M. BRAUNER and B. NICOLAENKO, Perturbation singulière, solutions multiples et hystérésis dans un problème de biochimie, C.R. Acad. Sc. Paris, Série A, 283, p. 775-778 (1976).

[9] C.M. BRAUNER and B. NICOLAENKO, Singular perturbation, multiple solutions and hysteresis in a nonlinear problem, Lect. Notes in Math. 594, Springer-Verlag, p. 50-76 (1977).

[10] C.M. BRAUNER and B. NICOLAENKO, Sur une classe de problèmes elliptiques non linéaires, C.R. Acad. Sc. Paris, Série A, 286, p. 1007-1010 (1978).

[11] C.M. BRAUNER and B. NICOLAENKO, Sur des problèmes aux valeurs propres non linéaires qui se prolongent en problèmes à frontière libre, C.R. Acad. Sci. Paris, Série A, 287, p. 1105-1108 (1978), and 288, p. 125-127 (1979).

[12] C.M. BRAUNER and B. NICOLAENKO, To appear.

[13] C.M. BRAUNER, B. GAY, and B. NICOLAENKO, Colloque d'Analyse Numérique, Giens (1978).

[14] H. BREZIS (Private communication).

[15] A.J. CALLEGARI, H.B. KELLER and E.L. REISS, Membrane buckling : a study of solution multiplicity, C.P.A.M., 24, p. 499-527 (1971).

[16] M.G. CRANDALL and P.H. RABINOWITZ, Bifurcation from simple eigenvalues, J. Funct. Anal., 8, p. 321-340 (1971).

[17] M.G. CRANDALL and P.H. RABINOWITZ, Bifurcation, perturbation of simple eigenvalues and linearized stability, Arch. Rat. Mech. Anal., 52, p. 161-180 (1973).

[18] M.G. CRANDALL and P.H. RABINOWITZ, Some continuation and variational methods for positive solutions of nonlinear elliptic eigenvaluc problems, Arch. Rat. Mech. Anal., 58, p. 207-218 (1975).

[19] J.P. KEENER and H.B. KELLER, Positive solutions of convex nonlinear eigenvalue problems, J. Diff. Equ., 16, p. 103-125 (1974).

[20] J. LERAY, Thèse, Paris (1934), J. Math. Pures et appl., 12, p. 1-80 (1933).

[21] J. LERAY and J. SCHAUDER, Topologie et équations fonctionnelles, Ann. Sci. Ecole Norm. Sup., 3, vol. 51, p. 45-78 (1934).

[22] J.L. LIONS, Quelques méthodes de résolution des problèmes aux limites non linéaires, Dunod (1969).

[23] J.L. LIONS, Perturbations singulières dans les problèmes aux limites et en contrôle optimal, Lect. Notes in Math. 323, Springer-Verlag (1973).

[24] J.L. LIONS (Private communication).

[25] F. MIGNOT and J.P. PUEL, Sur une classe de problèmes non linéaires avec non linéarité positive croissante, convexe, Colloque d'Analyse non linéaire, Rome, Mai 1978.

[26] J.P. PUEL, Existence, comportement à l'infini et stabilité dans certains problèmes quasilinéaires elliptiques et parabolique d'ordre 2, Ann. Sc. Norm. Pisa, 3, p. 85-119 (1976).

[27] P.H. RABINOWITZ, Some global results for nonlinear eigenvalue problems, J. Funct. Anal., 7, p. 487-513 (1971).

[28] D.H. SATTINGER, Topics in stability and bifurcation theory, Lect. Notes in Math. 309, Springer-Verlag (1973).

[29] K. STEWARTSON, Further solutions of the Falkner-Skan equation, Proc. Camb. Phil. Soc, 50, p. 454-465 (1954).

[30] A.N. IL'IN, A.S. KALASHNIKOV, and O.A. OLEINIK, Linear equations of the second order of parabolic type, Russian Math. Surveys, 17, N° 3, p. 1 - 143 (1962).

[31] J. FAVARD, Cours de Géométrie, Gauthier - Villars, Paris (1957).

INTRODUCTION AUX THEORIES STATISTIQUES DE LA TURBULENCE PLEINEMENT DEVELOPPEE

Marcel LESIEUR et Jean-Pierre CHOLLET
Institut de Mécanique de Grenoble

RESUME

Une des principales caractéristiques de la turbulence pleinement développée est l'accroissement considérable des coefficients de transport du fluide. L'importance relative de la diffusion turbulente et de la diffusion moléculaire se mesure par un paramètre appelé le nombre de Reynolds, qui caractérise également le rapport entre les termes non linéaires et le terme de dissipation visqueuse dans les équations de Navier-Stokes. Une simulation numérique directe d'un écoulement turbulent n'est possible que pour des nombres de Reynolds modérés. A grand nombre de Reynolds, une prédiction déterministe des grandes échelles suppose une paramétrisation correcte des petites échelles, et est rendue impossible à long terme à cause de la propagation dans les grandes échelles de l'incertitude initiale sur les petites échelles (imprédicibilité). On rappelle la théorie phénoménologique de KOLMOGOROV et la loi de RICHARDSON. On présente les théories statistiques dérivées de la théorie quasi-normale (théorie quasi-normale markovianisée avec amortissement turbulent EDQNM). Ces théories conduisent, à viscosité nulle, à une divergence de l'enstrophie (vorticité carrée moyenne) au bout d'un temps fini, ainsi qu'à une dissipation finie d'énergie cinétique. On interprète ces résultats comme une transition entre écoulement laminaire et écoulement turbulent. On propose enfin, dans le cas instationnaire du problème d'évolution, une méthode de paramétrisation compatible avec l'apparition d'un spectre de KOLMOGOROV au bout d'un temps fini.

ABSTRACT

One of the main features of fully developed turbulence is a considerable increase of the value of transport coefficients. The relative importance of eddy diffusion and molecular diffusion can be measured by a parameter called the Reynolds number, which characterizes also the ratio between the nonlinear terms and the dissipative term in the Navier-Stokes equations. A direct numerical simulation of a turbulent flow is only possible at moderate Reynolds numbers. At high Reynolds numbers, a deterministic prediction of large scales encounters the problem of the parameterization of small scales (subgridscale modeling), and is for long times made impossible because of the propagation towards large scales of the initial uncertainty in small scales (unpredictability). The phenomenological theory of KOLMOGOROV and the RICHARDSON law are recalled. Statistical theories derived from the quasi-normal approximation (eddy-damped quasi-normal markovian theory - EDQNM) are presented.

At zero viscosity these theories yield a divergence of the enstrophy (mean square vorticity) at a finite time, and a finite dissipation of kinetic energy. These results are interpreted as a transition between laminar and turbulent flow. Finally we propose a method of parameterization which, in the non-stationary problem of evolution, is compatible with the appearance of a KOLMOGOROV spectrum at a finite time.

1. INTRODUCTION

Il est extrêmement difficile de donner une définition précise de la turbulence dans un fluide. On emploie ce terme pour désigner des écoulements très désordonnés, d'aspect chaotique, que l'on rencontre dans un grand nombre de situations : en laboratoire, dans des installations industrielles, en aéronautique, dans l'atmosphère, l'océan, en astrophysique, etc...[+] *On peut montrer expérimentalement que cet état se manifeste lorsqu'un nombre sans dimension lié à l'écoulement, appelé le nombre de Reynolds, dépasse une certaine valeur critique. Ce nombre de Reynolds*

$$R = \frac{UL}{\nu} \tag{1-1}$$

est construit à partir d'une vitesse caractéristique U *de l'écoulement, d'une longueur caractéristique* L *(par exemple la dimension de l'obstacle induisant la turbulence), et de la viscosité cinématique* ν *. La signification physique du nombre de Reynolds peut être comprise comme suit : considérons un "tourbillon" de vitesse* U *et de dimension* L *; le temps mis par une particule fluide pour parcourir ce tourbillon est de l'ordre de* $T_U = L/U$ *, alors que le temps caractéristique pour que la particule fluide perde sa quantité de mouvement par diffusion moléculaire est de l'ordre de* $T_\nu = L^2/\nu$ *(* ν *est en effet le coefficient de diffusion moléculaire de la quantité de mouvement). Le nombre de Reynolds apparaît donc comme le rapport* T_ν / T_U *: pour un nombre de Reynolds petit devant* 1 *, une perturbation n'aura pas la possibilité de se développer et sera immédiatement amortie par la viscosité. La turbulence sera d'autant plus développée que le nombre de Reynolds est grand devant* 1 *. Nous allons dans ce qui suit nous intéresser uniquement aux écoulements à densité constante. Nous excluons donc à priori l'étude de problèmes tels que la turbulence supersonique ou les problèmes liés à la convection. Nous supposerons enfin que l'écoulement s'étend sur un domaine infini, ce qui est la situation la plus favorable pour que la turbulence puisse se développer sans rencontrer de frontières qui pourraient perturber l'évolution de l'écoulement. Soit* $\vec{u}(\vec{x},t)$ *la vitesse eulérienne*

[+] *on pourra consulter à ce sujet "le monde vu par un dynamicien des fluides" (MOFFATT, 1973).*

du fluide (vitesse de la particule fluide située en $\vec{x}$ à l'instant t), $p(\vec{x},t)$ sa pression, $\rho(\vec{x},t)$ sa masse volumique. Les équations de l'écoulement (équations de Navier-Stokes) s'écrivent alors :

$$\begin{cases} \dfrac{\partial \vec{u}}{\partial t} + (\vec{u}.\vec{\nabla})\vec{u} = -\dfrac{1}{\rho}\vec{\nabla}p + \nu\Delta\vec{u} & (1\text{-}2) \\ \vec{\nabla}.\vec{u} = 0 & (1\text{-}3) \end{cases}$$

L'équation (1-3) exprime l'incompressibilité, et l'équation (1-2) le bilan de quantité de mouvement pour une particule fluide. Dans l'équation (1-2), les termes $(\vec{u}.\vec{\nabla})\vec{u}$ et $\frac{1}{\rho}\vec{\nabla}p$ sont nonlinéaires quadratiques par rapport à $\vec{u}$. En effet, prenant la divergence de (1-2), et compte-tenu de (1-3), il vient :

$$-\frac{1}{\rho}\Delta p = \frac{\partial^2 u_i u_j}{\partial x_i \partial x_j} \qquad (1\text{-}4)$$

Pour un tourbillon de taille L et de vitesse U, les termes nonlinéaires sont de l'ordre de U^2/L, alors que le terme visqueux est de l'ordre de $\nu U/L^2$. Le nombre de Reynolds construit sur U et L caractérise donc l'importance relative du terme non linéaire et du terme d'amortissement visqueux. On sera parfois amené, dans l'étude de certains problèmes stationnaires, à introduire des forces extérieures dans (1-2). Quand la viscosité est nulle(fluide parfait), les équations obtenues s'appellent les équations d'Euler. Dans le cas particulier de l'écoulement à densité constante, le problème thermique et le problème dynamique sont découplés : la température $T(\vec{x},t)$ d'une particule fluide diffuse alors par agitation moléculaire, et satisfait l'équation :

$$\frac{\partial T}{\partial t} + (\vec{u}.\vec{\nabla})T = \kappa\,\Delta T \qquad (1\text{-}5)$$

où κ est la conductibilité thermique (coefficient de diffusion thermique moléculaire).

Il n'est pas question de citer ici tous les travaux mathématiques récents concernant les équations de Navier-Stokes et d'Euler. Le lecteur pourra se référer à l'ouvrage "Turbulence and Navier-Stokes Equation" (édité par R. TEMAM, 1975), et à l'article de revue de ROSE et SULEM (1978). Dans le cas bidimensionnel, on peut considérer que les problèmes d'existence, d'unicité et de régularité des solutions sont résolus. Dans le cas tridimensionnel on sait démontrer l'existence, l'unicité et la régularité des solutions des équations d'Euler jusqu'à un temps fini t_* dépendant des conditions initiales (KATO, 1972). Une conjecture, qui s'appuie sur l'étude de *certains modèles liés aux théories statistiques exposées au paragraphe 4, expliquerait l'apparition de la turbulence dans un fluide parfait comme due à la perte de régularité des solutions des équations d'Euler au bout d'un temps fini : les gradients de vitesse, en étirant les lignes de tourbillon, provoqueraient une croissance catastrophique du tourbillon $\overrightarrow{rot\, u}$ qui localement deviendrait infini au bout d'un temps fini. Une autre conjecture concernant l'évolution ultérieure est que les singularités seraient sur un support de dimension fractionnaire (MANDELBROT, 1975). Nous reviendrons là-dessus à propos de l'intermittence.*

D'un point de vue physique , une des manifestations de la turbulence est l'augmentation considérable des coefficients de transport. Un polluant transporté par l'écoulement (fumée d'une cheminée d'usine par exemple) diffuse beaucoup plus rapidement dans un écoulement turbulent que dans un écoulement laminaire, et le brassage induit par la turbulence est beaucoup plus intense que la diffusion engendrée par l'agitation moléculaire. Les premières théories de la turbulence reposaient justement sur une analogie avec la théorie cinétique des gaz : de même que la viscosité moléculaire (coefficient de diffusion moléculaire de la quantité de mouvement) est proportionnelle au produit du libre parcours moyen des molécules par une vitesse caractéristique d'agitation, de même la diffusion turbulente peut être caractérisée par une "viscosité turbulente" (BOUSSINESQ, 1897) égale au produit d'une "longueur de mélange" L_M par une vitesse caractéristique de l'agitation turbulente $\mathcal{V}$ (PRANDTL, 1925). Nous voyons là une nouvelle interprétation physique du nombre de Reynolds, comme rapport entre la viscosité turbulente et la viscosité moléculaire.

2. DETERMINISME ET TURBULENCE

Les équations de Navier-Stokes sont des équations "déterministes" dans le sens que la connaissance des conditions initiales $\vec{u}(\vec{x},o)$ (et des conditions aux limites quand il y a des frontières) doit en principe permettre de déduire $\vec{u}(\vec{x},t)$ pour tout temps. Ceci suppose bien entendu que nous disposions d'un théorème d'unicité. Sous cette seule restriction, une prédiction déterministe de l'écoulement peut être envisagée à condition de connaître $\vec{u}(\vec{x},O)$ pour tout $\vec{x}$. La résolution numérique doit prendre en compte toutes les structures présentes dans l'écoulement, des plus petites jusqu'aux plus grandes : les plus petites échelles sont les échelles en dessous desquelles toute perturbation est immédiatement amortie par la viscosité moléculaire. Elles ont donc des vitesses et échelles caractéristiques u_D et ℓ_D telles que le nombre de Reynolds associé soit de l'ordre de 1 , c'est à dire :

$$u_D \sim \frac{\nu}{\ell_D} \qquad (2\text{-}1)$$

Nous verrons au paragraphe 3 que la théorie phénoménologique de KOLMOGOROV conduit à la relation :

$$\frac{U^3}{L} = \frac{u_D^3}{\ell_D} \qquad (2\text{-}2)$$

dont on déduit :

$$R = \frac{UL}{\nu} = \left(\frac{L}{\ell_D}\right)^{4/3} \qquad (2\text{-}3)$$

L'échelle ℓ_D est appelée "échelle de Kolmogorov". A l'heure actuelle, on ne peut pas, même avec les plus gros ordinateurs, simuler numériquement des écoulements tri-dimensionnels ayant des nombres de Reynolds supérieurs à plusieurs centaines (ORSZAG et PATTERSON, 1972). En général les écoulements qui intéressent les utilisateurs ont des nombres de Reynolds supérieurs à plusieurs dizaines ou centaines de milliers. Ces écoulements ne peuvent donc être simulés directement sur ordinateur, puisque la discrétisation choisie aurait une maille ℓ_c très supérieure à ℓ_D et

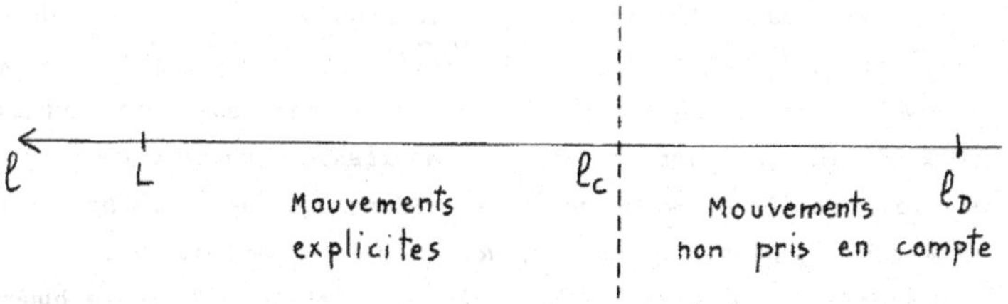

ne prendrait pas en compte les mouvements d'échelle inférieure à la maille. On est donc confronté au problème de la paramétrisation des petites échelles dans le calcul des grandes échelles, qui consiste à rajouter dans les équations d'évolution des modes explicites des termes tenant compte de l'interaction de ces échelles ($\ell > \ell_c$) avec les échelles de taille inférieure à la dimension de la maille ℓ_c (subgridscales). Jusqu'à présent les tentatives pour résoudre ce problème ont repris le vieux concept de viscosité turbulente, en assimilant l'effet des échelles $\ell < \ell_c$ à un "amortissement turbulent" sur les échelles $\ell > \ell_c$. Nous discuterons cette question au paragraphe 6.

Le problème de la paramétrisation des petites échelles en contient un autre, celui de la prédicibilité de l'écoulement : on peut montrer, sur la base des modèles statistiques développés au paragraphe 4 , que deux écoulements statistiquement identiques, initialement complètement corrélés sauf dans les petites échelles, vont progressivement se décorréler, jusqu'à devenir complètement indépendants (LORENZ, 1969, LEITH et KRAICHNAN, 1972). La généralisation de ces résultats aux équations de Navier-Stokes semble ne pas faire de doute dans le cas bidimensionnel mais peut se discuter dans le cas tridimensionnel. L'imprédicibilité de l'écoulement aurait pour conséquence l'impossibilité de la prédiction numérique déterministe à long terme d'un écoulement dont les conditions initiales ne sont pas connues dans les petites échelles[+].

Une conséquence du résultat d'imprédicibilité bidimensionnelle serait par exemple que l'écoulement atmosphérique à grande échelle ne serait pas prédicible numériquement au-delà de 10∿15 jours, donc qu'aucune prévision météorologique déterministe n'est possible au-delà de cette période. Signalons cependant que, dans le cas tridimensionnel, des contraintes extérieures imposées aux grandes échelles de l'écoulement (frontières, forces extérieures ...) pourraient avoir pour effet de ralentir, voire même d'inverser, ce "flux d'erreur" des petites échelles vers les grandes échelles.

3. PHENOMENOLOGIE DE LA TURBULENCE HOMOGENE ET ISOTROPE TRIDIMENSIONNELLE

Nous considérons donc le champ de vitesse $\vec{u}(\vec{x},t)$ *comme une fonction aléatoire. Nous noterons par* $\langle\ \rangle$ *l'opérateur de moyenne. Cette moyenne porte sur un ensemble de réalisations de l'écoulement qui correspond à un ensemble de conditions initiales défini par la fonction aléatoire* $\vec{u}(\vec{x},0)$. *Nous supposerons que la turbulence est homogène, c'est à dire que ses propriétés statistiques sont invariantes par translation. Ceci implique en particulier que la vitesse moyenne est constante, et donc nulle dans un repère lié au mouvement moyen. Signalons que certains auteurs utilisent le mot "homogène" pour une turbulence dont les fluctuations autour de la vitesse moyenne sont statistiquement invariantes par translation. Une telle hypothèse implique des gradients moyens constants (CRAYA, 1958). On peut mentionner à ce propos les expériences de MARECHAL (1972) et les travaux de COURSEAU et LOISEAU (1978); JEANDEL, BRISON et MATHIEU (1978) ; CAMBON et JEANDEL (1979). Nous ne considérerons pas ce type de turbulence par la suite, et supposerons que la vitesse moyenne est nulle.*

Le tenseur des corrélations

$$U_{ij}(\vec{r},t) = \langle u_i(\vec{x},t)\, u_j(\vec{x}+\vec{r},t)\rangle \qquad (3\text{-}1)$$

est alors indépendant de $\vec{x}$. *Soit* $\hat{u}_i(\vec{k},t)$ *la transformée de Fourier au sens des distributions de* $u_i(\vec{x},t)$

$$\hat{u}_i(\vec{k},t) = \left(\frac{1}{2\pi}\right)^3 \int e^{-i\vec{k}.\vec{x}}\, u_i(\vec{x},t)\, d^3x \qquad (3\text{-}2)$$

[+] *Le problème de l'imprédicibilité d'un écoulement turbulent n'est qu'un des aspects de la discussion sur la notion de déterminisme en mécanique, discussion reprise par exemple par PRIGOGINE et al. (1979).*

et $\hat{U}_{ij}(\vec{k},t)$ le tenseur spectral, transformée de Fourier de $U_{ij}(\vec{r},t)$:

$$\hat{U}_{ij}(\vec{k},t) = \left(\frac{1}{2\pi}\right)^3 \int e^{-i\vec{k}.\vec{r}}\, U_{ij}(\vec{r},t)\, d^3r \qquad (3\text{-}3)$$

Le tenseur spectral apparait également comme un tenseur de corrélations dans l'espace de Fourier. On peut en effet montrer que :

$$< \hat{u}_i(\vec{k'},t).\hat{u}_j(\vec{k},t) > = \hat{U}_{ij}(\vec{k},t)\ \delta(\vec{k}+\vec{k'}) \qquad (3\text{-}4)$$

Nous supposerons également que la turbulence est <u>isotrope</u>, c'est à dire statistiquement invariante par rotation autour d'un axe quelconque. Enfin, nous supposerons que la turbulence est sans hélicité, c'est à dire statistiquement invariante par symétrie plane [+]. Pour une telle turbulence, le tenseur spectral se met sous la forme (BATCHELOR 1953)

$$\hat{U}_{ij}(\vec{k},t) = \left(\delta_{ij} - \frac{k_i k_j}{k^2}\right) \frac{E(k,t)}{4\pi k^2} \qquad (3\text{-}5)$$

où $E(k,t)$, appelé "spectre d'énergie tridimensionnel", ne dépend que du nombre d'onde $k = |\vec{k}|$. L'énergie cinétique moyenne par unité de masse

$$\frac{1}{2} < u^2(\vec{x},t) > = \frac{1}{2}\ v^2(t) \qquad (3\text{-}6)$$

vérifie évidemment

$$\frac{1}{2}\ v^2(t) = \int_0^\infty E(k,t)\, dk \qquad (3\text{-}7)$$

en sorte que $E(k,t)$ appparait comme la densité d'énergie cinétique au nombre d'onde k.

[+] L'hélicité joue un rôle fondamental dans la dynamique de la turbulence magnétohydrodynamique (turbulence dans un fluide conducteur de l'électricité). Elle permet en particulier l'amplification par la turbulence d'un champ magnétique infinitésimal (effet dynamo : cf. POUQUET, FRISCH et LEORAT 1976 ; MOFFATT 1978).

On peut enfin écrire l'équation de Navier-Stokes dans l'espace de Fourier. On obtient :

$$\left(\frac{\partial}{\partial t} + \nu k^2\right)\hat{u}_i(\vec{k},t) = -i\, k_m \left(\delta_{ij} - \frac{k_i k_j}{k^2}\right) \int_{\vec{p}+\vec{q}=\vec{k}} \hat{u}_j(\vec{p},t)\, \hat{u}_m(\vec{q},t)\, d^3p \qquad (3\text{-}8)$$

La condition d'incompressibilité s'écrivant :

$$k_i \cdot \hat{u}_i(\vec{k},t) = 0 \qquad (3\text{-}9)$$

On remarque que les interactions non linéaires se font par triades de vecteurs d'onde $\vec{k}, \vec{p}, \vec{q}$ *telles que* $(\vec{k},\vec{p},\vec{q})$ *forment un triangle.*

Il est facile de montrer que les termes non linéaires des équations de Navier-Stokes conservent l'énergie cinétique moyenne. La dissipation d'énergie est donc due à la viscosité, et on a :

$$\bar{\varepsilon} = -\frac{d}{dt}\frac{1}{2}v^2 = \nu\langle \overrightarrow{\mathrm{rot}\, u}^2 \rangle \qquad (3\text{-}10)$$

La théorie de KOLMOGOROV (1941) suppose qu'il existe dans le spectre d'énergie une zone appelée "zone inertielle" où le spectre $E(k)$ *ne dépend que de* $\bar{\varepsilon}$ *et de* k *. Une analyse dimensionnelle montre alors que :*

$$E(k) = C_E\, \bar{\varepsilon}^{2/3}\, k^{-5/3} \qquad (3\text{-}11)$$

où C_E *est une constante universelle appelée constante de Kolmogorov.*

Une telle théorie suppose en fait que les interactions sont "locales" dans l'espace de Fourier, c'est à dire que n'interagissent que des triades dont les trois nombres d'onde sont du même ordre. Dans la zone inertielle l'énergie cascade à un taux constant $\bar{\varepsilon}$ *vers des structures de plus en plus petites qui sont finalement dissipées par la viscosité.*

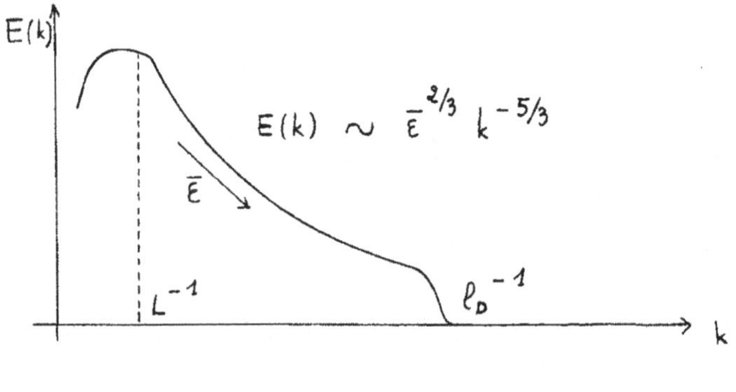

Dans la zone inertielle, $k . \left[kE(k)\right]^{3/2}$ est constant et égal à $\bar{\varepsilon}$. Si l'on associe au nombre d'onde k un "tourbillon" de vitesse $u = \left[kE(k)\right]^{1/2}$ et de dimension $\ell = k^{-1}$, on en déduit que, pour les tourbillons de taille inertielle, le produit u^3/ℓ est égal à $\bar{\varepsilon}$. En supposant que la zone inertielle est bornée par $1/\ell_D$ et $1/L$, il vient la relation (2-2). On remarquera en particulier que le nombre d'onde de Kolmogorov $1/\ell_D$ est de l'ordre de $\left[\bar{\varepsilon}/\nu^3\right]^{1/4}$. Pour plus de détails sur une telle approche de la théorie de KOLMOGOROV dans l'espace des configurations (espace physique), consulter ROSE et SULEM (1978).

Cette théorie est relativement bien vérifiée par les expériences (cf. par exemple GRANT, STEWART et MOLLIET 1962). Elle ne peut cependant être rigoureusement exacte puisqu'elle ne prend pas en compte les fluctuations de

$$\varepsilon = \nu \; \overrightarrow{\text{rot}\, u}^{\,2}$$

autour de sa valeur moyenne $\bar{\varepsilon}$. Ces fluctuations sont importantes et l'on constate expérimentalement qu'un écoulement turbulent est fortement intermittent, c'est à dire que les fluctuations de vitesse ne sont pas réparties spatialement de façon homogène pour une réalisation de l'écoulement. Cette notion d'intermittence est longuement discutée dans l'ouvrage de MONIN et YAGLOM (1975). Des mesures expérimentales (cf. par exemple KOLMYANSKII 1972, GAGNE et al. 1978) ont montré que les corrélations spatiales $\langle \varepsilon(\vec{x}) . \varepsilon(\vec{x}+\vec{r}) \rangle$ suivaient une loi en $r^{-\mu}$, μ étant de l'ordre de 0.5. D'un point de vue théorique, KOLMOGOROV (1962) a proposé pour ε une loi lognormale conduisant à un spectre d'énergie en $k^{-5/3-\mu/9}$. FRISCH, SULEM et NELKIN (1978) on proposé la théorie dite du " β - modèle", conduisant à un spectre d'énergie en $k^{-5/3\;-\mu/3}$. Dans cette dernière théorie, on peut montrer que la dissipation se fait sur un ensemble de dimension $3 - \mu$. Une détermination expérimentale de la correction apportée par l'intermittence à l'exposant $-5/3$ est extrêmement difficile, en particulier à cause des erreurs introduites dans le spectre d'énergie aux grands nombres d'onde par l'utilisation de l'hypothèse de Taylor (cf. par exemple CHAMPAGNE 1978).

L'analyse phénoménologique de KOLMOGOROV peut être appliquée au problème de la diffusion de paires de particules transportées par la turbulence. Soit $r^2 = \langle MM'^2 \rangle$ la distance quadratique moyenne entre deux particules M et M'. Si r est de l'ordre de grandeur des tourbillons inertiels (c'est à dire des tourbillons caractérisés par un nombre d'onde situé dans la zone inertielle), dr/dt est de l'ordre de la vitesse u correspondant au nombre d'onde $1/r$, c'est à dire $(\bar{\varepsilon}\, r)^{1/3}$. Le coefficient de diffusion des paires de traceurs est alors donné par

$$\frac{d}{dt}\, r^2 \sim \bar{\varepsilon}^{1/3}\, r^{4/3} \qquad (3\text{-}12)$$

et correspond à une loi

$$r(t) \propto t^{3/2} \tag{3-13}$$

La loi (3-12) a été proposée par RICHARDSON (1926) sur des bases expérimentales, et fait de celui-ci le précurseur de la théorie de KOLMOGOROV (1941). Il est difficile à l'heure actuelle de savoir comment cette loi peut être modifiée par l'intermittence.

4. THEORIES ANALYTIQUES

En turbulence homogène, la vitesse moyenne $\vec{u}(\vec{x},t)$ *est indépendante de* $\vec{x}$ *. On peut donc se ramener à un problème de vitesse moyenne nulle en se plaçant dans un repère lié au mouvement moyen. De l'équation de Navier-Stokes, écrite symboliquement sous la forme :*

$$\left(\frac{\partial}{\partial t} + \nu\right) u = u\, u \tag{4-1}$$

on peut déduire des équations liant les moments du champ de vitesse d'ordre n *aux moments d'ordre* $(n+1)$

$$\left(\frac{\partial}{\partial t} + \nu\right) \underbrace{\langle u u \cdots u \rangle}_{n \text{ termes}} = \underbrace{\langle u u \cdots u u \rangle}_{(n+1) \text{ termes}} \tag{4-2}$$

Pour n *fixé, le système de* n *équations ainsi obtenu contient* $(n + 1)$ *inconnues, et n'est donc pas "fermé". C'est ce que l'on appelle le problème de fermeture. On est obligé d'introduire arbitrairement une équation supplémentaire entre certains moments. L'approximation quasi-normale, introduite par MILLIONSCHTCHIKOV (1941) supposait que les moments d'ordre quatre étaient liés aux moments d'ordre deux comme si la fonction aléatoire* $\vec{u}(\vec{x},t)$ *était gaussienne. En fait il a été montré par O'BRIEN et FRANCIS (1962) et OGURA (1963) que cette approximation conduisait à des spectres d'énergie négatifs et était donc inacceptable. Ce comportement aberrant est dû à une croissance démesurée des corrélations triples dans cette approximation. En introduisant une relaxation linéaire supplémentaire ("eddy-damping") des corrélations triples, il est possible d'obtenir une approximation physiquement plus réaliste. C'est l'approximation Quasi-Normale Markovianisée avec amortissement turbulent (EDQNM), introduite par ORSZAG (1970, 1973) et LEITH (1971). On obtient alors pour le spectre d'énergie l'équation d'évolution suivante*

$$\left(\frac{\partial}{\partial t} + 2\nu k^2\right) E(k,t) = T(k,t) \tag{4-3}$$

$$T(k,t) = \iint_{\Delta_k} \theta_{kpq}\, b_{kpq}\left(\frac{k}{pq}\right)\left[k^2 E(p)E(q) - p^2 E(q)E(k)\right] dp\, dq \tag{4-4}$$

où Δ_k est le domaine du plan (p,q) tel que k,p,q soient les côtés d'un triangle ; b_{kpq} est donné par

$$b_{kpq} = \frac{p}{k}\,(xy + z^3)$$

où x,y,z sont les cosinus des angles intérieurs au triangle k,p,q. θ_{kpq} est le temps de relaxation des corrélations triples, donné par :

$$\theta_{kpq} = \frac{1 - \exp -[\mu(k,t) + \mu(p,t) + \mu(q,t)]\,t}{\mu(k,t) + \mu(p,t) + \mu(q,t)} \qquad (4\text{-}5)$$

avec :

$$\mu(k,t) = \nu k^2 + \lambda \left[\int_0^k p^2 E(p,t)\,dp\right]^{1/2} \qquad (4\text{-}6)$$

La forme particulière de l'approximation EDQNM avec un taux de relaxation des corrélations triples donné par (4-6) a été étudiée par SULEM et al.(1975), POUQUET et al. (1975), ANDRE et LESIEUR (1977), LESIEUR et SCHERTZER (1978). Cette approximation conduit à un spectre de KOLMOGOROV en $k^{-5/3}$ et coïncide dans la zone inerticlle avec les résultats d'une approche beaucoup plus sophistiquée due à KRAICHNAN (Test-Field-Model 1971 a), à condition de prendre la valeur de la constante λ intervenant dans (4-6) égale à 0.360.[+] La valeur de la constante de Kolmogorov C_E correspondante est égale à 1.4, ce qui est en bon accord avec les expériences. Les résultats mathématiques et numériques obtenus sur des équations du type (4-4) sont exposés dans les références ci-dessus ainsi que dans BRISSAUD et al. (1973), LESIEUR et SULEM (1975), PENEL(1975), BARDOS et al. (197) et ROSE-SULEM (1978). Les conclusions essentielles dans le cas tridimensionnel sont les suivantes : lorsque la viscosité tend vers 0 et pour un spectre initial $E(k,0)$ à décroissance rapide, il existe un temps fini

$$t_* \sim \left\{\int_0^\infty p^2 E(p,0)\,dp\right\}^{1/2} \qquad \text{tel que}$$

(i) pour $t < t_*$ $$\lim_{\nu \to 0} \nu \langle \overrightarrow{\mathrm{rot}\, u}^{\,2} \rangle = 0$$

+ cette constante doit être prise égale à $0.360/\sqrt{2} = 0.255$ si l'on définit le spectre d'énergie comme densité de $\langle u^2 \rangle$ (et non $\frac{1}{2}\langle u^2 \rangle$)au nombre d'onde k.

(ii) pour $t > t_*$

$$\lim_{\nu \to 0} \nu \langle \overrightarrow{\text{rot } u}^2 \rangle = \bar{\varepsilon} \neq 0$$

$$E(k) \sim \bar{\varepsilon}^{2/3} k^{-5/3} \quad \text{pour} \quad \begin{matrix} k \to \infty \\ \nu \to 0 \end{matrix}$$

Pendant la phase (i), le spectre reste à décroissance rapide et l'énergie est conservée. Au temps $t = t_*$, il y a perte de régularité et "l'enstrophie" $\frac{1}{2} \langle \overrightarrow{\text{rot } u}^2 \rangle$ devient infinie, tandis qu'apparait un spectre de KOLMOGOROV s'étendant jusqu'à l'infini.

Les modèles du type EDQNM suppriment évidemment l'intermittence puisqu'ils conduisent à un spectre en $k^{-5/3}$. Mais ils sont extêmement utiles pour étudier le sens et l'importance des transferts d'énergie entre les différentes échelles de l'écoulement. Ils ont permis en particulier de vérifier l'existence d'une cascade inverse d'énergie en turbulence bidimensionnelle, et d'une cascade inverse de magnétohélicité en turbulence MHD tridimensionnelle (POUQUET et al. 1976).

5. A PROPOS DE LA TRANSITION "ECOULEMENT LAMINAIRE - ECOULEMENT TURBULENT"

Les résultats ci-dessus conjecturant l'existence d'un temps critique t_* correspondant à l'apparition d'une singularité dans un fluide parfait (viscosité nulle) et au-delà duquel il y aurait perte d'énergie, permettent d'envisager une nouvelle interprétation de la transition laminaire - turbulent. Cette transition est généralement considérée comme un changement de l'état de l'écoulement consécutif à un certain nombre de bifurcations lorsque le nombre de Reynolds dépasse une valeur critique.
En fait il est possible d'envisager cette transition dans le cadre d'un fluide parfait : l'état laminaire est alors caractérisé par une superposition de gros tourbillons ; cette superposition peut être aléatoire, en sorte qu'il n'y a pas identité entre la notion d'écoulement laminaire et la notion d'ordre. On pourrait définir l'état laminaire comme un état de désordre à grande échelle qui conserve l'énergie. Avec l'état turbulent au contraire apparaissent l'irréversibilité (perte d'énergie) et les singularités.

Il est tentant d'interpréter ces conjectures concernant la turbulence dans un fluide parfait à la lumière des travaux épistémologiques de SERRES (1977) à propos du texte "De natura rerum" de LUCRECE. Le modèle de LUCRECE du flux laminaire d'atomes tombant en chute libre et caractérisant cependant le plus grand désordre ("turba") représente alors tout simplement les deux volets (écoulement bien organisé — désordre conservatif à grande échelle) de la notion d'écoulement laminaire. Aucune intervention extérieure n'est alors requise pour créer le mystérieux "clinamen" (déclinaison) de

LUCRECE, qui correspond alors à l'apparition au bout d'un temps fini de singularités nées des interactions non linéaires (étirement des lignes de tourbillon par les gradients de vitesse). Le "désordre indifférent" des gros tourbillons cède alors la place au "désordre organisé" de la cascade d'énergie où s'ordonnent tous les tourbillons jusqu'aux plus petites échelles.

6. QUELQUES IDEES SUR LA PARAMETRISATION DES PETITES ECHELLES

Le problème de la paramétrisation des petites échelles dans une simulation numérique des grandes échelles peut se poser dans l'espace de Fourier : si k_c *est un nombre d'onde de coupure, on cherche comment les modes implicites* $k > k_c$ *interagissent avec les modes explicites* $k < k_c$. *Dans un premier temps, on peut chercher à prévoir simplement l'évolution du spectre d'énergie des modes explicites, ce qui permet d'écarter les difficultés dues à l'imprédicibilité de l'écoulement. Une telle démarche a permis dans le cas bidimensionnel d'obtenir un spectre d'énergie en* k^{-3} *correspondant à la cascade d'enstrophie et s'étendant jusqu'au nombre d'onde de coupure* k_c *(BASDEVANT et al. 1978). Une telle paramétrisation ne peut cependant prétendre prévoir correctement le champ de vitesse dans les échelles explicites d'une réalisation particulière de l'écoulement.*

Commençons par considérer le modèle de spectre suivant :

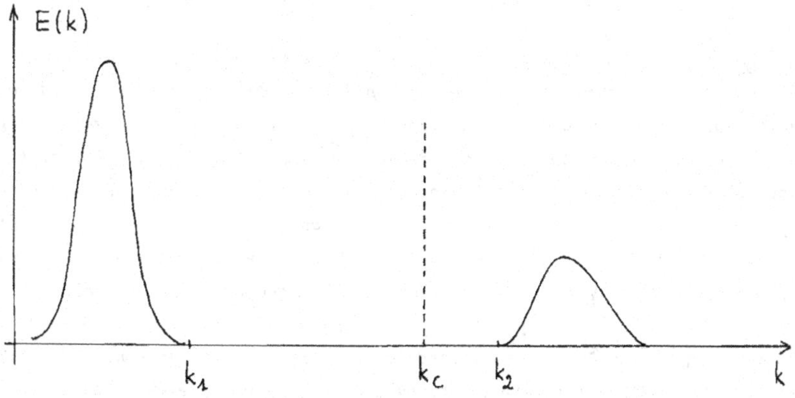

dans lequel les "petites" échelles $(k > k_2)$ *sont séparées des "grandes" échelles* $(k < k_1)$ *par un "fossé" (gap) où le spectre est nul. Nous supposerons* $k_1 \ll k_2$, *et*

$$\sup_{k > k_2} E(k) \ll \sup_{k < k_1} E(k) \qquad (6\text{-}1)$$

Il est alors possible de développer le terme de transfert $T(k,t)$ de l'équation (4-3) en puissances du petit paramètre k_1/k_2, selon des techniques mises en oeuvre par KRAICHNAN (1966, 1971 b, 1976), POUQUET et al. (1975), BASDEVANT et al. (1978), LESIEUR et SCHERTZER (1978). On trouve alors que, dans le cas tridimensionnel, l'effet des petites échelles sur les grandes échelles est un effet de viscosité turbulente, c'est à dire que les grandes échelles $(k < k_c)$ satisfont l'équation

$$\left[\frac{\partial}{\partial t} + 2(\nu + \nu_t) k^2\right] E(k,t) = T_{<k_c}(k,t) \tag{6-2}$$

où $T_{<k_c}$ est le transfert "explicite ", c'est à dire correspondant aux interactions $k < k_c$, $p < k_c$, $q < k_c$. ν_t est une viscosité turbulente donnée par

$$\nu_t = \frac{1}{15} \int_{k_c}^{\infty} \theta_{opp} \left[5\, E(p) + p \frac{\partial E}{\partial p}\right] dp \tag{6-3}$$

cette intégrale peut être évaluée en prenant un modèle de spectre $E(k) \propto k^{-5/3}$ pour $k > k_2$. On obtient alors :

$$\nu_t = \frac{1}{9\lambda}\, \tilde{v}\, k_2^{-1} \tag{6-4}$$

où $\frac{1}{2}\tilde{v}^2 = \int_{k_2}^{\infty} E(k)\, dk$ est l'énergie cinétique "turbulente", c'est à dire l'énergie cinétique des modes implicites. On retrouve dans (6-4) une formulation analogue à celle de la longueur de mélange, mais avec une valeur explicite de la constante numérique figurant devant $\tilde{v}\, k_2^{-1}$. Pour $\lambda = 0.360$, cette constante est égale à 0.31 . Si le spectre $E(k)$ dans les structures explicites $(k < k_c)$ est un pic très étroit situé au voisinage de k_1, ce gros tourbillon, en interagissant avec les petites échelles $(k > k_2)$, verra son énergie décroître en $\exp\{-2\nu_t k_1^2 t\}$, donc aura un temps de vie de l'ordre de $1/2\nu_t k_1^2$ soit $1.61\ \tilde{v}^{-1} k_2/k_1^2$. Ce temps est grand devant le temps propre d'évolution $\tilde{v}^{-1} k_2^{-1}$ de la turbulence petite échelle, et pour que le calcul soit valable, il faut supposer que la turbulence petite échelle est entretenue par des forces extérieures.

Ce modèle de spectre avec une séparation marquée entre les petites échelles et les grandes échelles n'est cependant pas réaliste, puisqu'en général le spectre d'énergie décroît continûment des grandes échelles vers les petites échelles. Le transfert $T_{>k_c}(k,t)$ correspondant aux interactions entre échelles explicites et implicites peut encore (pour $k \ll k_c$) être mis sous la forme $-2\tilde{\nu}_t\, k^2 E(k,t)$ où $\tilde{\nu}_t$ est une viscosité turbulente toujours donnée par (6-3). La seule différence avec le problème du spectre avec "gap" est que dans $\tilde{\nu}_t$ interviennent maintenant par l'intermédiaire de θ_{opp}, des nombres d'onde $p < k_c$. On trouve alors :

$$\tilde{\nu}_t = \frac{\sqrt{3}}{18\lambda} E(k_c)^{1/2} k_c^{-1/2} \qquad (6\text{-}5)$$

ou encore :

$$\tilde{\nu}_t = \frac{\sqrt{3}}{18\lambda} C_E^{1/2} \bar{\varepsilon}^{1/3} k_c^{-4/3} \qquad (6\text{-}6)$$

De l'expression (6-6), on déduit que k_c est proportionnel à $\left[\bar{\varepsilon}/\tilde{\nu}_t^3\right]^{1/4}$.
k_c apparaît donc comme un "nombre d'onde de Kolmogorov turbulent". Il serait hasardeux d'utiliser la viscosité turbulente (6-6), calculée dans la limite $k \ll k_c$, au voisinage de k_c. En effet, le flux d'énergie $-\int_0^{k_c} 2\tilde{\nu}_t k^2 E(k)\, dk$ sortant du système est, pour une telle paramétrisation, égal (pour un spectre en $k^{-5/3}$) à $\frac{0.239}{\lambda}\bar{\varepsilon}$. Pour $\lambda = 0.360$, on obtient un flux égal à $0.664\,\bar{\varepsilon}$.

KRAICHNAN (1976) a étudié de façon plus précise à l'aide du Test-Field-Model le transfert $T_{>k_c}(k,t)$ dans la zone inertielle. On trouve alors que le rapport $\nu_t(k) = -T_{>k_c}/2k^2E$ est égal à $\Lambda\left(\frac{k}{k_c}\right) C_E^{1/2} \bar{\varepsilon}^{1/3} k_c^{-4/3}$ où $\Lambda(\lambda)$ est une fonction croissant au voisinage de 1, qui est représentée sur la figure 1.

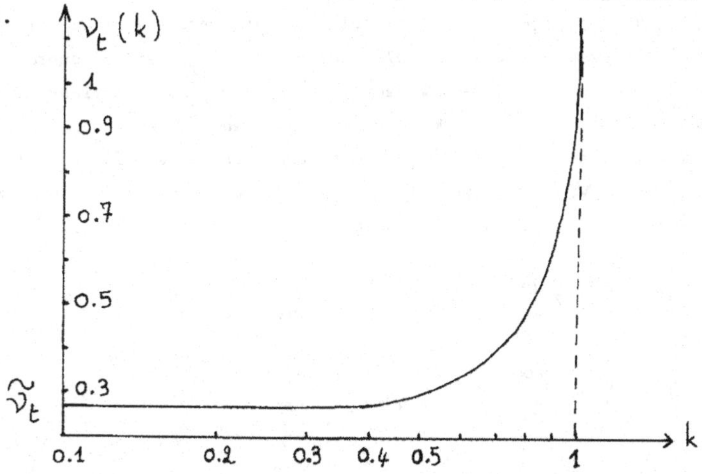

Fig.1 - Variation de la viscosité turbulente en fonction de k. Les unités sont k_c pour k, et $C_E^{1/2}\ \bar{\varepsilon}^{1/3}\ k_c^{-4/3}$ pour $\nu_t(k)$ (d'après KRAICHNAN, 1976).

Une telle viscosité turbulente garantit alors un flux sortant du système égal à $\bar{\varepsilon}$ *pour un spectre stationnaire en* $k^{-5/3}$.

A priori, les unités $E(k_c)^{1/2}\, k_c^{-1/2}$ *et* $C_E^{1/2}\, \bar{\varepsilon}^{1/3}\, k_c^{-4/3}$ *sont équivalentes et correspondent à un spectre d'énergie en* $k^{-5/3}$. *La viscosité turbulente* $\nu_t(k)$ *ne serait donc utilisable que pour des calculs stationnaires et dans la zone inertielle. Cependant, si l'on prend pour unité de viscosité turbulente* $E(k_c)^{1/2}\, k_c^{-1/2}$ *plutôt que* $C_E^{1/2}\, \bar{\varepsilon}^{1/3}\, k_c^{-4/3}$, *le calcul peut être étendu aux problèmes instationnaires où la valeur de* $\bar{\varepsilon}$ *est inconnue: pour un spectre initial confiné dans les grosses structures, et tant que l'excitation n'a pas atteint le nombre d'onde de coupure* k_c, *la viscosité turbulente est nulle et donc inopérante. En outre, il semble alors possible d'éviter soit une accumulation d'énergie soit une zone de dissipation artificielle au voisinage de* k_c. *L'accumulation d'énergie semble exclue à cause de la brusque croissance de* $\nu_t(k)$ *au voisinage de* k_c, *un accroissement de* $E(k_c)$ *augmentant la viscosité turbulente et provoquant donc un amortissement de* $E(k_c)$. *De même une zone de dissipation artificielle ne peut se développer, les faibles valeurs de* $E(k_c)$ *engendrant une faible valeur de la viscosité.*

Nous avons testé cette méthode de paramétrisation pour l'équation spectrale EDQNM, en utilisant d'une part la viscosité turbulente $\nu_t(k) = \Lambda\left(\frac{k}{k_c}\right) E(k_c)^{1/2}\, k_c^{-1/2}$, *d'autre part la viscosité turbulente* $\tilde{\nu}_t$.

La figure 2 représente la comparaison entre ces deux paramétrisations dans lesquelles le nombre d'onde de coupure a été pris égal à 64 k_I *(où* k_I *est un nombre d'onde caractéristique des gros tourbillons initiaux), et un calcul complet (sans paramétrisation) à nombre de Reynolds* $R = 32800$. *Les résultats peuvent être considérés comme satisfaisants, quoique la viscosité* $\nu_t(k)$ *soit trop forte au voisinage de la coupure, alors que la viscosité* $\tilde{\nu}_t$ *provoque une légère tendance à l'accumulation d'énergie sur* k_c. *Le tableau suivant représente, aux instants* $t = 6$ *et* $t = 10$, *les valeurs respectives de :*

$$\bar{\varepsilon} = 2\nu \int_0^\infty k^2 E(k,t)\,dk \quad \text{dans le calcul "exact"}$$

$$\Pi(k_c) = \int_{k_c}^\infty T(k,t)\,dk \quad \text{dans le calcul "exact"}$$

$$\Pi^{(1)}(k_c) = 2\int_0^k \nu_t(k)\, k^2 E(k,t)\,dk \quad \text{dans le calcul paramétrisé par } \nu_t(k)$$

$$\Pi^{(2)}(k_c) = 2\int_0^k \tilde{\nu}_t\, k^2 E(k,t)\,dk \quad \text{dans le calcul paramétrisé par } \tilde{\nu}_t .$$

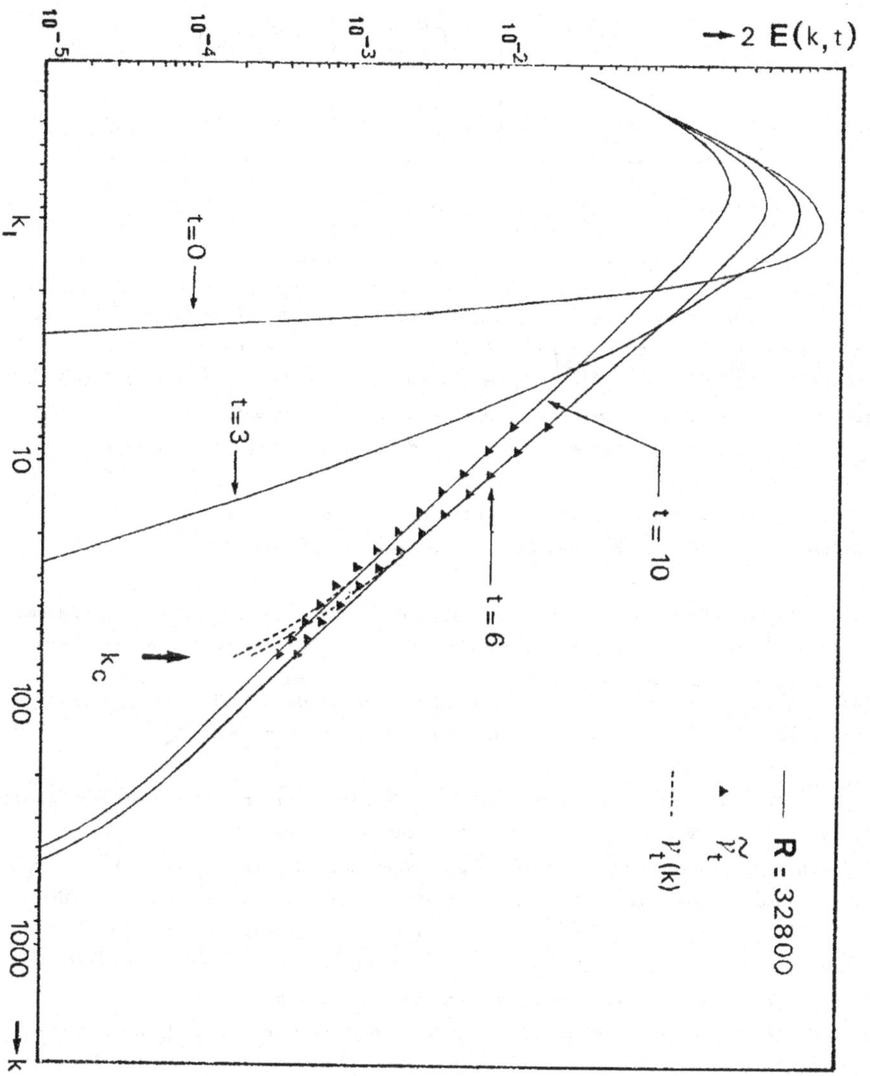

Fig.2 - Evolution dans le temps du spectre $E(k,t)$ pour les temps $t = 0, 3, 6, 10$. Trait plein : spectre "exact" ; tirets : spectre "paramétrisé" par $\nu_t(k)$. ▲ : spectre paramétrisé par $\tilde{\nu}_t$. Le spectre initial est $E(k,0) \sim \exp\{-2k^2\}$. L'unité de nombre d'onde est k_I.

	$\bar{\varepsilon}$	$\Pi(k_c)$	$\Pi^{(1)}(k_c)$	$\Pi^{(2)}(k_c)$
t = 6	0.097	0.101	0.095	0.097
t = 10	0.070	0.063	0.064	0.062

Ce tableau montre que les deux méthodes de paramétrisation proposées donnent de bons résultats pour les calculs instationnaires, puisque les flux paramétrisés $\Pi^{(1)}(k_c)$ et $\Pi^{(2)}(k_c)$ sont très proches des flux $\Pi(k_c)$. Il semble cependant au vue de ces résultats, que la viscosité turbulente proposée par KRAICHNAN soit trop importante pour ce type de calcul instationnaire ; par contre la viscosité asymptotique $\tilde{\nu}_t$ serait trop faible. Un ajustement empirique de la viscosité turbulente peut alors être envisagé de façon à assurer une parfaite coïncidence entre spectre exact et spectre paramétrisé.

A cet effet, il serait intéressant de tester les viscosités turbulentes généralisées proposées par LESLIE et QUARINI (1979). Cet ajustement de la fonction $\Lambda(k/k_c)$ qui module $E(k_c)^{1/2}\,k_c^{-1/2}$ dans l'expression de la viscosité turbulente ν_t pourrait se faire dans le cadre de calculs stationnaires où des forces extérieures injectent de l'énergie au voisinage de k_I.

Il resterait à tester ces méthodes de paramétrisation sur des simulations numériques directes instationnaires des équations de Navier-Stokes. SIGGIA et PATTERSON (1978) ont utilisé une viscosité turbulente n'agissant que sur les nombres d'onde au voisinage de k_c, et ajustée empiriquement pour produire des spectres stationnaires en $k^{-5/3}$. Sur la base de ces calculs on peut penser que la généralisation des méthodes de paramétrisation que nous proposons est envisageable pour une simulation numérique directe. Mais ces méthodes ne peuvent à priori prétendre qu'à une prédiction statistique et non déterministe de l'écoulement.

R E F E R E N C E S

ANDRE, J.C. et LESIEUR, M. 1977 : "Influence of helicity on the evolution of Isotropic Turbulence at high Reynolds number" J. Fluid. Mech.81 p. 187.

BARDOS, C. , PENEL, P.,FRISCH U, et SULEM, PL. 1979 , "Modified dissipativity for a non-linear evolution equation arising in turbulence". Arch. Rat. . Mech. Anal., to appear.

BASDEVANT, C. , LESIEUR M. et SADOURNY , R. 1978, "Subgridscale modeling of enstrophy transfer in two-dimensional turbulence" J. Atm . Sci. 35, P 1028.

BATCHELOR, G.K. 1953 : "The theory of homogeneous turbulence". Cambridge University Press.

BOUSSINESQ, J. 1897 : "Théorie de l'écoulement tourbillonnant et tumultueux des liquides dans des lits rectilignes de grande section". Gauthier-Villars.

BRISSAUD, A., FRISCH, U., LEORAT, J., LESIEUR, M. , MAZURE, A.,POUQUET, A.,SADOURNY R. et SULEM , P.L. : 1973 "Catastrophe énergétique et nature de la turbulence". Ann. Géophys. 29, P. 539

CAMBON, C. et JEANDEL, D. 1979 : "Spectral Modeling of three-dimensional homogeneous Turbulence". Ecole Centrale Lyonnaise - Prépublication.

CHAMPAGNE, F.H. , 1978 : "The fine-scale structure of the turbulent velocity field" J. Fluid, Mech. 86 , p 67.

COURSEAU, P.A. , LOISEAU M. 1978 : "Contribution à l'analyse de la turbulence homogène anisotrope". Journal de Mécanique . 17, n° 2, p 245.

CRAYA, A : "Contribution à l'analyse de la turbulence associée à des vitesses moyennes " 1958 P.S.T. n° 345, SDST PARIS.

FRISCH, U. , SULEM, P.L. et NELKIN, M. 1978 : " A simple dynamical model of intermittent fully developed turbulence" . J. Fluid Mech. 87 P. 719.

GAGNE, Y., HOPFINGER, E. et MARECHAL, J 1978 :"Measurements of internal intermittency and dissipation correlations in fully developed turbulence" Publié dans les Proceedings de "Dynamics flow conferences", MARSEILLE, BALTIMORE septembre 1978.

GRANT, H.L., STEWART , R.W. et MOILLIET, A. 1962 : "Turbulence spectra from a tidal channel". J. Fluid. Mech. 12, p 241.

JEANDEL , D., BRISON, J.F. et MATHIEU, J. 1978 : "Modeling methods in physical and spectral space" . Phys. Fluids 21, p 169.

KATO, T. 1972 " Non stationary flows of viscous and ideal fluids in R^3", . J. Funct. Anal. 9, p 296.

KHOLMYANSKII, M.Z. , 1972, Izv. Akad. Nank. SSSR . Fiz. Atmosf. i Okeana 8,p 818.

KOLMOGOROV, A.N. , 1941 "the local structure of turbulence in incompressible viscous fluid for very large Reynolds number ". C.R. Acad. Sci. URSS. 30 p. 301

KOLMOGOROV, A.N. , 1962 "A refinement of previous hypotheses concerning the local structure of turbulence in a viscous incompressible fluid at high Reynolds number". J. Fluid. Mech. 13, p 82.

KRAICHNAN , R.H. 1966 "Isotropic turbulence and inertial-range structure". Phys. Fluids, 9, p. 1728.

KRAICHNAN, R.H. 1971 a "An almost-Markovian Galilean-invariant turbulence model" J. Fluid. Mech. 47, p 513

KRAICHNAN , R.H. 1971 b "Inertial-Range transfer in two and three dimensional turbulence". J. Fluid. Mech. 47, p 525.

KRAICHNAN , R.H. 1976, "Eddy viscosity in two and three dimensions" J. Atm. Sci. 33, p 1521.

LEITH , C.E. 1971 : "Atmospheric predictability and two-dimensional turbulence" J. Atm . Sci.28, p 145.

LESLIE, D.C. et QUARINI, G.L. 1979 "The application of turbulence theory to the formulation of subgrid modelling procedures" J. Fluid. Mech 91, p. 65.

LEITH, C.E. et KRAICHNAN, R.H. 1972 "Predictability of turbulent flows". J. Atm. Sci., 29, p. 1041.

LESIEUR, M. et SCHERTZER, D. 1978. "Amortissement antosimilaire d'une turbulence à grand nombre de Reynolds". Journal de Mécanique 17, p 609.

LESIEUR, M. et SULEM, P.L. 1975 : "Les équations spectrales en turbulence homogène et isotrope. Quelques résultats théoriques et numériques".
Dans "Turbulence and Navier Stokes Equation". p 113 (cf ref. Terman 1975).

LORENZ, E.N. 1969 "The predictability of a flow which possesses many scales of motion." Tellus , 21 , p 289.

MANDELBROT, B.B. . 1975. "Intermittent turbulence and fractal dimension : kurtosis and the spectral exponen t 5/3 + β " , dans "Turbulence and Navier - stokes equation " p 121. Voir réf. Temam.

MARECHAL , J. 1972. "Etude expérimentale de la déformation plane d'une turbulence homogène ". Journal de Mécanique 11, 1972 , p 263.

MILLIONSCHTCHIKOV, M. 1941 "on the theory of homogeneous isotropic turbulence". CR. Acad. Sci. URSS. 32 , P 615.

MOFFATT, H.K. 1973 "six lectures on general fluid dynamics", dans "Fluid dynamics", Ecole d'E té de Physique des Houches 1973, édité par R. BALIAN et J.L. PEUBE.

MOFFATT, H.K. 1978 "Magnetic field generation in electrically conducting fluids". Cambridge University Press.

MONIN,A . et YAGLOM, A.M. 1975. : "Statistical fluid Mechanics". t2. Edition en langue anglaise par J.L. Lumley . M I T Press (Cambridge).

O'BRIEN, E.E. et FRANCIS, G.C. 1962 "A consequence of the zero fourth cumulant approximation ", J. Fluid . Mech. , 13, p. 369. :e'.

OGURA, Y. 1963 "A consequence of the zero fourth cumulant approximation in the decay of isotropic turbulence", J. Fluid . Mech, 16, p. 38.

ORSZAG, S.A. 1970 "Analytical Theories of turbulence" J. Fluid. Mech. 41, p 363.

ORSZAG, S.A. 1973. "Lectures on the statistical theory of turbulence" dans "Fluid Dynamics" (voir réf. Moffatt 1973).

ORSZAG, S.A. et PATTERSON , G.S. 1972 "Numerical simulation of the three dimensional homogeneous turbulence". In "statistical Models and Turbulence" p 127. Edited by M. ROSENBLATT and C. VAN ATTA, Springer-Verlag.

PENEL , P. 1975 "Sur une équation d'évolution non linéaire liée à la théorie de la Turbulence". Thèse d'Etat. Université de PARIS-SUD.

POUQUET, A. , LESIEUR, M. , ANDRE, J.C. et BASDEVANT, C. 1975 "Evolution of high Reynolds number two-dimensional turbulence". J. Fluid. Mech. 72,p 305.

POUQUET, A. , FRISCH, U. et LEORAT , J. 1976. "Strong MHD helical turbulence and the non-linear dynamo effect". J. Fluid. Mech 77 p 321.

PRIGOGINE, I. , STENGERS, I. et PAHAUT , S. 1979 "La dynamique de Leibniz à Lucrèce" (Editions de Minuit). Critique 35, n° 380, p 35.

PRANDTL, L. 1925 . Z.A. Math. Mech. 5, P 136

RICHARDSON, L.F. 1926 "Atmospher ic diffusion shown on a distance- neighbour graph" Proc. roy. Soc. A, 110, p 709

ROSE, H.A. et SULEM , P.L. 1978 "Fully developed turbulence and statistical Mechanics" journal de Physique 39, p 441.

SERRES, M. 1977 " La naissance de la physique dans le texte de Lucrèce. Fleuves et Turbulences". Editions de Minuit Coll. "Critique ".

SIGGIA, E.D. et PATTERSON, G.S. 1978 "Intermittency effects in a numerical simulation of stationary three-dimensional turbulence". J. Fluid Mech. 86 p 567 .

TEMAM, R. 1975 "Turbulence and Navier-Stokes Equation". Proceedings of the conference held at the University of PARIS-SUD Orsay, 12 -13 juin 1975. Springer-Verlag.

SULEM, P.L. , LESIEUR, M. et FRISCH, U. 1975 "Le test-field-model interprété comme méthode de fermeture des équations de la turbulence". Ann. Géophys. 31 p 487.

EXPERIMENTAL STUDY OF THE MECHANISM OF A NEW HYDRODYNAMICAL INSTABILITY OBSERVED AT SOME INTERFACES BETWEEN IMMISCIBLE LIQUIDS.

M. DUPEYRAT
Université Pierre et Marie Curie
Laboratoire de Chimie Physique
11, rue Pierre et Marie Curie
75 231 PARIS CEDEX 05 - FRANCE -

Description of the instability - It is a MARANGONI effect

The hydrodynamical instabilities reported in this paper occur as an interfacial turbulence, at the interface between two liquid immiscible phases containing charged species, one of which is surface active. They were described, for the first time in 1969 by Dupeyrat et Michel (1) and found out, by chance, from a rather complicated experiment. A system consisting of a solution in nitrobenzene of octadecyltrimethyl ammonium picrate ($C_{18}Pi$ 5 10^{-3}M), a surface active compound, in contact with an aqueous solution of potassium chloride (KCl 10^{-2}M) was brought far from equilibrium by means of an electrolysis. During the relaxation of the system, after the electrolysis is over, instability appears as waves arising along the wall of the vessel used, which are about 1 cm in amplitude and which induce a deformation of the whole interface.(2,3) The motion can last one hour or more.

The problem was to determine the physical parameters accountable for the phenomenon and to propose a model consistent with the experimental results.

Due to the requirement of a surface active compound, we studied the influence of the interfacial tension (4). We observed that this motions is related to the curvature of the interface which, at first convex for example, becomes concave, imparting a movement. We could show, indeed, that the contact angle between the wall and the tangent to the interface at the contact point with the wall must be about 90 degrees for inducing movements. We attributed the motion to the local variation in the interfacial tension (4) and described it as a MARANGONI effect (5,6). The experimental conditions, in such a system are not clearly defined and in order to go further we looked for another simpler one. We succeeded to obtain motion from a system initially far from equilibrium, made from a saturated solution in nitrobenzene of potassium iodide KI, a compound very soluble in water and an aqueous solution of octadecyltrimethylammonium chloride, C_{18} Cl, a very hydrophobic compound. Besides the previously described interfacial turbulence, we observed movements in the interface plane, appearing as local contractions or expansions of the interface made visible by a whitish emulsion arising during the experiment which we attributed to local variation of the interfacial tension (2,3). We chose, for the sake of simplicity, to eliminate the wall effect by coating the glass with paraffin, which suppresses the waves.

It is different from the instabilities described up to now

From this rather simple system, we examined wether the explanations proposed up to now for the various hydrodynamical instabilities are suitable or not.

It does not seem probable that the phenomenon is concerned with a heat transfer as in Bénard's cells (7) because the thermal energy exchange involved in such a system is about 10^{-2} calorie/cm^3. We measured the density of each phase before and after the movements and observed a modification within the third decimal place. Therefore, it is neither a density effect as in Rayleigh Taylor effect.(8), nor the spontaneous agitation of the interface between two unequilibrated liquids since the solvents are mutually saturated (9). Moreover, the instability criteria derived by Sternling and Scriven for example (10) from an analysis of the interfacial turbulence as a manifestation of hydrodynamic instability, taking into account diffusion, convection and interfacial tension, are not sufficient to explain our experiments.

The occurence of the instability required a chemical reaction

Contrary to LINDE, SCHWARTZ and WILKE (11), we observed movements only for two components each separately dissolved in one of the phases and at concentrations far from equilibrium. But we failed to get instability, whatever the concentration range, using a solution of KI very pure and fresh and we noticed that the more aged the solution of KI is, the more important the movements were. Finally we checked that it is iodine - an oxidation product, present as an impurity in the solution - which is responsible for the phenomenon. Our observations schematically summarized in the fig.1 show that not only a transfer of each compound in the other phase is required to get motion, but also a chemical reaction must occur. Indeed, we could observe from spectroscopic and pH metric measurements (12) that a set of more or less coupled redox reactions involving various oxidized derivatives of iodine occurs just as in Bray's reaction, a chemical oscillating reaction occuring in one phase observed for the first time in 1921 (13) and very much studied since then, both by the experimental and the theoretical workers.(14) But such a system still is too complicated to be described without speculative assumptions by a modem suitable for a mathematical treatement and we had to find out a simpler system.

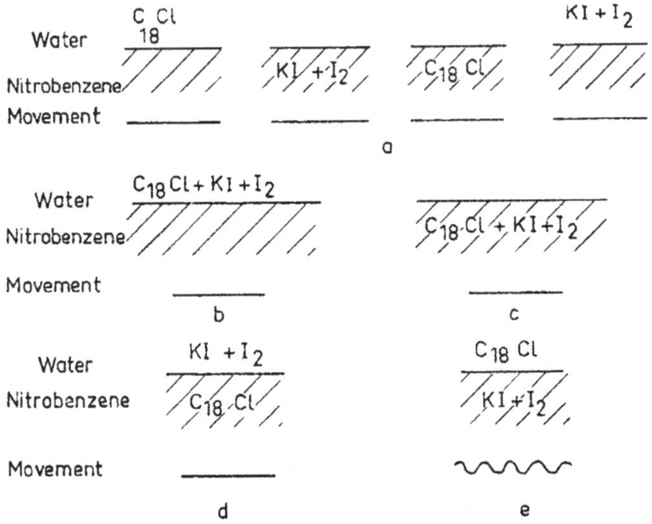

Fig.1 Experimental conditions under the movements occur

Which is the mechanism by which a chemical reaction induces an instability ?

We succeeded in getting the same instabilities as previously described from an aqueous solution of a long chain alkyltrimethyl ammonium chloride or bromide (RCl or Br) in contact with a solution of picric acid (HPi) in nitrobenzene or nitroethane in a given concentration range (15). The period and the amplitude of the movements depend on the concentration and the lenght of the hydrocarbon chain. The total reaction would be

$$R^+_w\ Cl^-_w \ + \ H\ Pi_n \longrightarrow R^+_n\ Pi^-_n \ + \ H^+_w\ Cl^-_w$$

the question is to know there the non linearity, required to explain instability, can appear in a such simple system.

SANFELD and coworkers (16,17,18,19),have studied for several years the coupling between chemical reactions and hydrodynamics from a theoretical point of view, using NAVIER-STOKES equations and the balance of the interfacial matter involving diffusion, convection and chemical production. They recently applied their treatement (20) to the system we have just described. They assume that the adsorption - desorption process, during the transfer , is the determining step and come to the conclusion that the non linearity should appear in the interfacial chemical reaction. For this, they proposed either a solvation - desolvation process or the formation of triplet ions (two ammonium and one picrate for example). These reactions seem to us not very probable and we decided to make some measurements to experimentally determine the controlling parameter and the nature of the interfacial reaction in order to support a model on some hypothesis consistent with the experiments.

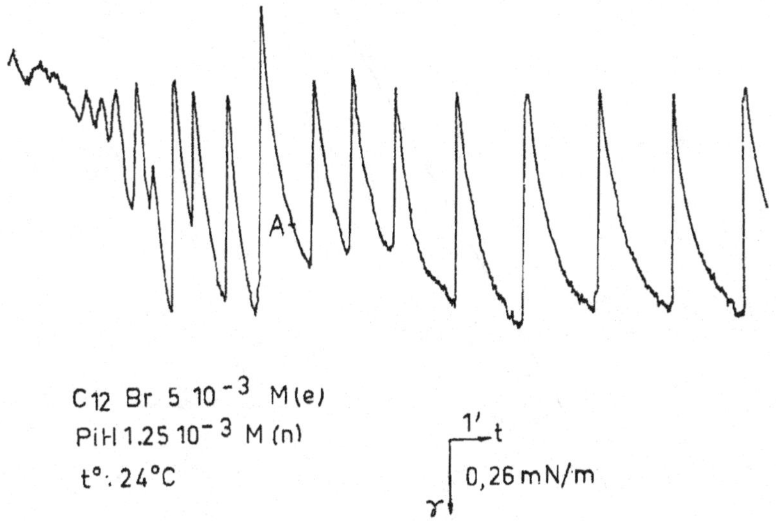

Fig.2 Variation of the interfacial tension as a function of time

We studied a system made from $C_{12}Br$ ($5\ 10^{-3}M$) in water and HPi ($1.25\ 10^{-3}M$) in nitroethane in such a device as we can only obtain motion in the interface plane (21). We analysed the movements from the variations ,as a function of time ,of the interfacial tension measured by the stirrup method. The fig.2 shows somewhat regular oscillations with a approximately one minute pseudo period. It can be noticed that, during each oscillation, the interfacial tension gradually increases at the beginning, then abruptly falls. This fall can be correlated with an "explosion" at the interface, that is, with a massive arrival of surface active molecules at the interface.

Now, we found from another type of experiments involving very analogous systems (22) that the observed variations of interfacial tension and potential are in very good agreement with the calculated

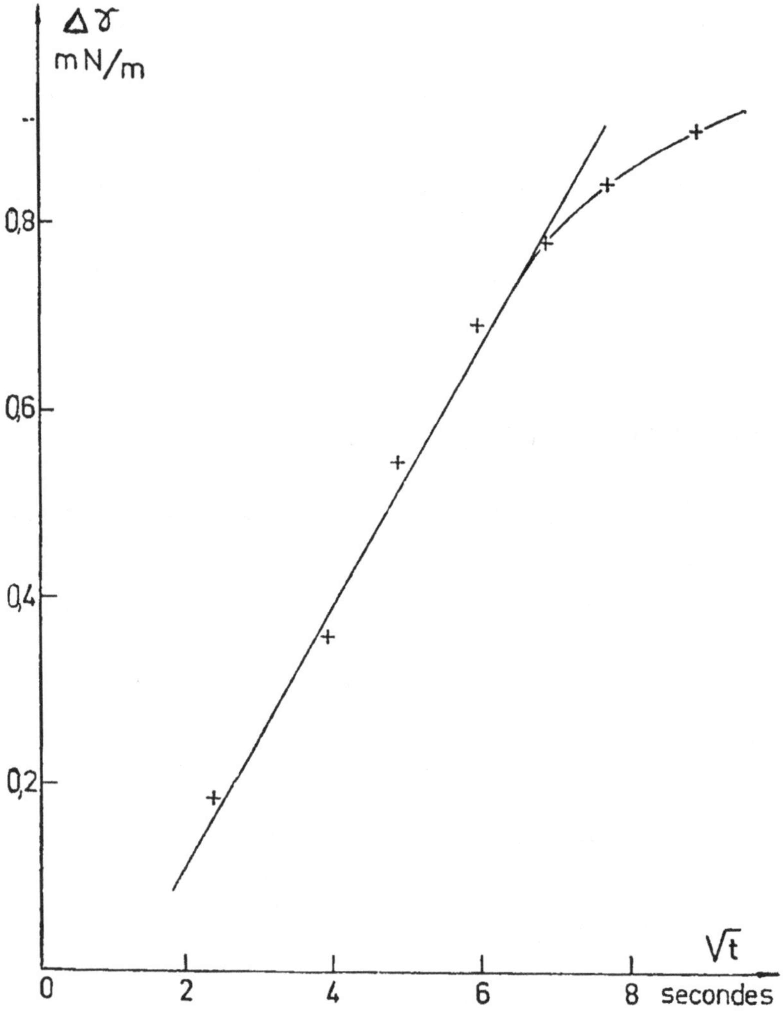

Fig.3 $\Delta\gamma$ as a function of $\sqrt{t}$

results if we admit that the adsorption - desorption process is fast with respect to the diffusion - convection and that the interfacial reaction is only an exchange between the counter-ions of the surface active ions according to the equation :

$$RCl_{surf} + Pi^-_{sol} \rightleftharpoons RPi_{surf} + Cl^-_{sol} \qquad (1)$$

If the diffusion - convection process is the determining step, it is expected that, at the beginning of the oscillation when the diffusion is more important, the interfacial tension variation is linear as a function of $\sqrt{t}$. The fig.3 shows the results of the fig.2 so plotted for one oscillation. The curve is a straight line on the climbing part of the oscillation. Therefore, we come to the hypothesis that the adsorption desorption process is fast and that, at every time, an equilibrium between the interface and the subjacent layers is established. Therefore, the interfacial reaction accountable for the variation of the interfacial tension depends on the composition of the subjacent layers.

The interfacial reaction (1) would consist of a simple exchange of $C_{12}Cl$ in $C_{12}Pi$. This supposes the transfer of C_{12}^+ and Pi^- ions, through the interface, in opposite direction, following the chemical potential gradient, which is in agreement with the experiments.

It is easy to check that, at the same bulk concentration, $C_{12}Pi$ reduces the tension less than $C_{12}Cl$ does, because the number of C_{12}^+ ions present at the interface in equilibrium with the subjacent layers is smaller when these ions are associated with Pi^-. The replacement of choride ions by picrate ions in the bulk near the interface is sufficient to explain the variation of the tension. Therefore the interfacial reaction would only result in a variation of the number of C_{12}^+ ions present at the interface. The non linearity could not come from the interfacial reaction. Where does it come from ? Obviously, the variation of the concentration of Pi^- ions in the subjacent layers is gouverned by diffusion and convection from the core of the solution but also by a chemical reaction in the bulk of the subjacent layers, which is :

$$H\,Pi_n \rightleftharpoons H^+_n + Pi^-_n$$

For this reason we proposed the following mechanism (23).

$$
\begin{array}{ccccccc}
H^+_e & + & Pi^-_e & \longrightarrow & Pi^-_e & C^+_{12e} & Cl^-_e \\
\Big\uparrow & & & & \Big\downarrow & \Big\downarrow & \Big\downarrow \\
HPi_n & \rightleftharpoons & H^+_n & + & Pi^-_n & C^-_{12n} & Cl^-_n
\end{array}
$$

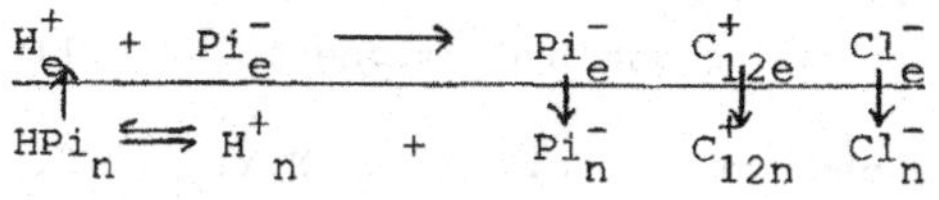

HPi would diffuse from nitroethane towards water and react with C_{12} ions giving $C_{12}Pi$. This compound, which has a large affinity for nitroethane would pass into this solvent in which it is entirely dissociated. So the Pi^- ion would come back in the organic solution where they could partially combine again with the H^+ ions, because of the high formation constant of HPi - So a feed back would be settled which would bring the non linearity if the kinetic constants of the various transfers and reactions are suitable. This mechanism could be supported by the fact that, repla-cing HPi by KPi; salt for which the formation constant in nitroethane is much more smaller than for HPi, no movement occurs.

In conclusion, an interfacial chemical reaction with a non linear kinetics is not absolutely required to get instability. The cou-pling between the interfacial transfer and the chemical reactions <u>in the bulk</u> of the layers subjacent to the interface is sufficient to ex-plain the observed phenomenon. So it would be confirmed that the inter-face is closely related to the subjacent phases and exists only in their presence.

The question is open to know what is the right mathema-tical approach to solve this kind of problems which are very important for several fields. Indeed, this mechanism could be used as a model as well for the movements related to the biological organisms as for the liquid-liquid extraction in chemical engineering or as an example of PRIGOGINE and GLANSDORFF's dissipative structures (24).

(1) M. DUPEYRAT and J. MICHEL - XX réunion CITCE Strasbourg p.346, 1969.

(2) M. DUPEYRAT and E. NAKACHE - "Mouvements spontanés à l'interface de deux phases liquides non miscible. Film SERDDAV CNRS, Paris 1976.

(3) M. DUPEYRAT and E. NAKACHE in "Electrical phenomena at the biolo-gical membrane level" edit by E. Roux, Elsevier Sc. Company Amsterdam, p.377, 1977.

(4) M. DUPEYRAT and J. MICHEL - Experientia Suppl. <u>18</u>, 269-73, 1971.

(5) J.J. BIKERMAN - Surface Chemistry Acad. Press. N.Y. 1958.

(6) C.V. STERNLING,L.E.SCRIVEN- Nature <u>187</u>, 186, 1960.

(7) M.G. VELARDE in "Fluids Dynamics" ed by R. BALIAN and J.L. PEUBE GORDON and BREACH, N.Y., p.469, 1976.

(8) S. CHANDRASEKHAR, "Hydrodynamic and hydromagnetic stability" Oxford, University Press, 1961.

(9) J.T. DAVIES "Turbulence phenomena" Ac. Press. N.Y., 1972.

(10) C.V. STERNLING,L.E.SCRIVEN A.I.Ch.E.J., <u>5</u>, 514, 1959.

(11) H. LINDE, P. SCHWARTZ and H. WILKE - Symposium on "Dynamics and Instability of fluid Interface". LYNGBY Copenhagen, May 1978.

(12) M. DUPEYRAT and E. NAKACHE - Proceedings of the 7th Intern Congress CID, Moscou, 1976.

(13) W.C. BRAY - J. Amerc. Chem Soc, 43, 1262, 1921.

(14) G. SCHMITZ and H. ROOZE - Colloque 'Loin de l'équilibre. Instabilités Structures". Bordeaux, Sept.1978. to be published

(15) M. DUPEYRAT and E. NAKACHE - Physicochemical hydrodynamics - Edit by B. SPALDING , Advance Publications limited, London , 1977, Vol.II, p.591.

(16) A. SANFELD and A. STEINCHEN - Biophys. Chem., 3, 99, 1975.

(17) M. HENNENBERG, T.S. SØRENSEN, A. STEINCHEN and A. SANFELD - J. Chim. Phys. 72, 1202, 1975.

(18) T.S. SØRENSEN, M. HENNENBERG, A. STEINCHEN and A. SANFELD - J. Coll. Int. Sc. 56, 191, 1976.

(19) M.G. VELARDE, J.L. IBANEZ, T.S. SØRENSEN, A. SANFELD and M. HENNENBERG - Proc. Levich Conference, Oxford p.483, 1977.

(20) W. DALLE-VEDOVE, P.M. BISCH, A. SANFELD and A. STEINCHEN - C.R Acad. Sc., Paris 287 C 297, 1978.

(21) M. DUPEYRAT and E. NAKACHE - Colloque "Loin de l'équilibre Instabilités et Structures" Bordeaux, Sept.1978 to be published

(22) M. DUPEYRAT and E. NAKACHE - to be published.

(23) M. DUPEYRAT and E. NAKACHE - Biochemistry and Bioenergetics 5, 134, 1978.

(24) P. GLANSDORFF and I. PRIGOGINE. Structure, Stabilité et Fluctuations Masson, Paris 1971.

Remarques sur un problème de valeurs propres non linéaires faisant intervenir des fonctions non différentiables.

C. GUILLOPÉ.

Analyse Numérique et Fonctionnelle,

C.N.R.S et Université de Paris-Sud

Bâtiment 425, 91405 ORSAY, (France).

0. INTRODUCTION.

Dans cette conférence, nous étudions l'unicité (ou la non-unicité) des solutions d'un problème de valeur propre non linéaire issu de la physique des plasmas.

Notre problème est de déterminer un réel λ et une fonction réelle u, définie sur un ouvert Ω de $\mathbb{R}^n$ $(n \leq 3)$, tels que

(0.1) $$-\Delta u + \lambda u_- = 0 \quad \text{dans} \quad \Omega,$$

(0.2) $$u = \text{constante (inconnue) sur} \quad \Gamma = \partial\Omega,$$

(0.3) $$\int_\Gamma \frac{\partial u}{\partial n} \, d\ell = I,$$

(0.4) $$|u_-|_{L^2(\Omega)} = C,$$

où $I > 0$, $C > 0$ sont deux constantes données. L'ouvert Ω est supposé borné, convexe et régulier. Les équations peuvent modéliser la position d'équilibre d'un plasma confiné dans une cavité toroïdale axisymétrique : u désigne la fonction de flux du champ magnétique méridien, Ω la section du tore, et le plasma occupe la région $\Omega_- = \{x \in \Omega \mid u(x) < 0\}$. (cf C. Mercier [4]).

Nous savons que ce problème admet des solutions régulières (cf R. Temam [6]). Par contre, le problème de l'unicité des solutions n'avait pas encore été abordé : nous en donnons ici une réponse partielle. Précisément, nous montrons que si la constante I/C est assez petite, alors le problème (0.1)-(0.4) admet une solution unique ; cette solution vérifie, de plus,

$$0 < \lambda < \lambda_1, \quad u = -u_-, \quad \Omega_- = \Omega,$$

où λ_1 désigne la première valeur propre de l'opérateur Λ sur Ω. En outre, nous montrons l'existence d'un réel $\bar{\rho} > 0$, tel que le problème (0.1)-(0.4) admet (au moins) une solution (u,λ) vérifiant $\lambda \in]0,\lambda_1]$ si et seulement si $\frac{I}{C} \in]0,\bar{\rho}]$.

Pour obtenir ce résultat, nous abordons le problème de deux manières très différentes. Tout d'abord, nous montrons un certain nombre d'estimations a priori, d'où nous déduisons l'unicité de la solution pour I / C assez petit : c'est l'objet du paragraphe 2. Puis, nous étudions l'ensemble des solutions (u,λ) vérifiant $\lambda \in]0,\lambda_1]$: les propriétés de ces solutions sont développées dans le paragraphe 3.

Enfin, en annexe de l'étude précédente, nous nous intéressons au cas particulier où Ω est un ouvert de $\mathbb{R}$, qui ne vérifiera pas nécessairement l'hypothèse de connexité (donc de convexité) : nous vérifions que si Ω est un intervalle de $\mathbb{R}$, la solution est unique; par contre, si $\Omega \subset \mathbb{R}$ est non connexe, nous exhibons plusieurs branches de solutions : les résultats obtenus explicitement pour $\Omega \subset \mathbb{R}$, non connexe, sont à rapprocher de ceux obtenus, numériquement, par M. Sermange [5] dans le cas où Ω est un ouvert de $\mathbb{R}^2$, en forme de sablier (en particulier non convexe).

Les espaces fonctionnels utilisés sont les espaces usuels $L^p(\Omega)$, les espaces de Sobolev $W^{m,p}(\Omega)$ et $H^m(\Omega)$ (cf. J.L. Lions-E. Magenes [3]).

1. Enoncé des résultats.

En effectuant le changement de fonction inconnue $u \longmapsto u/I$, nous ramenons la résolution de (0.1)-(0.4) à la résolution de

(1.1) $$-\Delta u + \lambda u_- = 0 \quad \text{dans } \Omega,$$

(1.2) $$u = \text{constante sur } \Gamma,$$

(1.3) $$\int_\Gamma \frac{\partial u}{\partial n}\, d\ell = 1,$$

(1.4) $$|u_-| = \sigma,$$

où nous avons posé $\sigma = 1/\rho = C/I$. Par ailleurs, en intégrant (1.1) sur Ω, et en utilisant (1.3), nous remarquons que

(1.5) $$\lambda\left(\int_\Omega u_- dx\right) = 1$$

et donc, nécessairement, $\lambda > 0$. En outre, d'après le principe du maximum, nous avons les propriétés (cf R. Temam [7])

$$\begin{cases} u(\Gamma) < 0 \iff \lambda < \lambda_1, \\ u(\Gamma) = 0 \iff \lambda = \lambda_1, \\ u(\Gamma) > 0 \iff \lambda > \lambda_1, \end{cases}$$

où $u(\Gamma)$ dénote la valeur constante de u sur Γ.

Nous déduisons le résultat, énoncé dans l'introduction, de deux types de résultats. D'une part, par une méthode d'estimations a priori, nous montrons :

Théorème 1.1. *Si le réel* $\rho = 1/\sigma$ *est assez petit, le Problème* (1.1)-(1.4) *admet une solution unique.*

D'autre part, par une étude directe du cas où $u(\Gamma)$ est négatif, nous montrons le résultat suivant.

Théorème 1.2. *Soit* $\bar{w}$ *un vecteur propre associé à la première valeur propre* λ_1 *de l'opérateur* Δ *sur* Ω. *Alors*

(i) *Il existe un réel* $\bar{\rho} \geq \lambda_1 \dfrac{|\int_\Omega \bar{w}dx|}{|\bar{w}|}$ *tel que le Problème* (1.1)-(1.4) *admet* (au moins) une solution (u,λ), vérifiant $0 < \lambda \leq \lambda_1$, si et seulement si $0 < \rho \leq \bar{\rho}$.

(ii) Si le réel ρ est assez petit, la solution (u,λ) vérifiant $0 < \lambda \leq \lambda_1$ est unique.

Ces résultats sont démontrés dans les paragraphes suivants.

En remarque préliminaire, montrons l'existence d'un unique $\sigma > 0$ tel que le Problème (1.1)-(1.4) admette une solution de la forme (u,λ_1) ; cette solution est alors unique. En effet, d'après (1.5)-(1.6), u vérifie les relations

(1.7) $$-\Delta u = \lambda_1 u \quad \text{dans} \quad \Omega,$$

(1.8) $$u = 0 \quad \text{sur} \quad \Gamma,$$

(1.9) $$\int_\Omega u\,dx = -1/\lambda_1,$$

(1.10) $$|u_-| = \sigma.$$

D'après (1.7)-(1.8), la fonction u est fonction propre de l'opérateur Δ associé à la valeur propre (simple) λ_1 et est déterminée de façon unique par la relation (1.9) : la valeur de σ, telle qu'il existe une solution (u,λ_1), est donc déterminée, de façon unique, par la relation (1.10).

2. Quelques estimations à priori.

Utilisant (1.5), nous mettons le Problème (1.1)-(1.4) sous la forme

(2.1) $$-\Delta u + \frac{u_-}{\int_\Omega u_- dx} = 0 \quad \text{sur} \quad \Omega,$$

(2.2) $$u = \text{constante sur} \quad \Gamma,$$

(2.3) $$|u_-| = \sigma.$$

2.1. Démonstration du Théorème 1.1.

Soient u et v deux solutions régulières de (2.1)-(2.4). Alors $\Phi = u-v$ vérifie les relations

(2.4) $$\Delta\Phi = \frac{u_-}{\int_\Omega u_- dx} - \frac{v_-}{\int_\Omega v_- dx},$$

(2.5) $$\Phi = \text{constante sur} \quad \Gamma,$$

En multipliant (2.4) par Φ et en intégrant sur Ω, nous obtenons

$$|\mathrm{grad}\ \Phi|^2 = -\frac{(u_- - v_-, u - v)}{\int_\Omega u_- dx} + \left(\frac{1}{\int_\Omega u_- dx} - \frac{1}{\int_\Omega v_- dx}\right)(v_-, u - v),$$

d'où nous déduisons la majoration

$$|\mathrm{grad}\ \Phi|^2 \leqslant \frac{|\Phi|^2}{\int_\Omega u_- dx} + \frac{\left|\int_\Omega (u_- - v_-) dx\right|}{\left(\int_\Omega u_- dx\right)\left(\int_\Omega v_- dx\right.} |v_-||\Phi|,$$

soit, en utilisant l'inégalité de Cauchy-Schwarz et la relation (2.3),

$$(2.6) \qquad |\mathrm{grad}\ \Phi|^2 \leqslant \frac{1}{\int_\Omega u_- dx} \left(1 + \frac{|\Omega|^{\frac{1}{2}} \sigma}{\int_\Omega v_- dx}\right) |\Phi|^2,$$

où $|\Omega|$ désigne la mesure de l'ouvert Ω.

Les lemmes ci-dessous donnent des minorations de l'intégrale $\int_\Omega u_- dx$ et du rapport $|\mathrm{grad}\ \Phi|^2 / |\Phi|^2$.

<u>Lemme</u> 2.1. <u>Il existe une fonction</u> $\delta : \mathbb{R}_+^* \longrightarrow \mathbb{R}_+^*$, <u>décroissante, telle que, pour toute solution régulière</u> (2.1)-(2.3), <u>on ait</u>

$$(2.7) \qquad \boxed{\int_\Omega u_- dx \geqslant \sigma\delta(\tfrac{1}{\sigma}).}$$

<u>Lemme</u> 2.2. <u>Il existe une fonction</u> $\delta_1 : \mathbb{R}_+ \longrightarrow \mathbb{R}_+$, <u>croissante, telle que si</u> Φ <u>est la différence de deux solutions régulières de</u> (2.1)-(2.3), <u>on ait</u>

$$(2.8) \qquad \boxed{|\Phi|^2 \leqslant \delta_1(\tfrac{1}{\sigma})|\mathrm{grad}\ \Phi|^2.}$$

Utilisant ces deux résultats, nous terminons la démonstration du Théorème 1.1. En effet, d'après (2.7), la relation (2.6) s'écrit

$$|\mathrm{grad}\ \Phi|^2 \leqslant \delta_2(\tfrac{1}{\sigma})|\Phi|^2,$$

où la fonction δ_2, définie pour $t \in \mathbb{R}_+^*$ par $\delta_2(t) = \frac{t}{\delta(t)}\left(1 + \frac{|\Omega|^{\frac{1}{2}}}{\delta(t)}\right)$, est croissante. Utilisant (2.8), nous en déduisons

$$(1 - \delta_1(\tfrac{1}{\sigma})\delta_2(\tfrac{1}{\sigma}))|\Phi|^2 \leqslant 0,$$

et donc $\Phi = 0$, dès que $\delta_1(\frac{1}{\sigma})\delta_2(\frac{1}{\sigma}) < 1$. Or, la fonction $t \longmapsto \delta_1(t)\delta_2(t)$ décroît vers 0, quand t décroît vers 0, car si $t < t_0$, nous avons

$$0 < \delta_1(t)\delta_2(t) < \frac{t}{t_0}\,\delta_1(t_0)\delta_2(t_0).$$

□

2.2. Démonstration du Lemme 2.1.

Nous utiliserons une inégalité, introduite par R. Temam dans [6], et énoncée sous une forme plus précise par H. Brézis dans [8]. Nous aurons besoin, ici, de ce résultat dans le cas particulier $r = 2$ (cf [8]).

Si Ω est un ouvert borné, régulier de $\mathbb{R}^n$, et p, q, p' sont des réels vérifiant

$$1 \leqslant p < n, \qquad q = \frac{pn}{n-p}, \qquad \frac{1}{p} + \frac{1}{p'} = \frac{n+2}{2n},$$

alors, il existe une constante k, ne dépendant que de Ω et p, telle que

$$(2.9) \qquad |u|_{L^q(\Omega)} \leqslant k\{|\mathrm{grad}\, u|_{\mathbb{L}^p(\Omega)} + \max\left[|u_-|, \frac{|\mathrm{grad}\, u|^{p'/2}_{\mathbb{L}^p(\Omega)}}{|u_-|^{p'/2-1}}\right]\}$$

pour toute fonction u de $W^{1,p}(\Omega)$, telle que $|u_-| \neq 0$.

Soit u une solution régulière du Problème (2.1)-(2.3). Intégrant (2.1) sur Ω, nous remarquons que

$$\int_\Omega \Delta u \, dx = 1, \quad \text{avec} \quad \Delta u \geqslant 0 \quad \text{dans} \quad \Omega,$$

et donc que la fonction Δu appartient à un borné de $L^1(\Omega)$, et par conséquent, d'après les théorèmes d'injection de Sobolev, à un borné de $W^{-1,p}(\Omega)$, avec $1 \leqslant p < \frac{n}{n-1}$. Les résultats de régularité pour les problèmes elliptiques (cf Agmon-Douglis-Nirenberg [1]) montrent alors l'existence d'une constante $c_0(\Omega,p)$ telle que

$$\|u - u(\Gamma)\|_{W^{1,p}(\Omega)} \leqslant c_0(\Omega,p)|\Delta u|_{L^1(\Omega)} \leqslant c_0(\Omega,p) \ ;$$

en particulier, nous avons

$$(2.10) \qquad |\mathrm{grad}\, u|_{\mathbb{L}^p(\Omega)} \leqslant c_0.$$

Le lemme 2.1 se déduira immédiatement du résultat suivant

Lemme 2.3. *Soit* Ω *un ouvert borné de* $\mathbb{R}^n$ (n=2 *ou* 3), *de frontière régulière, et soit* p *un réel tel que* $\frac{2n}{n+2} < p < \frac{n}{n-1}$.

Soit $B(t)$ *le compact de* $L^2(\Omega)$ *défini, pour* $t \in \mathbb{R}_+^*$, *par*

$$B(t) = \{\Phi \in W^{1,p}(\Omega) \mid |\Phi_-| = 1, \quad |\text{grad } \Phi|_{\mathbb{L}^p(\Omega)} \leq c_0 t\}$$

où la constante c_0 *est définie dans la relation* (2.10).

Alors, la fonction δ, *définie sur* $\mathbb{R}_+$ *par*

$$\delta(t) = \operatorname*{Inf}_{\Phi \in B(t)} \int_\Omega \sigma \Phi_- dx,$$

est décroissante et strictement positive.

Démonstration. (i) L'ensemble $B(t)$ est un compact non vide de $L^2(\Omega)$.

En effet, c'est un borné de l'espace $W^{1,p}(\Omega)$: en appliquant la relation (2.9), nous obtenons, pour toute fonction Φ de $B(t)$,

(2.11) $$|\Phi|_{L^p(\Omega)} \leq |\Phi|_{L^q(\Omega)} |\Omega|^{\frac{1}{n}} \leq k(c_0 t + 1 + c_0^{p'/2} t^{p'/2}) |\Omega|^{\frac{1}{n}}$$

où $\frac{1}{p} = \frac{1}{q} + \frac{1}{n}$ et $\frac{1}{p} + \frac{1}{p'} = \frac{n+2}{2n}$. Or, comme $\frac{2n}{n+2} < p < \frac{n}{n-1}$, l'injection de $W^{1,p}(\Omega)$ dans $L^2(\Omega)$ est compacte : donc, l'ensemble $B(t)$ est relativement compact dans $L^2(\Omega)$. Enfin, $B(t)$ est un fermé de $L^2(\Omega)$, car la fonctionnelle $\Phi \longmapsto \int_\Omega (\Phi_-)^2 dx$ est continue sur $L^2(\Omega)$, et la semi-norme $\Phi \longrightarrow |\text{grad } \Phi|_{\mathbb{L}^p(\Omega)}$ est faiblement semi-continue sur $L^2(\Omega)$.

(ii) Nous minimisons la fonctionelle continue $\Phi \longrightarrow \int_\Omega \Phi_- dx$ sur le compact $B(t)$ de $L^2(\Omega)$: le minimum est atteint, et donc strictement positif. □

Pour obtenir le résultat du Lemme 2.1, il suffit de fixer p ($p = \frac{7}{5}$, par exemple), et de minimiser la fonctionnelle $\Phi \longmapsto \int_\Omega \Phi_- dx$ sur l'ensemble

$$B = \{u \in W^{1,p}(\Omega) \mid |u_-| = \sigma, \quad |\text{grad } u|_{\mathbb{L}^p(\Omega)} \leq c_0\}. \quad \square$$

2.3. Démonstration du Lemme 2.2.

Par les mêmes techniques que précédemment (théorèmes d'injection de Sobolev, résultats de régularité dans les équations elliptiques) nous montrons le résultat suivant.

Lemme 2.4. *Il existe une fonction* $\delta_3 : \mathbb{R}_+^* \longrightarrow \mathbb{R}_+^*$, *croissante, telle que, pour toute solution régulière* u *de* (2.1)-(2.3), *on ait*

(2.12) $$\|u\|_{H^3(\Omega)} \leq \sigma\delta_3(\tfrac{1}{\sigma}).$$

Démonstration. Soit u une solution régulière de (2.1)-(2.3). Les réels p, q, p' sont fixés comme précédemment.

(i) Dans l'inégalité (2.11), nous posons $\Phi = \frac{u}{\sigma} = tu$, et nous obtenons

$$|u| \leq |u|_{L^q(\Omega)} |\Omega|^{\frac{1}{2}-\frac{1}{q}} \leq k\sigma(\frac{c_0}{\sigma} + 1 + c_0^{p'/2}(\frac{1}{\sigma})^{p'/2})|\Omega|^{\frac{1}{2}-\frac{1}{q}},$$

ce qui montre l'existence d'une fonction $\delta_4 : \mathbb{R}_+^* \longrightarrow \mathbb{R}_+^*$, croissante, telle que

(2.13) $$|u| \leq \sigma\delta_4(\tfrac{1}{\sigma}).$$

(ii) Nous majorons maintenant $|\text{grad } u_-|_{\mathbb{L}^2(\Omega)}$.

Nous avons

$$|u_-|_{L^p(\Omega)} \leq |\Omega|^{\frac{1}{p}-\frac{1}{2}}|u_-|,$$

et, donc, d'après les relations (2.3) et (2.10), il existe une constante c, telle que

(2.14) $$\|u_-\|_{W^{1,p}(\Omega)} \leq c(1+\sigma).$$

Les résultats de régularité des problèmes elliptiques montrent que la fonction $u - u(\Gamma)$ est bornée dans $W^{3,p}(\Omega)$, et donc que grad u est bornée dans $W^{2,p}(\Omega)$, espace qui s'injecte de façon continue dans $L^2(\Omega)$. Donc, il existe une constante, notée encore c, telle que

$$|\text{grad } u_-|_{\mathbb{L}^2(\Omega)} \leqslant |\text{grad } u|_{\mathbb{L}^2(\Omega)} \leqslant c \frac{1}{\int u_- dx} \|u_-\|_{W^{1,p}(\Omega)}$$

ce qui, d'après les relations (2.7) et (2.14), montre l'existence d'une fonction $\delta_5 : \mathbb{R}_+^* \longrightarrow \mathbb{R}_+^*$, croissante, telle que

(2.15) $$|\text{grad } u_-|_{\mathbb{L}^2(\Omega)} \leqslant \sigma\delta_5(\tfrac{1}{\sigma}),$$

δ_5 est définie par $\delta_5(t) = c\,\dfrac{t(1+t)}{\delta(t)}$.

(iii) Nous en déduisons une majoration de $\|\text{grad } u\|_{\mathbb{H}^2(\Omega)}$.

D'après les théorèmes de régularité pour les problèmes elliptiques, la fonction $u - u(\Gamma)$ est bornée dans l'espace $H^3(\Omega)$: donc la fonction grad u est bornée dans $H^2(\Omega)$ et il existe une constante notée c, telle que

$$\|\text{grad } u\|_{H^2(\Omega)} \leqslant c \frac{1}{\int_\Omega u_- dx} \|u_-\|_{H^1(\Omega)},$$

ce qui, d'après les relations (2.3), (2.7) et (2.15), montre l'existence d'une fonction $\delta_6 : \mathbb{R}_+^* \longrightarrow \mathbb{R}_+^*$ croissante, telle que

(2.16) $$\|\text{grad } u\|_{\mathbb{H}^2(\Omega)} \leqslant \sigma\delta_6(\tfrac{1}{\sigma}) ;$$

δ_6 est définie par $\delta_6(t) = ct\,\dfrac{1+\delta_5(t)}{\delta(t)}$.

(iv) Le résultat du Lemme 2.4 en découle. En effet

$$\|u\|_{H^3(\Omega)} \leqslant \|\text{grad } u\|_{\mathbb{H}^2(\Omega)} + |u| ;$$

d'après (2.13) et (2.16), nous obtenons la relation (2.12), en posant $\delta_3 = \delta_4 + \delta_6$. □

Le Lemme 2.2 se déduira immédiatement du résultat suivant.

Lemme 2.5. Soit $B_3(t)$ le compact de $L^2(\Omega)$ défini, pour $t \in \mathbb{R}_+^*$, par

$$B_3(t) = \{\phi \in H^3(\Omega) \mid |\phi_-| = 1, \|\phi\|_{H^3(\Omega)} \leqslant \delta_3(t)\}.$$

Alors, la fonction α définie, pour $t \in \mathbb{R}_+^*$, par

$$\alpha(t) = \inf_{\substack{\Phi,\psi \in B_3(t) \\ \Phi \neq \psi}} \frac{|\mathrm{grad}\,(\Phi - \psi)|}{|\Phi - \psi|}$$

est décroissante, et strictement positive.

Démonstration. (i) L'ensemble $B_3(t)$ est un compact non vide de $L^2(\Omega)$. En effet, par définition, c'est un ensemble borné dans $H^3(\Omega)$. Or, comme l'injection de $H^3(\Omega)$ dans $\mathcal{C}^0(\bar{\Omega})$ est compacte (Ω est borné), l'ensemble $B_3(t)$ est relativement compact dans $L^2(\Omega)$. Enfin, $B_3(t)$ est fermé dans $L^2(\Omega)$.

(ii) Montrons que $\alpha(t)$ est strictement positif pour tout $t > 0$. Fixons $t \in \mathbb{R}_+^*$ et supposons que $\alpha(t) = 0$. Soit $(\Phi_i,\psi_i)_{i \in \mathbb{N}}$ une suite minimisante : la suite $(F_i)_{i \in \mathbb{N}}$, définie par $F_i = \Phi_i - \psi_i$, vérifie

(2.18) $$\lim_{i \to +\infty} \frac{|\mathrm{grad}\,F_i|}{|F_i|} = 0.$$

L'ensemble $B_3(t)$ étant compact dans $L^2(\Omega)$, nous pouvons de la suite (Φ_i,ψ_i) extraire une sous-suite, toujours notée (Φ_i,ψ_i) telle que

(2.19) $$\left.\begin{array}{l} \Phi_i \longrightarrow \Phi \\ \psi_i \longrightarrow \Psi \end{array}\right\} \text{ dans } H^2(\Omega) \text{ fort et } H^3(\Omega) \text{ faible.}$$

Nous distinguons alors deux cas :

(i) soit $\Phi \neq \psi$. Les relations (2.18) et (2.19) impliquent

$$|\mathrm{grad}\,F_i| \longrightarrow 0, \quad |F_i| \longrightarrow |\Phi - \psi| \quad (\neq 0),$$

si bien qu'il existe un réel non nul d, que nous pouvons supposer négatif, tel que $\Phi = \psi + d$. Les fonctions Φ et ψ appartiennent à $B_3(t)$, donc ψ vérifie

$$\int_\Omega \{[(\Psi + d)_-]^2 - [\psi_-]^2\}dx = 0,$$

c'est-à-dire

$$\int_{\{0 \leq \psi \leq -d\}} (\psi + d)^2 dx + \int_{\{\psi \leq 0\}} d(2\psi + d)dx = 0 \;;$$

cette relation n'est possible que si $\psi(x) \geq -d > 0$, pour presque tout x de Ω, ce qui contredit l'hypothèse $|\psi_-| = 1$.

(ii) Soit $\Phi = \psi$. Les fonctions de $B_3(t)$ sont continues sur $\bar{\Omega}$, et donc les suites (ϕ_i) et (ψ_i) convergent uniformément vers Φ. Nous posons

(2.20) $$\varepsilon_i = |F_i| \quad (\neq 0), \qquad G_i = F_i / \varepsilon_i \ ;$$

nous avons alors

$$|G_i| = 1, \qquad \lim_{i \to +\infty} |\text{grad } G_i| = 0.$$

Donc, la suite (G_i) converge, dans $H^1(\Omega)$, vers la fonction constante $G = |\Omega|^{-\frac{1}{2}}$. Les fonctions ϕ_i et ψ_i, liées par la relation $\phi_i = \psi_i + \varepsilon_i G_i$ d'après (2.20), appartiennent à $B_3(t)$, c'est-à-dire

$$\int_\Omega \{[(\psi_i + \varepsilon_i G_i)_-]^2 - [(\psi_i)_-]^2\} dx - 0$$

soit, aussi

(2.21) $$\int_\Omega \frac{[(\psi_i + \varepsilon_i G_i)_-]^2 - [(\psi_i + \varepsilon_i G)_-]^2}{\varepsilon_i} dx + \int_\Omega \frac{[(\psi_i + \varepsilon_i G)_-]^2 - [(\psi_i)_-]^2}{\varepsilon_i} dx = 0.$$

Nous notons I_1 et I_2 ces intégrales, et nous passons à la limite dans chacune d'elle.

D'une part nous avons

$$I_1 \leq |(\psi_i + \varepsilon_i G_i)_- + (\psi_i + \varepsilon_i G)_-| \ \frac{|(\psi_i + \varepsilon_i G_i)_- - (\psi_i + \varepsilon_i G)_-|}{\varepsilon_i},$$

soit

$$I_1 \leq (|\phi_i| + |\psi_i| + \varepsilon_i |\Omega|^{-\frac{1}{2}}) |G_i - G|$$

et donc, comme les fonctions ϕ_i et ψ_i sont bornées, que la suite (ε_i) tend vers 0, et que la suite G_i tend vers G dans $L^2(\Omega)$, l'intégrale I_1 converge vers 0.

Pour l'intégrale I_2, nous remarquons que la fonction $A_i = \{[(\psi_i + \varepsilon_i G)_-]^2 - [(\psi_i)_-]^2\} / \varepsilon_i$ est majorée, presque partout dans Ω, par $(2|\psi|_i + \varepsilon_i G)G$, ce qui est majoré par une constante ; d'autre part, $A_i(x)$ converge presque partout vers $-2(\psi(x))_- G(x)$; d'après le théorème de Lebesgue, l'intégrale I_2 converge vers $-2 \int_\Omega \psi_- G dx$. La relation (2.21) implique alors

$$\int_\Omega \psi_- G \, dx = 0$$

ce qui contredit l'hypothèse $|\psi_-| = 1$ □

Pour obtenir le résultat du Lemme 2.2, il suffit de poser $\phi = \sigma\tilde{\Phi}$ et $\psi = \sigma\tilde{\psi}$ et nous obtenons $\delta_1(t) = \frac{1}{\alpha(t)}$.

3. Etude des solutions u telles que $u(\Gamma) < 0$.

3.1. Résolution du Problème (1.1)-(1.4).

Soit (u,λ) est une solution de (1.1)-(1.4) vérifiant $\lambda \in]0,\lambda_1[$: nous posons

$$c = -u(\Gamma) \quad \text{et} \quad v = -u + u(\Gamma) \ ;$$

les réels λ et c, la fonction $v : \Omega \longrightarrow \mathbb{R}$ sont alors les solutions de

(3.1) $$-\Delta v - \lambda v = \lambda c \quad \text{dans } \Omega,$$

(3.2) $$v = 0 \quad \text{sur } \Gamma,$$

(3.3) $$\lambda\left(\int_\Omega (v + c)\, dx\right) = 1$$

(3.4) $$\int_\Omega (v + c)^2 dx = \sigma^2,$$

vérifiant, de plus,

(3.5) $$\lambda \in]0,\lambda_1[, \quad c > 0, \quad v + c \geq 0.$$

En fait, il suffit de chercher les solutions de (3.1)-(3.4) vérifiant

(3.6) $$\lambda \in]0,\lambda_1[, \quad c > 0 \ ;$$

en effet, si (3.6) et (3.1) sont vérifiées, nous avons

$$-\Delta v - \lambda v > 0 \quad \text{dans } \Omega, \quad \text{avec } v = 0 \text{ sur } \Gamma,$$

et donc, par le principe du maximum, la fonction v est positive dans Ω.

Les réels λ et c étant fixés et vérifiant (3.6), le Problème (3.1)-(3.2) admet une solution unique, soit $v(\lambda,c)$ dans l'espace $H_0^1(\Omega) \cap H^2(\Omega)$; de plus, l'application $c \longmapsto v(\lambda,c)$ est linéaire, et donc $v(\lambda,c)$ se met sous la forme

$$v(\lambda,c) = cv_1(\lambda)$$

où v_1 est défini par $v_1 = v(.,1)$.

D'autre part, la fonction v vérifie la relation (3.3), c'est-à-dire

$$(3.7) \qquad \lambda c\left(\int_\Omega (v_1(\lambda)+1)dx\right) = -1 \ ;$$

la fonction v_1 étant positive sur Ω, il existe un unique réel c, fonction de λ, tel que $v(\lambda,c) = cv_1(\lambda)$ soit l'unique solution de (3.1)-(3.3). De plus, c est strictement positif.

Dans la suite, nous notons, pour $\lambda \in]0,\lambda_1[$, par $v(\lambda) = c(\lambda)v_1(\lambda)$ la solution de (3.1)-(3.3), où c est donné par

$$(3.8) \qquad c(\lambda) = \frac{1}{\lambda\left(\int_\Omega (v_1(\lambda)+1)dx\right)}$$

et $v_1(\lambda)$ est la solution de

$$(3.9) \qquad \begin{cases} -\Delta v_1 - \lambda v_1 = \lambda & \text{dans } \Omega, \\ v_1 \in H_0^1(\Omega). \end{cases}$$

Il reste maintenant à déterminer les réels λ de $]0,\lambda_1[$ tels que la fonction $v(\lambda)$ vérifie, de plus, la relation (3.4), c'est-à-dire tels que nous ayons

$$(3.10) \qquad \int_\Omega [v_1(\lambda)+1]^2 dx = \frac{\sigma^2}{c^2(\lambda)}.$$

Nous montrons qu'en fait, la condition (3.10) se met sous la forme d'une relation vérifiée par la dérivée de la fonction $\lambda \longmapsto c(\lambda)$.

<u>Lemme</u> 3.1. <u>Les applications</u> $v_1 :]0,\lambda_1[\longrightarrow H_0^1(\Omega)$, $c :]0,\lambda_1[\longrightarrow \mathbb{R}_+$, <u>et</u> $v :]0,\lambda_1[\longrightarrow H_0^1(\Omega)$ <u>sont de classe</u> $\mathscr{C}^\infty$.

<u>Démonstration</u>. La fonction $v_1(\lambda)$ est la solution du problème (3.9), que nous pouvons aussi mettre sous la forme

$$(3.11) \qquad \mathcal{F}(\lambda,\phi) = 0, \qquad \phi \in H_0^1(\Omega)$$

où $\mathcal{F} : \begin{cases}]0,\lambda_1[\times H_0^1(\Omega) \longrightarrow H^{-1}(\Omega) \\ (\lambda,\phi) \longmapsto -\Delta\phi - \lambda\phi - \lambda \end{cases}$ est de classe $\mathscr{C}^\infty$.

Pour tout $(\lambda_0,u_0) \in]0,\lambda_1[\times H_0^1(\Omega)$, la dérivée $\mathcal{F}_u(\lambda_0,u_0) = -\Delta - \lambda_0$ est un isomorphisme de $H_0^1(\Omega)$ sur $H^{-1}(\Omega)$. Le théorème des fonctions implicites appliqué au Problème (3.11) montre qu'au voisinage de toute solution (λ_0,u_0), il existe un arc unique $\lambda \longrightarrow u(\lambda)$ de classe $\mathscr{C}^\infty$ d'un voisinage $\mathcal{V}_{\lambda_0}$ dans $]0,\lambda_1[$ dans $H_0^1(\Omega)$

tel que

$$\mathcal{F}(\lambda,u(\lambda))=0 \qquad \forall\,\lambda\in \mathcal{V}_{\lambda_0}.$$

Comme $v_1(\lambda)$ est la solution unique du problème (3.9) et que $u(\lambda)$ est également une solution de ce Problème, nous avons $v_1(\lambda)=u(\lambda)$: cela montre que $\lambda \longrightarrow v_1(\lambda)$ est de classe $\mathcal{C}^\infty$.

L'application $u \longrightarrow \int_\Omega u dx$ étant linéaire continue sur $H_0^1(\Omega)$, est de classe $\mathcal{C}^\infty$; donc, les applications c et v, définies par la relation (3.8), sont aussi de classe $\mathcal{C}^\infty$. □

Remarque 3.1. La fonction $\lambda \longrightarrow v_1(\lambda)$ est croissante et convexe, pp dans Ω.

En effet, la fonction $w_1(\lambda)=\dfrac{\partial v_1(\lambda)}{\partial\lambda}$ est la solution de

(3.12)
$$\begin{cases}(-\Delta-\lambda)w_1=v_1+1 & \text{dans } \Omega,\\ w_1=0 & \text{sur } \Gamma\end{cases}$$

et, d'après le principe du maximum, $w_1(\lambda)>0$ sur Ω.

De la même façon, $\dfrac{\partial^2 v_1}{\partial\lambda^2}$ est une fonction positive sur Ω. □

Proposition 3.2. Une condition nécessaire et suffisante pour que $\lambda\in]0,\lambda_1[$, vérifie la relation (3.10) est que λ vérifie

(3.13)
$$\boxed{c'(\lambda)=-\sigma^2.}$$

Démonstration. Multipliant les relations (3.8) par $w_1=\dfrac{\partial v_1}{\partial\lambda}$ et (3.9) par v_1, nous obtenons en intégrant par parties sur Ω

$$((v_1,w_1))-\lambda(v_1,w_1)=\lambda(\int_\Omega w_1 dx),$$

$$((v_1,w_1))-\lambda(w_1,v_1)=\int_\Omega v_1(v_1+1)dx,$$

c'est-à-dire

$$\lambda(\int_\Omega w_1 dx)=\int_\Omega v_1(v_1+1)dx,$$

que nous mettons sous la forme

$$(3.14) \qquad \int_\Omega (v_1+1)^2 dx = \lambda\left(\int_\Omega w_1 dx\right) + \int_\Omega (v_1+1)dx.$$

D'autre part, dérivant la relation (3.7), nous obtenons

$$(3.15) \qquad (\lambda c' + c)\left(\int_\Omega (v_1+1)dx + \lambda c\left(\int_\Omega w_1 dx\right) = 0,$$

ce qui, joint aux relations (3.7) et (3.14), s'écrit

$$(3.16) \qquad \int_\Omega (v_1+1)^2 dx = -\frac{c'}{c^2}. \qquad \square$$

Remarque 3.2. La fonction $\lambda \longmapsto v_1(\lambda)$ étant croissante, la fonction $\lambda \longmapsto c(\lambda)$ est décroissante : donc, si $(-\sigma^2) \in \operatorname{Im} c'$, l'équation $c'(\lambda) = -\sigma^2$ admet (au moins) une solution λ.

Remarque 3.3. L'équation $c'(\lambda) = -\sigma^2$ admet une solution unique λ si et seulement si la fonction c est convexe. En effet, nous verrons plus loin que, au voisinage de $\lambda = 0$, c'' est positif, et donc que c' est monotone, si et seulement si c est convexe.

La nature de la fonction c semble dépendre de la convexité ou non-convexité du domaine Ω. Notons que, si Ω est un intervalle borné de $\mathbb{R}$, un calcul explicite de c montre que c est convexe (cf section 4). $\square$

3.2. Etude de la fonction c.

3.2.1. Etude au voisinage de $\lambda = 0$.

Proposition 3.3. Au voisinage de $\lambda = 0$, nous avons

$$(3.17) \qquad \boxed{\text{(i)}\ c(\lambda) \sim \frac{1}{\lambda|\Omega|}, \quad \text{(ii)}\ c'(\lambda) \sim -\frac{1}{\lambda^2|\Omega|}, \quad \text{(iii)}\ c''(\lambda) \sim \frac{2}{\lambda^3|\Omega|}.}$$

Démonstration. a) La solution v_1 de (3.9) vérifie

$$\lim_{\lambda \to 0} v_1(\lambda) = 0 \qquad \text{dans } H_0^1(\Omega).$$

En effet, multipliant (3.9) par v_1 et intégrant par parties sur Ω, nous obtenons

$$\|v_1\|^2 = \lambda\left(\int_\Omega v_1 dx\right) + \lambda |v_1|^2,$$

d'où

$$\|v_1\|^2 \leq \lambda|\Omega|^{\frac{1}{2}}|v_1| + \lambda|v_1|^2 \leq \lambda_1^{-\frac{1}{2}}|\Omega|^{\frac{1}{2}}\|v_1\| + \frac{\lambda}{\lambda_1}\|v_1\|^2,$$

l'injection de H_0^1 dans L^2 étant continue, de norme $(\lambda_1)^{-\frac{1}{2}}$; donc

$$\|v_1\| \leq \lambda_1^{-\frac{1}{2}}|\Omega|^{\frac{1}{2}} \frac{\lambda}{\lambda_1 - \lambda},$$

ce qui montre que $v_1(\lambda)$ converge vers 0 dans $H_0^1(\Omega)$.

Des relations (3.8) et (3.16), nous déduisons alors (3.17.i) et (3.17.ii).

b) De la même façon que précédemment, nous montrons que

$$\lim_{\lambda \to 0} w_1(\lambda) = w_{1o} \quad \text{dans } H_0^1,$$

où w_{1o} est la solution de

$$\begin{cases} -\Delta w_{1o} = 1 & \text{dans } \Omega, \\ w_{1o} = 0 & \text{sur } \Gamma. \end{cases}$$

Une expression de la fonction c'' est obtenue en dérivant la relation (3.16).

$$-\frac{c''}{c^2} = -\frac{2c'^2}{c^3} + 2\int_\Omega (v_1 + 1) w_1 dx.$$

Or, d'après les résultats obtenus précédemment, nous avons

$$\lim_{\lambda \to 0} \int_\Omega (v_1 + 1) w_1 dx = \int_\Omega w_{1o} dx.$$

Utilisant de plus (3.17.i) et (3.17.ii), nous en déduisons (3.17.iii) □

3.2.2. <u>Etude au voisinage de</u> $\lambda = \lambda_1$.

Soit $\bar{w}$ le vecteur propre positif, associé à la valeur propre simple λ_1 de l'opérateur Δ sur Ω, vérifiant de plus

$$|\bar{w}| = 1.$$

Nous notons par $\{\bar{w}\}^\perp$ le sous-espace orthogonal à $\bar{w}$ dans $H_0^1(\Omega)$.

<u>Lemme</u> 3.4. <u>Il existe un voisinage</u> $\mathcal{V}_{\lambda_1}$ <u>de</u> λ_1 <u>et une fonction</u> $\theta : \mathcal{V}_{\lambda_1} \longrightarrow \{\bar{w}\}^\perp$,

<u>de classe</u> $\mathcal{C}^\infty$, <u>tels que</u>

$$(3.18)\qquad v_1(\lambda) = \frac{\lambda\bar{w}}{\lambda_1-\lambda}\left(\int_\Omega \bar{w}dx\right) + \theta(\lambda), \qquad \forall\,\lambda\in \mathcal{V}_{\lambda_1}\in\,]0,\lambda_1[$$

<u>Démonstration</u>. Nous décomposons la fonction $v_1(\lambda)$ sur le sous-espace propre engendré par $\bar{w}$ et sur $\{\bar{w}\}^\perp$:

$$v_1(\lambda) = a\bar{w} + \theta, \qquad \lambda\in\,]0,\lambda_1[,$$

où a est une constante réelle et θ une fonction de $\{\bar{w}\}^\perp$, que nous allons exprimer en fonction de λ. La fonction $v_1(\lambda)$ vérifiant (3.9), la fonction θ et réel a vérifient

$$(-\Lambda-\lambda_1)\theta + (\lambda_1-\lambda)(\theta + a\bar{w}) = \lambda \quad \text{dans} \quad \Omega.$$

En multipliant cette relation par $\bar{w}$, et en intégrant sur Ω, nous obtenons a comme fonction de λ :

$$a(\lambda) = \frac{\lambda}{\lambda_1-\lambda}\left(\int_\Omega \bar{w}dx\right), \qquad \lambda\in\,]0,\lambda_1[,$$

La fonction θ est alors solution de

$$(3.19)\qquad \begin{cases} \theta\in\{\bar{w}\}^\perp, \\ (-\Delta-\lambda_1)\theta + (\lambda_1-\lambda)\theta = -\lambda\bar{w}\left(\int_\Omega \bar{w}dx\right) + \lambda \quad \text{dans} \quad \Omega, \end{cases}$$

où $\lambda\in\,]0,\lambda_1[$. Nous appliquons le théorème des fonctions implicites à la fonction

$$\mathcal{F} : \begin{cases} \mathbb{R}\times\{\bar{w}^\perp\} \longrightarrow (\{\bar{w}^\perp\})' \\ (\lambda,\theta) \longmapsto (-\Delta-\lambda)\theta + \bar{w}\left(\int_\Omega \bar{w}dx\right) - \lambda \end{cases}$$

au point (λ_1,θ_1), où θ_1 est la solution dans $\{\bar{w}\}^\perp$ (obtenue par l'alternative de Fredholm) de $\mathcal{F}(\lambda_1,\theta_1) = 0$. □

<u>Lemme</u> 3.5. <u>Nous avons, pour</u> $\lambda<\lambda_1$,

$$\lim_{\lambda\to\lambda_1} \theta(\lambda) = \theta_1 \quad \underline{\text{dans}} \quad L^2(\Omega) \quad \underline{\text{fort, et}} \quad H_0^1(\Omega) \quad \underline{\text{faible}}.$$

<u>Démonstration</u>. Soit $\lambda<\lambda_1$: notons $\varepsilon = \lambda_1-\lambda$ et $u_\varepsilon = \theta_1-\theta(\lambda)$.

D'après (3.19), la fonction u_ε vérifie

$$(3.20)\qquad \begin{cases} u_\varepsilon \in \{\bar{w}\}^\perp, \\ (-\Delta - \lambda_1)u_\varepsilon + \varepsilon u_\varepsilon = \varepsilon(\theta_1 + 1 - \bar{w}(\int_\Omega \bar{w}dx)) \quad \text{dans} \quad \Omega. \end{cases}$$

En multipliant cette relation par u_ε et en intégrant sur Ω, nous obtenons

$$\|u_\varepsilon\|^2 = (\lambda_1 - \varepsilon)|u_\varepsilon|^2 + \varepsilon(\int_\Omega (\theta_1 + 1)u_\varepsilon dx) \ ;$$

par suite

$$\|u_\varepsilon\|^2 \leqslant (\lambda_1 - \varepsilon)|u_\varepsilon|^2 + \varepsilon|\theta_1 + 1||u_\varepsilon|,$$

soit, comme l'injection de $H_0^1(\Omega)$ dans $L^2(\Omega)$ est continue, de norme $(\frac{1}{\lambda_1})^{\frac{1}{2}}$,

$$\|u_\varepsilon\| \leqslant (\lambda_1)^{\frac{1}{2}}|\theta_1 + 1|.$$

Il existe donc une sous-suite, notée $(u_\varepsilon)_\varepsilon$, et une fonction u^* de $\{\bar{w}\}^\perp$ tels que

$$\lim_{\varepsilon \to 0} u_\varepsilon = u^* \quad \text{dans } L^2(\Omega) \text{ fort et } H_0^1(\Omega) \text{ faible.}$$

Enfin, d'après (3.20), la fonction u_ε vérifie

$$((u_\varepsilon, v)) - \lambda_1(u_\varepsilon, v) + \varepsilon(u_\varepsilon, v) = \varepsilon(\theta_1 + 1, v), \quad \forall\, v \in \{\bar{w}\}^\perp,$$

et donc, en passant à la limite, $u^* \in \{\bar{w}\}^\perp$ vérifie

$$((u^*, v)) - \lambda_1(u^*, v) = 0, \qquad \forall\, v \in \{\bar{w}\}^\perp,$$

ce qui montre que $u^* \equiv 0$ dans $H_0^1(\Omega)$, et c'est donc toute la suite $(u_\varepsilon)_\varepsilon$ qui converge vers 0 dans $H_0^1(\Omega)$ faible. □

Nous déduisons alors des estimations, au voisinage de $\lambda = \lambda_1$, des fonctions c et c'.

<u>Proposition</u> 3.6. <u>Nous avons, pour</u> $\lambda < \lambda_1$,

(i)
$$c(\lambda) \sim \frac{|\bar{w}|^2}{\lambda_1^2(\int_\Omega \bar{w}dx)^2}(\lambda_1 - \lambda), \text{ \underline{au voisinage de} } \lambda = \lambda_1,$$

(ii)
$$\lim_{\lambda \to \lambda_1} c'(\lambda) = -\frac{|\bar{w}|^2}{\lambda_1^2(\int_\Omega \bar{w}dx)^2}.$$

Démonstration. D'après (3.2), nous avons

$$\int_\Omega v_1(\lambda)dx = \frac{\lambda}{\lambda_1-\lambda}\left(\int_\Omega \bar{w}dx\right)^2 + \int_\Omega \theta(\lambda)dx$$

et

$$|v_1(\lambda)+1|^2 = \frac{\lambda(2\lambda_1-\lambda)}{(\lambda_1-\lambda)^2}\left(\int_\Omega \bar{w}dx\right) + |\theta(\lambda)+1|^2.$$

En faisant tendre λ vers λ_1 dans les relations (3.8) et (3.16) et en utilisant la convergence, dans $L^2(\Omega)$, de $\theta(\lambda)$ vers θ_1, nous obtenons (i) et (ii). □

3.3. Démonstration du Théorème 2.2.

Au paragraphe 3.1, nous avons ramené l'étude du Problème (1.1)-(1.4) à la résolution de l'équation

$$(3.13)\qquad \begin{cases} \lambda \in]0,\lambda_1[, \\ c'(\lambda) = -\sigma^2. \end{cases}$$

D'après l'étude précédente, nous savons que la fonction c est strictement décroissante sur $]0,\lambda_1[$. Par contre, comme $\lim_{\lambda\to 0} c''(\lambda) = +\infty$, la fonction c', prenant la valeur $-\infty$ en $\lambda = 0$, est strictement croissante sur un voisinage $]0,\lambda^*[$ de 0 ; mais, nous n'avons pas obtenu de renseignements plus précis sur le sens de variation de la fonction c'. Les résultats sont représentés dans le tableau suivant

λ	0	λ^*		λ_1
$c(\lambda)$	$+\infty$	↘	↘	0
$c'(\lambda)$	$-\infty$	↗	?	$-\dfrac{\lvert\bar{w}\rvert^2}{\lambda_1^2\left(\int_\Omega \bar{w}dx\right)^2}$

La fonction c' étant continue sur $]0,\lambda_1[$ et strictement négative, le réel $\bar{\rho} = \left(-\max_{\lambda\in]0,\lambda_1[} c'(\lambda)\right)^{-\frac{1}{2}}$ existe et vérifie

$$\bar{\rho} \geqslant \lambda_1 \frac{\left|\int_\Omega \bar{w}dx\right|}{|\bar{w}|}.$$

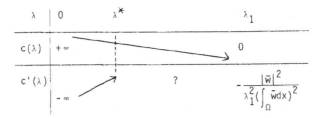

Le Problème (1.1)-(1.4) admet donc au moins une solution $(u,\lambda) = (u(\lambda),\lambda)$ vérifiant $\lambda \in]0,\lambda_1[$, si et seulement si

$$0 < \rho \leq \bar{\rho}.$$

C'est la partie (i) du Théorème 2.2.

De plus, comme c' ne prend la valeur $-\infty$ qu'en $\lambda = 0$, et est strictement croissante dans un voisinage de $\lambda = 0$, alors, pour σ assez grand, c'est-à-dire pour ρ assez petit, le Problème (1.1)-(1.4) admet une unique solution $(u,\lambda) = (u(\lambda),\lambda)$ vérifiant $0 < \lambda < \lambda_1$. □

Nous terminons ce paragraphe par l'énoncé d'une condition suffisante pour l'unicité de la solution $(u(\lambda),\lambda)$, avec $\lambda \in]0,\lambda_1[$, du Problème (1.1)-(1.4). Nous avons remarqué que l'équation (3.13) admet une solution unique si et seulement si la fonction c est convexe (cf Remarque 3.2).

Pour $\lambda \in]0,\lambda_1[$, notons $u(\lambda)$ la solution de

(3.21) $$-\Delta u - \lambda u = 0 \quad \text{dans } \Omega,$$

(3.22) $$\int_\Gamma \frac{\partial u}{\partial n}\, d\ell = 1,$$

(3.23) $$u = \text{constante sur } \Gamma.$$

D'après (3.1)-(3.3), nous avons $u(\lambda) = -c(\lambda)[v_1(\lambda) + 1]$, et nous remarquons que la fonction $\lambda \longrightarrow u(\lambda)|_\Gamma$ est croissante.

<u>Proposition</u> 3.7. <u>Si la fonction</u> $\lambda \longrightarrow u(\lambda)$ <u>est croissante, alors la fonction</u> c <u>est convexe</u>.

<u>Démonstration</u>. Soient $(\lambda,u(\lambda))$ et $(\mu,u(\mu))$ deux solutions distinctes de (3.21)-(3.23) ; soit $\theta \in]0,1[$: le barycentre (α,w), défini par

$$\alpha = \theta\lambda + (1-\theta)\mu, \qquad w = \theta u(\lambda) + (1-\theta)u(\mu),$$

vérifie alors l'équation

$$-\Delta w - \alpha w = \theta(1-\theta)(\lambda-\mu)(u(\lambda) - (\mu)).$$

D'après l'hypothèse, (α,w) est une "sur-solution" de (3.21)-(3.23), car vérifie

$$\begin{cases} -\Delta w - \alpha w \geq 0 & \text{dans } \Omega, \\ \displaystyle\int_\Gamma \frac{\partial w}{\partial n}\, d\ell = 1, \end{cases}$$

$$\Big\{ w = \text{constante sur } \Gamma .$$

Utilisant le Lemme 3.8 ci-dessous, nous obtenons

$$w_{|\Gamma} \leqslant u(\alpha)_{|\Gamma}$$

c'est-à-dire $\theta u(\lambda)_{|\Gamma} + (1-\theta)u(\mu)_{|\Gamma} \leqslant u(\lambda\theta + (1-\theta)\mu)_{|\Gamma}$,

ce qui montre que $c(\lambda) = -u(\lambda)_{|\Gamma}$ est convexe.

Remarquons que si la fonction $\lambda \longrightarrow u(\lambda)$ est strictement croissante, alors la fonction c est strictement convexe. □

Lemme 3.8. Soit $\alpha \in]0,\lambda_1[$. Si w est une sur-solution stricte (respectivement sur-solution) du Problème (3.21)-(3.23), c'est-à-dire vérifie

$$\begin{cases} -\Delta w - \alpha w > 0 & (\text{respectivement} : \geqslant) \text{ dans } \Omega, \quad (3.24) \\ \displaystyle\int_\Gamma \frac{\partial w}{\partial n}\, d\ell \geqslant 1, & (3.25) \\ w = \text{constante sur } \Gamma, & (3.26) \end{cases}$$

alors

$$w_{|\Gamma} < u(\alpha)_{|\Gamma} \qquad (\text{respectivement} : \leqslant).$$

Démonstration. Soit w une sur-solution stricte. Supposons que $w_{|\Gamma} \geqslant u(\alpha)_{|\Gamma}$; alors la fonction $\phi = w - u(\alpha) - (w - u(\alpha))_{|\Gamma}$ vérifie

$$-\Delta\phi - \alpha\phi = -\Delta w - \alpha w + \alpha(w - u(\alpha))_{|\Gamma}$$

c'est-à-dire

$$-\Delta\phi - \alpha\phi > 0 \quad \text{dans } \Omega. \tag{3.27}$$

Comme, de plus, ϕ est nulle sur Γ, le principe du maximum montre que $\phi > 0$ dans Ω.

Or, intégrant (3.27) sur Ω, nous obtenons

$$-\int_\Gamma \frac{\partial\phi}{\partial n}\, d\ell - \alpha\Big(\int_\Omega \phi dx\Big) > 0,$$

c'est-à-dire, d'après l'expression de la fonction ϕ,

$$-\int_\Gamma \frac{\partial w}{\partial n}\, d\ell + 1 - \alpha\Big(\int_\Omega \phi dx\Big) > 0 \;;$$

comme $\phi > 0$ sur Ω, cette relation contredit la relation (3.25). □

4. Calcul des solutions, dans le cas "$\Omega \subset \mathbb{R}$".

Nous cherchons à déterminer le nombre de solutions du Problème (1.1)-(1.4). Pour cela, comme d'après (1.5), λ est strictement positif, nous posons $\lambda = \mu^2$ où $\mu > 0$; μ étant fixé, nous calculons explicitement les solutions $u(\mu)$ de

(4.1) $$-u'' + \mu^2 u_- = 0 \quad \text{dans} \quad \Omega,$$

(4.2) $$u = \text{constante (inconnue) sur} \quad \Gamma,$$

(4.3) $$\int_\Gamma \frac{\partial u}{\partial n}\, d\ell = 1 \ ;$$

puis nous calculons l'application f (multivoque en général) défini par

(4.4) $$f(\mu) = 1 / (\int_\Omega [u(\mu)_-]^2 dx) \ ;$$

Enfin, le réel positif ρ étant donné, nous résolvons l'équation en μ

(4.5) $$f(\mu) = \rho^2.$$

Dans la suite, nous étudions les trois cas suivants :

(i) $\Omega =]0,\alpha[$, $\alpha > 0$ (il y a unicité de la solution)

(ii) $\Omega =]-2,-1[\cup]0,1[$,

(iii) $\Omega =]-2,-1[\cup]0,\alpha[$ avec $\alpha > 1$.

Les résultats obtenus sont très différents dans les cas (ii) et (iii).

Nous notons X l'espace des fonctions continues sur Ω, muni de la topologie de la convergence uniforme.

4.1. Cas où Ω est convexe.

Soit $\Omega =]0,\alpha[$, avec $\alpha > 0$: la première valeur propre du problème de Dirichlet sur Ω est π^2/α^2. Le Problème (4.1)-(4.3) admet une solution unique (cf D. Goldman-Hilhorst [2]), notée $y_\alpha(\mu)$:

(i) si $\mu \in]0,\frac{\pi}{\alpha}[$, $\quad y_\alpha(\mu)(x) = -\frac{1}{2\mu} \dfrac{\sin(\mu x) + \sin[\mu(\alpha - x)]}{1 - \cos(\alpha\mu)}$;

(ii) si $\mu = \frac{\pi}{\alpha}$, $\quad y_\alpha(\frac{\pi}{\alpha})(x) = -\frac{\alpha}{2\pi} \sin(\frac{\pi}{\alpha} x)$;

(iii) si $\mu \in]\frac{\pi}{\alpha}, +\infty[$,

$$y_\alpha(\mu)(x) = \begin{cases} -\frac{1}{2}(x - \frac{\alpha}{2} + \frac{\pi}{2\mu}), & \text{pour } x \in [0, \frac{\alpha}{2} - \frac{\pi}{2\mu}], \\ \frac{1}{2\mu}\sin(\frac{\alpha\mu}{2} - \frac{\pi}{2} - \mu x), & \text{pour } x \in]\frac{\alpha}{2} - \frac{\pi}{2\mu}, \frac{\alpha}{2} + \frac{\pi}{2\mu}[, \\ \frac{1}{2}(x - \frac{\alpha}{2} - \frac{\pi}{2\mu}), & \text{pour } x \in [\frac{\alpha}{2} + \frac{\pi}{2\mu}, \alpha]. \end{cases}$$

De plus, la fonction f, définie par la relation (4.4), est de la forme

$$f(\mu) = g(\alpha\mu)/\alpha^3 \qquad \text{pour } \mu > 0,$$

où

$$g(\mu) = \begin{cases} \frac{4\mu^3(1 - \cos\mu)}{\mu + \sin\mu}, & \text{si } \mu \in]0, \pi[, \\ 8\mu^3/\pi, & \text{si } \mu \in [\pi, +\infty[. \end{cases}$$

La fonction f est de classe $\mathscr{C}^1$, strictement croissante, applique $\mathbf{R}_+$ sur $\mathbf{R}_+$ (cf Figure 4.1 (1)) et donc l'équation (4.5), où $\rho > 0$ est fixé, admet une solution unique μ. En outre, quand μ tend vers π/α par valeurs inférieures ou supérieures, la fonction $y_\alpha(\mu)$ converge dans X vers $y_\alpha(\pi/\alpha)$.
Nous avons donc vérifé le résultat suivant.

<u>Proposition</u> 4.1. <u>Soit</u> Ω <u>un intervalle de</u> $\mathbf{R}$. <u>Alors le Problème</u> (<u>1.1</u>)-(<u>1.4</u>) <u>admet une unique branche de solutions dans</u> $X \times \mathbf{R}$, <u>définie pour tout</u> $\rho > 0$.

4.2. <u>Cas où Ω est réunion disjointe de deux intervalles égaux.</u>

Soit $\Omega =]-2,-1[\cup]0,1[$: la première valeur propre du problème de Dirichlet est double et égale à π^2. Nous distinguons trois cas, suivant la valeur de μ par rapport à π.

<u>(i) $\mu \in]0,\pi[$</u>. Le Problème (4.1)-(4.3) admet une solution unique, notée $u^0(\mu)$:

$$u^0(\mu)(x) = \begin{cases} \frac{1}{2} y_1(\mu)(x+2), & \text{pour } x \in]-2,-1[, \\ \frac{1}{2} y_1(\mu)(x), & \text{pour } x \in]0,1[. \end{cases}$$

Le calcul de la fonction f nous donne

$$f(\mu) = 2g(\mu), \qquad \text{pour} \quad \mu \in]0,\pi[.$$

Donc l'application f est $\mathscr{C}^\infty$ sur $]0,\pi[$, strictement croissante de $]0,\pi[$ sur $]0,16\pi^2[$. Donc, si $\rho \in]0,16\pi^2[$, le Problème (1.1)-(1.4) admet une solution unique, de la forme $(\lambda, u^o(\sqrt{\lambda}))$, où $\sqrt{\lambda} = g^{-1}(\frac{\rho^2}{2})$ est compris entre 0 et π.

(ii) $\mu = \pi$. Le Problème (4.1)-(4.3) admet une infinité de solutions de la forme

$$u(x) = \begin{cases} a y_1(\pi)(x+2), & \text{pour } x \in]-2,-1[, \\ b y_1(\pi)(x), & \text{pour } x \in]0,1[, \end{cases}$$

avec $a \geq 0$, $b \geq 0$ et $a+b=1$. De plus, l'application f est multivoque et vaut

$$f(\pi) = \frac{8\pi^2}{a^2+b^2}.$$

Résoudre l'équation (4.5) est équivalent à trouver les couples (a,b) tels que

$$\begin{cases} a \geq 0, \quad b \geq 0, \quad a+b=1 \\ a^2+b^2 = \dfrac{8\pi^2}{\rho^2}. \end{cases}$$

Ce problème n'est résoluble que pour $\rho^2 \in [8\pi^2, 16\pi^2]$, et admet deux solutions (a,b), de la forme $(a,1-a)$ et $(1-a,a)$, où

(4.6) $$0 \leq a \leq \frac{1}{2}, \qquad a^2+(1-a)^2 = 8\pi^2/\rho^2.$$

Donc, si $\rho \in [2\sqrt{2}\pi, 4\pi[$, le Problème (1.1)-(1.4) admet deux solutions distinctes, soient $(u^1(\pi),\pi^2)$ et $(u^2(\pi),\pi^2)$, tandis que si $\rho = 4\pi$, il admet une seule solution de ce type, soit $(u^o(\pi),\pi^2)$: ce sont les seules valeurs de ρ pour lesquelles il existe une solution (u,λ), vérifiant de plus $\lambda = \pi^2$.

(iii) $\mu \in [\pi, +\infty[$. Le Problème (4.1)-(4.3) admet trois solutions distinctes, soient

$$u^o(\mu)(x) = \begin{cases} \frac{1}{2}\, y_1(\mu)(x+2), & \text{pour } x \in]-2,-1[, \\ \frac{1}{2}\, y_1(\mu)(x), & \text{pour } x \in]0,1[, \end{cases}$$

$$u^1(\mu)(x) = \begin{cases} \frac{1}{2}\, y_1(\mu)(x+2) & , \text{ pour } x \in]-2,-1[, \\ \frac{1}{2}\,(\frac{1}{2} - \frac{\pi}{2\mu}) & , \text{ pour } x \in]0,1[, \end{cases}$$

$$u^2(\mu)(x) = \begin{cases} \frac{1}{2}\,(\frac{1}{2} - \frac{\pi}{2\mu}) & , \text{pour } x \in]-2,-1[, \\ y_1(\mu)(x) & , \text{ pour } x \in]0,1[. \end{cases}$$

De plus, l'application f est multivoque et vaut

$$f = (f^o, f^1, f^2),$$

avec $f^o(\mu) = 2g(\mu)$ et $f^1(\mu) = f^2(\mu) = g(\mu)$, pour $\mu > \pi$. La fonction f^o est une bijection croissante de $]\pi, +\infty[$ sur $]16\pi^2, +\infty[$, tandis que f^1 et f^2 sont des bijections croissantes de $]\pi, +\infty[$ sur $]8\pi^2, +\infty[$.

De cette étude, nous déduisons la structure de l'ensemble des solution du Problème (1.1)-(1.4).

Proposition 4.2. *Soit Ω un ouvert de $\mathbb{R}$, réunion disjointe de deux intervalles de longeur égale. Alors l'ensemble des solutions du Problème (1.1)-(1.4) est constitué*

(i) d'une branche "triviale" de solutions dans $X \times \mathbb{R}$, dans le sens où elle est définie pour tout $\rho > 0$; c'est la branche des solutions symétriques ;

(ii) de deux branches bifurquant au même point de (ρ^*, u^*, λ^*) de $\mathbb{R} \times X \times \mathbb{R}$; en outre, les branches bifurquées possèdent un point de retournement.

Démonstration. Soit $\Omega =]-2,-1[\cup]0,1[$. Les propriétés de solutions du Problème (1.1)-(1.4) sont visualisées sur la Figure 4.1(2), donnant le graphe de l'application multivoque $\begin{cases} \rho^2 \longmapsto f^{-1}(\rho^2). \\ \mathbb{R}_+^* \longrightarrow \mathbb{R}_+^*. \end{cases}$

(i) La branche "triviale" de solutions $(u(\rho), \lambda(\rho))$ est définie, pour tout $\rho > 0$, par

$$\begin{cases} \lambda(\rho) = [\mu(\rho)]^2 \quad \text{et} \quad u(\rho)(x) = \begin{cases} \frac{1}{2}\, y_1(\mu(\rho))(x+2) & \text{pour } x \in]-2,-1[, \\ \frac{1}{2}\, y_1(\mu(\rho))(x) & \text{pour } x \in]0,1[, \end{cases} \\ \text{où } \mu(\rho) = g^{-1}(\rho^2/2). \end{cases}$$

(ii) Le point $(\rho^*, u^*, \lambda^*) = (4\pi, u(4\pi), \pi^2)$ est un point de bifurcation. Il suffit de vérifier qu'en ce point, les solutions $u(\rho)$, correspondantes à chacune des branches mises en évidence sur la fonction $\rho^2 \longrightarrow f^{-1}(\rho^2)$, se raccordent. Or, quand $\rho \longrightarrow 4\pi$, le réel a défini par (4.6) tend vers 1/2, et donc les solution $(u^1(\pi), \pi^2)$ et $(u^2(\pi), \pi^2)$, définies dans 4.2.ii convergent uniformément vers $(u(4\pi), \pi^2)$.

En outre, au point $\rho = 2\sqrt{2}\pi$, les fonctions u_1 et u_2, définies pour $\mu = \pi$ dans 4.2.ii d'une part, et pour $\mu > \pi$ dans 4.2.iii d'autre part, se raccordent. Nous avons donc bien deux branches de solutions bifurquées. Enfin, le point $(2\sqrt{2}\pi, u^1(\pi), \pi^2)$ (respectivement $(2\sqrt{2}\pi, u^2(\pi), \pi^2)$) est un point de retournement sur la branche bifurquée car pour $\rho < 2\sqrt{2}\pi$, il n'existe pas de solution sur cette branche.

4.3. Cas où Ω est réunion disjointe de deux intervalles non égaux.

Soit $\Omega =]-2,-1[\cup]0,\alpha[$, avec $\alpha > 1$. Les premières valeurs propres de l'opérateur Δ sur $]-2,-1[$ et $]0,\alpha[$ sont respectivement π^2 et $\frac{\pi^2}{\alpha^2}$. Nous distinguons cinq cas, suivant la place de μ par rapport à π/α et π.

(i) $\mu \in]0, \frac{\pi}{\alpha}[$. Le Problème (4.1)-(4.3) admet une solution unique, soit $u_\alpha^0(\mu)$

$$u_\alpha^0(\mu)(x) = \begin{cases} a_\alpha(\mu) y_1(\mu)(x+2), & \text{pour } x \in]-2,-1[, \\ b_\alpha(\mu) y_\alpha(\mu)(x), & \text{pour } x \in]0,\alpha[\ ; \end{cases}$$

où
$$a_\alpha(\mu) = \frac{\sin(\mu/2)\ \cos(\alpha\mu/2)}{\sin[(\alpha+1)\mu/2]}, \quad b_\alpha(\mu) = \frac{\sin(\alpha\mu/2)\ \cos(\mu/2)}{\sin[(\alpha+1)\mu/2]}.$$

De plus, la fonction f est donnée par

$$f(\mu) = \frac{g(\mu)\ g(\alpha\mu)}{[a_\alpha(\mu)]^2 g(\alpha\mu) + [b_\alpha(\mu)]^2 \alpha^3 g(\mu)}, \quad \forall\ \mu \in]0, \tfrac{\pi}{\alpha}[.$$

(ii) $\mu = \frac{\pi}{\alpha}$. Le Problème (4.1)-(4.3) admet une solution unique, soit

$$u_\alpha^0(\tfrac{\pi}{\alpha})(x) = \begin{cases} 0 & , \text{pour } x \in]-2,-1[, \\ y_\alpha(\frac{\pi}{\alpha})(x), & \text{pour } x \in]0,\alpha[. \end{cases}$$

De plus, la fonction f, au point $\frac{\pi}{\alpha}$, vaut $f(\frac{\pi}{\alpha}) = \frac{8\pi^2}{\alpha^3}$.

(iii) $\mu \in]\frac{\pi}{\alpha},\pi[$. Le Problème (4.1)-(4.3) admet une solution unique, soit

$$u_\alpha^0(\mu)(x) = \begin{cases} \frac{1}{2}(\frac{\alpha}{2} - \frac{\pi}{2\mu}), & \text{pour} \quad x \in]-2,-1[, \\ y_\alpha(\mu)(x), & \text{pour} \quad x \in]0,\alpha[. \end{cases}$$

De plus, la fonction f est donnée par

$$f(\mu) = \frac{g(\alpha\mu)}{\alpha^3}, \qquad \text{pour} \quad \mu \in]\frac{\pi}{\alpha},\pi[.$$

(iv) $\mu = \pi$. Le Problème (4.1)-(4.3) admet deux solutions, soient

$$u_\alpha^0(\mu)(x) = \begin{cases} \frac{1}{2}\ (\frac{\alpha}{2} - \frac{1}{2}), & \text{pour} \quad x \in]-2,-1[, \\ y_\alpha(\pi)(x), & \text{pour} \quad x \in]0,\alpha[, \end{cases}$$

$$u_\alpha^1(\mu)(x) = \begin{cases} y_1(\pi)(x+2), & \text{pour} \quad x \in]-2,-1[, \\ 0 & , \ \text{pour} \quad x \in]0,\alpha[. \end{cases}$$

De plus, la fonction f est multivoque et vaut

$$f(\pi) = (f^0(\pi), f^1(\pi))$$

avec $\qquad f^0(\pi) = f^1(\pi) = 8\pi^2.$

(v) $\mu \in]\pi,+\infty[$. Le Problème (4.1)-(4.3) admet trois solutions, soient

$$u_\alpha^0(\mu)(x) = \begin{cases} \frac{1}{2}\ (\frac{\alpha}{2} - \frac{\pi}{2\mu}), & \text{pour} \quad x \in]-2,-1[, \\ y_\alpha(\mu)(x), & \text{pour} \quad x \in]0,\alpha[, \end{cases}$$

$$u_\alpha^1(\mu)(x) = \begin{cases} y_1(\mu)(x+2), & \text{pour} \quad x \in]-2,-1[, \\ \frac{1}{2}\ (\frac{1}{2} - \frac{\pi}{2\mu}), & \text{pour} \quad x \in]0,\alpha[, \end{cases}$$

$$u_\alpha^2(\mu)(x) = \begin{cases} a_\alpha(\mu) y_1(\mu)(x+2), & \text{pour} \quad x \in]-2,-1[, \\ b_\alpha(\mu) y_\alpha(\mu)(x), & \text{pour} \quad x \in]0,\alpha[, \end{cases}$$

avec $\qquad a_\alpha(\mu) = \frac{\alpha\mu - \pi}{(\alpha+1)\mu - 2\pi} \qquad \text{et} \qquad b_\alpha(\mu) = \frac{\mu - \pi}{(\alpha+1)\mu - 2\pi}.$

La fonction f est donc ici multivoque et vaut, pour $\mu > \pi$

$$f(\mu) = (f^0(\mu), f^1(\mu), f^2(\mu)),$$

avec
$$\begin{cases} f^{o}(\mu) = f^{1}(\mu) = g(\mu), \\ f^{2}(\mu) = \dfrac{g(\mu)\ g(\alpha\mu)}{[a_{\alpha}(\mu)]^{2} g(\alpha\mu) + \alpha^{3}[b_{\alpha}(\mu)]^{2} g(\mu)}. \end{cases}$$

De cette étude, nous déduisons la structure de l'ensemble des solutions du Problème (1.1)-(1.4), structure essentiellement différente de celle obtenue dans le cas où $\Omega =]-2,-1[\cup]0,1[$.

Proposition 4.3. Soit Ω un ouvert de $\mathbb{R}$ réunion disjointe de deux intervalles de longueurs différentes. Alors l'ensemble des solutions du Problème (1.1)-(1.4) est constitué

(i) d'une branche triviale de solutions dans $X \times \mathbb{R}$;

(ii) d'une branche de solutions dans $X \times \mathbb{R}$, possédant un point de retournement c'est-à-dire définie seulement pour ρ supérieur à un certain $\rho^* > 0$.

Démonstration. Soit $\Omega =]-2,-1[\cup]0,\alpha[$. Sur la Figure 4.1 (3), est représenté le graphe de l'application multivoque $\begin{cases} \rho^2 \longrightarrow f^{-1}(\rho^2) \\ \mathbb{R}_+^* \longrightarrow \mathbb{R}_+^* \end{cases}$. Le point d'abscisse $\rho^2 = 8\pi^2$ n'est pas un point de bifurcation, comme le confirme le graphe de l'application $\mu \longmapsto u_{\alpha}(\mu)|_{\Gamma}$ représenté sur la figure 4.3.2.

(i) La branche triviale de solutions $(u(\rho),\lambda(\rho)) \in X \times \mathbb{R}$ est définie, pour tout $\rho > 0$, par
$$\lambda(\rho) = [\mu(\rho)]^2 \qquad \text{et} \qquad u(\rho) = u_{\alpha}^{o}(\mu(\rho))$$
où $\mu(\rho) = [f^{o}]^{-1}(\rho^2)$. D'autre part, il existe une valeur $\alpha > 0$ de α telle que, pour $\alpha < \alpha^*$, la branche triviale admet deux points de retournement, tandis que, pour $\alpha > \alpha^*$, la branche triviale est monotone : le réel α^* est obtenu en annulant la valeur au point $\frac{\pi}{\alpha}$ de la dérivée de la fonction f calculée pour $\mu \in]0,\frac{\pi}{\alpha}[$, soit
$$\alpha^* = \frac{\pi}{2}\ [\mathrm{Arctg}(\frac{3}{2\pi})]^{-1}.$$

(ii) La deuxième branche de solutions est celle correspondant aux solutions $u_{\alpha}^{1}(\mu)$ et $u_{\alpha}^{2}(\mu)$, définies seulement pour $\mu \geq \pi$, c'est-à-dire pour $\rho > 2\sqrt{2}\pi$ ($=\rho^*$). Nous vérifions que, lorsque μ tend vers π par valeurs supérieures, alors les fonctions $u_{\alpha}^{1}(\mu)$ et $u_{\alpha}^{2}(\mu)$ convergent dans X vers $u_{\alpha}^{1}(\pi)$. ▨

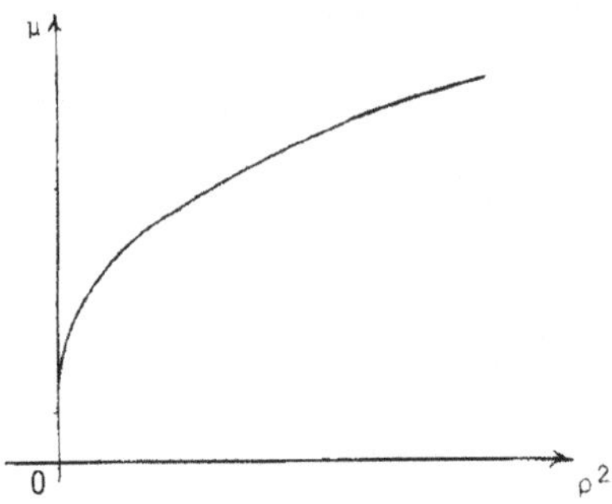

(1) $\Omega =]0,1[$

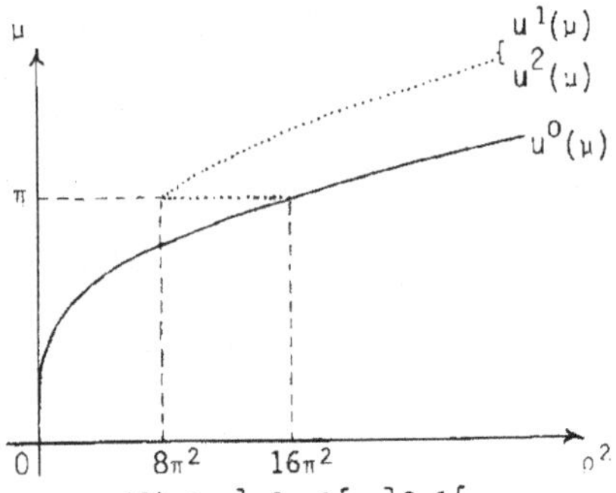

(2) $\Omega =]-2,-1[\cup]0,1[$

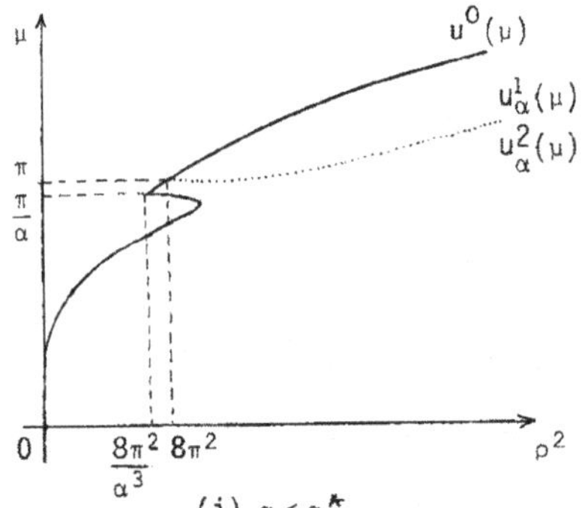

(i) $\alpha < \alpha^*$

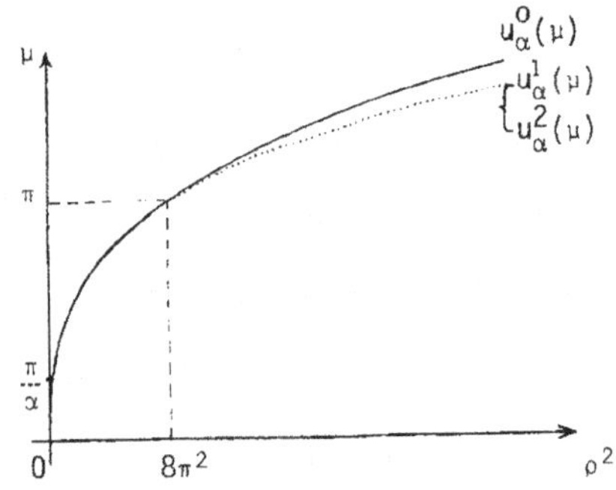

(ii) $\alpha \geq \alpha^*$

(3) $\Omega =]-2,-1[\cup]0,\alpha[,\ \alpha>1.$

Figure 4.1. Graphe de l'application $\rho^2 \longrightarrow f^{-1}(\rho^2)$, (f définie par (4.1)-(4.4)).

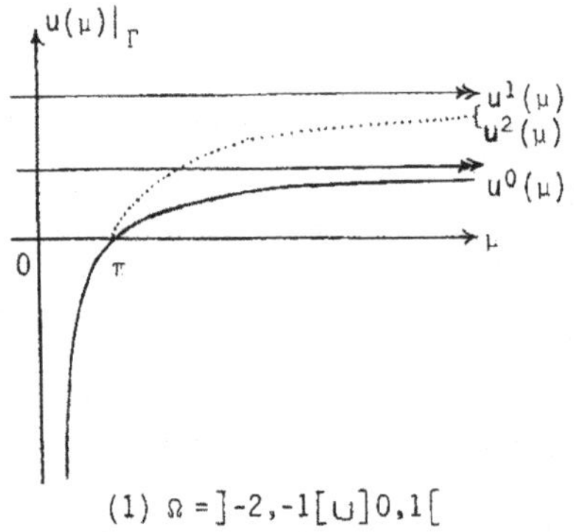

(1) $\Omega =]-2,-1[\cup]0,1[$

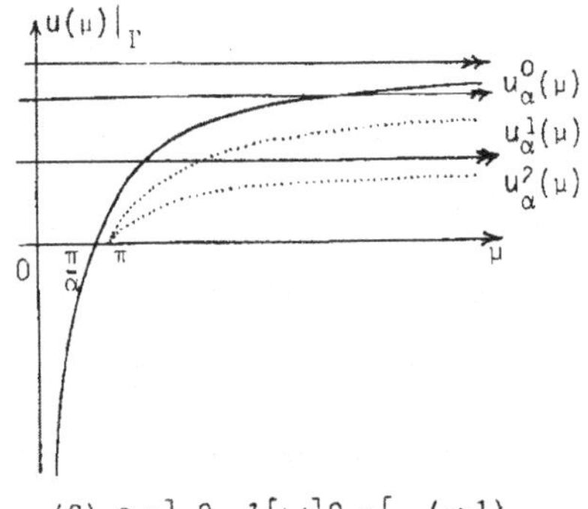

(2) $\Omega =]-2,-1[\cup]0,\alpha[,\ (\alpha>1)$.

Figure 4.2. Graphe de l'application $\mu \longrightarrow u(\mu)|_\Gamma$ ($u(\mu)$ solution de (4.1)-(4.3)).

- BIBLIOGRAPHIE -

[1] S. AGMON, A. DOUGLIS, L. NIRENBERG.
Estimates near the boundary for solutions of elliptic partial differential equations satisfying general boundary conditions I, Comm. Pure Appl. Math., 12, 623-727 (1959).

[2] D. GOLDMAN- HILHORST.
Communication personnelle.

[3] J.L. LIONS, E. MAGENES.
Non homogeneous boundary value problems and applications, Springer Verlag, Berlin-New-York 1972.

[4] C. MERCIER, The magneto-hydrodynamic approach to the problem of plasma confinment in closed magnetic configuration, Publ. of EURATOM-CEA, Luxembourg, 1974.

[5] M. SERMANGE, Une méthode numérique en bifurcation. Application à un problème à frontière libre de la physique des plasmas. Rapport n°310, IRIA, France, 1978.

[6] R. TEMAM, A non-linear eigenvalue problem : the shape at equilibrium of a confined plasma, Arch. Rat. Mech. Anal., 60, 51-73 (1975).

[7] R. TEMAM, Remarks on a free boundary value problem arising in plasma physics, Comm. in P.D.E., 2 (6), 563-585 (1977).

[8] R. TEMAM, Applications de l'analyse convexe au calcul des variations Proc. of the "Conference on non-linear operators and the calculus of variations", Bruxelles, sept. 1972, Lecture Notes in Mathematics, Berlin, 1975.

SOLAR FLARES : A NON LINEAR EIGENVALUE PROBLEM IN AN UNBOUNDED DOMAIN

by J. Heyvaerts [1]
J.M. Lasry [2]
M. Schatzman [3]
P. Witomski [4]

Abstract. Solar flares are considered, and a description of the astrophysical model is given; simplifications of the model make it possible to give mathematical results of two kinds : one part is devoted to theoretical results about a nonlinear eigenvalue problem in a half plane; the other one is concerned with numerical algorithms and results including turning points, for an approximating problem in a bounded domain.

(1) Observatoire de Meudon - 92190 Meudon,
and Université Paris 7, 75231 Paris Cedex 05

(2) CEREMADE, Université Paris 9,
75775 Paris Cedex 16

(3) LA 189, Tour 55-65, Université Paris 6,
75230 Paris Cedex 05

(4) Laboratoire IMAG, BP 53 X, 38041 Grenoble Cedex

I.- DESCRIPTION OF THE ASTROPHYSICAL MODEL

The solar flares are transient phenomena occuring in the so called active regions of the sun, near sunspots. They consist of the rapid conversion of energy stored in the form of extensive current systems flowing in part under the visible surface of the sun (the photosphere) (5000°K) and in part above it in a very tenuous but very hot medium (1.000,000°K) called the corona. Between the photosphere and corona exists an intermediate layer, the chromosphere. The flare energy release occurs in the corona, and appears immediately under various forms (motions, energetic particles, and radiation in various parts of the electromagnetic spectrum).

For a long time, the outstanding manifestation of a flare was the rapid and extreme brightening observed in a red line emission, characteristic of hydrogen ($H\alpha$). Two long and thin ribbons of H_α emission appear, extending parallel to a line where the magnetic polarity changes sign. Thanks to satellite space observations, the corona has now been observed against the visible disk of the sun by its X ray emission. X ray coronal emission is vigorously increased in the flaring region, and appears to consist of a system of many loops rooted in the $H\alpha$ emission ribbons, forming a long arcade which bridges above the line of magnetic polarity reversal. The system of electric current which suffers this catastrophe is built up prior to flare, progressively by the motion of the points where coronal lines of forces are rooted in the photosphere. The elongated structure of loop arcades and $H\alpha$ ribbons, which is systematic suggests that on a sufficiently coarse scale, the magnetic structure in the corona is endowed with translational symetry in one direction. This of course ignores the, also systematic, separation of the structure into individual loops, which may play an important role, but it is thought that as far as the global equilibrium is concerned, this translation-symetric modelization is safe enough.

The flare then occurs at the end of a very slow evolution of the active region, which justifies considering, instead of a truly dynamic problem, a succession of equilibria. Normally, these are the result of balancing several forces in the conducting gas (plasma) which constitues the medium pervading the corona : Laplace forces produced by electric currents flowing there ($\vec{J} \wedge \vec{B}$), gas pressure forces; and gravity forces. Fortunately, it can be shown that gravity and pressure forces at their maximum conceivable can only be a tiny fraction of any well developped Laplace force. This is because the coronal plasma is extremely tenuous, while the active region magnetic field is large (100 to 1000 times the terrestrial field). To a very good approximation, the magnetic structure and current flow in this gas must be free of Laplace forces.

$$\vec{J} \wedge \vec{B} = 0.$$

This approximation, termed force - free approximation, ceases to be valid near the visible surface of the sun. Our problem is then to calculate the current flow in the corona subject to boundary conditions on the photosphere, to be discussed below. There is no other natural boundary. Space observations have revealed recently that some flares make the magnetic loops, initially rather compact, expand very far in the interplanetary medium, forming a bubble - like structure which rapidly becomes larger than the sun itself. For this reason, it is believed that the semi-infinite nature of the domain in which the problem is to be solved may give rise to specific effects, and we shall see that this actually is the case. One would have missed these, if we had considered a finite "large", but bounded, domain.

Let us now see precisely how the problem appears mathematically. Because the build up phase is slow and gradual, one can safely consider that, up to the catastrophe, the evolution proceeds by a series of equilibria. In the translational symetry employed here, the magnetic field is a function of two variables only : x and y : y will be chosen to be the altitude above the solar surface (assumed to be plane), and $\vec{z}$ is the direction of translational symetry. The field is described by three components, B_x, B_y, B_z, but, because of the equation div $\vec{B} = 0$, only two functions are actually needed to precise it. Here it is convenient to introduce a potential function $u(x,y)$, such that :

$$B_x = \frac{\partial u}{\partial y} \quad ; \quad B_y = - \frac{\partial u}{\partial x} \quad ; \quad B_z = B_{11}(x,y) \ .$$

One immediately sees that u is constant along a given line of force $(\vec{B}.\vec{\nabla} u = 0)$, so that curves of constant u map the projection of actual lines of force parallel to the direction of translational symetry on the plane (x,y).

The electric current density $\vec{J}$ is, from Ampere's equation, given by :

$$\mu_o \vec{J} = \text{rot}\, \vec{B}$$

so :

$$J_x = \frac{1}{\mu_o} \frac{\partial B_{11}}{\partial y} \quad ; \quad J_y = - \frac{1}{\mu_o} \frac{\partial B_{11}}{\partial x} \quad ; \quad J_z = - \frac{1}{\mu_o} \left(\frac{\partial^2 u}{\partial x^2} + \frac{\partial^2 u}{\partial y^2}\right) .$$

The vector equation $\vec{J} \wedge \vec{B} = 0$ gives three equations, one of which expresses the fact that B_{11} is constant along a given line of force :

$$\frac{\partial B_{11}}{\partial x} \frac{\partial u}{\partial y} = \frac{\partial B_{11}}{\partial y} \frac{\partial u}{\partial x} \ .$$

Hence : $B_{11}(x,y) = B_{11}(u(x,y))$ provided there is only one topologically distinct line of force with a given value of u. We assume this to be the case. The other two equations are both equivalent to :

$$(1) \qquad \frac{\partial^2 u}{\partial x^2} + \frac{\partial^2 u}{\partial y^2} + \frac{d}{du}\left(\frac{B_{11}^2(u)}{2}\right) = 0 \ .$$

$B_{11}(u)$ is determined by the displacement of one with respect to the other of the feet of the line of force u : $\ell(u)$. This displacement is one of the quantities that the observation could give most easily, and it is the quantity most likely to be a monotonic function of time. B_{11} and ℓ are related by the line of force equation :

$$\frac{dz}{B_{11}} = \frac{dz}{B_x} \ ,$$

which can be integrated along a line of force to give :

$$(2) \qquad \ell(u) = B_{11}(u) \int_{\substack{x_A(u) \\ u = Cte}}^{x_B(u)} \frac{dx}{\left(\frac{\partial u}{\partial y}\right)}$$

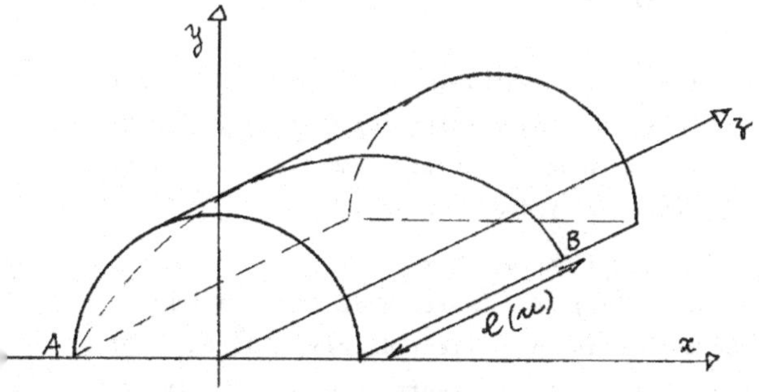

Fig. 1

The derivative under the integral sign has to be evaluated at the point x,y such that $u(x,y) = u$. Other considerations must be used to specify the value of $B_{11}(u)$ on those lines which do not connect to the boundary. Up to now, no precise statement concerning the choice to be made is available. We come back later to this question. The system of equations (1) and (2) should then in principle be solved for a set of functions $\ell(u,t)$ describing progressive shearing of the foot points. This has been considered up to now as too difficult, but eventually the problem (1) - (2) is the relevant one to solve. Facing with this difficulty, it has been imagined to consider a modified problem, for which the function $B_{11}(u)$ is known. The physical significance of this may be understood by noting that the vertical current J_y emitted by the boundary, $y = 0$, in the domain of interest is, at a point $(x,0)$:

$$J_y = -\frac{1}{\mu_0}\frac{dB_{11}}{du}\frac{\partial u}{\partial x} \ .$$

Now, u is known on the boundary, and so will be J_y provided dB_{11}/du or equivalently $\frac{d}{du}\frac{1}{2}B_{11}^2(u)$ is. We set from now on :

(3) $$\frac{1}{2}\frac{d}{du}B_{11}^2(u) = F(u) ,$$

and we shall consider, instead of problem (1), (2), with a set of functions $\ell(u,t)$ the problem (3) only with a set of functions $F(u)$, which we choose for simplicity as :

$$F(u) = \lambda(t)\ f(u) \qquad \lambda > 0$$

The "evolution" of the structure is modelized fact that λ be an increasing function of time. We recall that, up to a multiplicative factor $F(u)$ is the current density flowing on line u in the z direction. It can also be related to the other meaningful quantity $J_y(x,0)$, which represents the current density supplied by the interior of the sun to the external corona near point x. Actually dB_{11}/du may be obtained from $F(u)$ provided $B_{11}(u(\infty,0))$ beknown; here we assume it to be zero.

Knowing boundary conditions $u(x,0)$, which do not change during the deformation, J_y can be calculatd, and is equal to

$$J_y(x,0) = \sqrt{\lambda}\ j(x) ,$$

$j(x)$ being a known function of x.

Having this in mind, we can easily see that some reasonable conditions should be imposed on $f(u)$. First, one would like to exclude the possibility that $J_y(x,0)$ become infinite. This is easily obtained by imposing regularity conditions on the functions $f(u)$ and $u(x)$. We restrict the problem by demanding that

1) $u(x,0)$ be bounded and have continuous derivatives.

2) $\lim_{x \to \infty} x\ u(x,0) = 0$ (only derivatives of u are physically meaningful)

Moreover, the choice of $f(u)$ must be such that the current driven on line of force u be consistent with physics. Physics may prescribe currents driven by the interior of the sun into the corona. Up to now it has not been able to prescribe the current in field lines which do not connect to the boundary. A reasonable choice is to assume zero current in these regions. We then add the requirement : $f(u)$ vanishes for values of u refering to field lines which do not connect to the boundary. Let $[a,k]$ be the interval of values of $u(x,0)$, $x \in [-\infty, +\infty]$. The possibility that different field lines refer to the same value of u has abready been discarded. Our demand states then :

3) $f(u) = 0$ except for $u \in [0,k]$;

The fact that $k = \sup_x(u(x,0))$ does not seem to be of primary importance, and the mathematical developments ignored this; f was simply assumed to vanish outside some finite interval. We also, for the sake of simplicity and for no other reason, assumed $f(u)$ to be strictly positive. This is the case for J_z flowing always in the same direction, and corresponds to the simplest situation one could conceive of, one which represents a simple current bridge :

4) $f(u) \geq 0$

Physically different effects can be expected when this condition is not fulfilled.

II. THEORETICAL RESULTS IN AN UNBOUNDED DOMAIN

We consider the following boundary value problem, depending on the parameter λ :

$$(4)\qquad \begin{cases} -\ \Delta u = \lambda f(u) & \text{in } \Omega = R \times (0,\infty) \\ u(x,0) = g(x) & \forall\, x \in \mathbb{R}. \end{cases}$$

We seek continously differentiable, bounded, non-negative solutions under the following assumptions :

(5) The function g is non-negative, bounded, twice differentiable, with Hölder-continuous second derivative : $|g''(x) - g''(y) \leq \ell\ |x-y|^{\gamma}$ $\forall x,y \in \mathbb{R}$; $\gamma \in (0,1)$.

(6a) The function f is Lipschitz continous from $\mathbb{R}^+$ into itself, with Lipschitz constant k : $|f(\alpha) - f(\beta)| \leq k\ |\alpha-\beta|$ $\forall\, \alpha,\beta \in \mathbb{R}^+ = [0,\infty]$.

It also satisfies the supplementary assumptions :

(b) $f(\alpha) \sim C\,\alpha^s$, $s > 3$ is a neighborhood of zero

(6c) $f(\alpha) > 0$ if $\alpha \in (0, \alpha_o)$; $f(\alpha) = 0$ if $\alpha \geq \alpha_o$

where α_o is a given positive number.

The unusual feature of this problem is the unboundedness of the domain; moreover, bifurcation occurs in a Frechet space, which has been seldom considered.

Authors who worked on this kind of non-linear elliptic problem in unbounded domains are T. Küpper, C.A. Stuart, R. Chiapinelli, [1] - [8] who studied differential problems on the half-line, in the frame of Hilbert spaces, and K. Kirchgässner & J. Scheurle [9] who studied a problem in $\Omega \times \mathbb{R}$, where Ω is bounded. See the bibliography of these papers for other references.

We shall prove that (4) possesses (at least) two branches of solutions which have different asymptotic properties.

The proof comes in several stages :

- A priori estimates
- Summary of functional results
- Study of an associated ordinary differential equation
- Proof of existence by super- and subsolutions
- Behavior at infinity of the solutions
- Bifurcation in Frechet space.

II.1.A priori estimates.

Theorem 1. Any non-negative bounded solution of (13) satisfies the estimate :

$$(7)\qquad \sup_{(x,y) \in \Omega} u(x,y) \leq \max(\alpha_o, \sup g(x)).$$

Proof. Let m be an upper bound for u, and suppose $m \geq C = \max(\alpha_o, \sup g)$; let ϕ_a be a smooth function defined on $[-a,+a]$ such that

$$(8)\quad \begin{cases} \phi_a \text{ is even : } \phi_a(x) = \phi_a(-x), \ \forall x \in [-a,+a] \ ; \\ \phi_a(a) = m, \\ \phi_a(x) = C \text{ if } |x| \leq a-\varepsilon , \\ C \leq \phi_a(x) \leq m \quad \forall x \in [-a,+a] . \end{cases}$$

Let Ω_{ab} be the open rectangle $(-a,a) \times (0,b)$ and Γ_{ab} its boundary. The solution w_{ab} of the problem

$$(9)\quad \begin{cases} -\Delta w_{ab} = 0 \quad \text{in} \quad \Omega_{ab} \ ; \\ w_{ab}(x,0) = \phi_a(x) \quad \text{in} \quad (-a,a) \ ; \\ w_{ab}(x,y) = m \quad \text{elsewhere an} \quad \Gamma_{ab} \ ; \end{cases}$$

must be by the maximum principle larger than C on Ω_{ab}.

Set

(10) $v = w_{ab} - u.$

Then v satisfies

(11a) $-\Delta v = -\lambda f(u)$ in Ω_{ab} ;

(11b) $v \geq 0$ on Γ_{ab} .

Multiply (10) by $v^- = \max(-v,0)$, and integrate over Ω_{ab} :

$$(12)\quad \int_{\Omega ab} \nabla v . \nabla v^- - \int_{\Gamma ab} v^- \frac{\partial v}{\partial n} = -\lambda \int_{\Omega ab} v^- f(u).$$

The boundary term vanishes because v is positive on the boundary (11b), and the right hand side term vanishes also because, wherever v^- is non zero, u is larger than w_{ab} which is itself larger than C, and for such values of $u, f(u)$ vanishes. We deduce from (12) that

(13) $\nabla v^- = 0$ on Ω_{ab} ,

which implies that v^- must be constant. Actually, the constant must be zero, because we know from (11b) that v is non-negative on the boundary and so v^- vanishes there. We can conclude that :

(14) $u(x,y) \leq w_{ab}(x,y)$ on $\overline{\Omega}_{ab}$,

and it remains to estimate w_{ab} as a and b go to infinity. Define :

(15) $$v_{ab}(x,y) = w_{ab}(x,y) - \frac{m-c}{b}\; y - c$$

v_{ab} is the solution of the following boundary problem :

(16) $$\begin{cases} -\ \Delta v_{ab} = 0\ , \\ v_{ab}(x,0) = \phi_a(x,0) - C\ , \\ v_{ab}(x,b) = 0\ , \\ v_{ab}(a,y) = v_{ab}(a-y) = (m-c)(1-y/b)\,. \end{cases}$$

It can be shown by direct comparison, that, provided ϕ_a is suitably chosen, v_{ab} is smaller, on Γ_{ab}, than the function z_{ab} defined as follows :

(17) $$z_{ab}(x,y) = \frac{m-C}{a^2+b^2}\,(x^2-y^2+b^2)$$

which is also a solution of Laplace equation. By the maximum principle we see that :

(18) $$v_{ab}(x,y) \le z_{ab}(x,y) \quad \text{on} \quad \Omega_{ab}\ .$$

Therefore from (14) and (15)

(19) $$u(x,y) \le w_{ab}(x,y) = v_{ab}(x,y) + \frac{m-C}{b}\,y + C \le z_{ab} + \frac{m-C}{b}\,y + C\ ;$$

if a goes to infinity, then $z_{ab}(x,y) \to 0$ on every compact subset of $\mathbb{R} \times [0,b]$ and we deduce from (19) that :

(20) $$u(x,y) \le \frac{m-C}{b}\; y + C\ .$$

Letting b tend to infinity, we obtain the inequality (7). ■

II.2. Summary of functional results.

The functional results we need are mainly the Phragmen-Lindelöf principle, (Protter-Weinberger [10] which tells us that problem

$$\begin{cases} -\ \Delta u = h & \text{in} \quad \Omega = R \times R^+\ , \\ u_{|\Gamma} & \text{on} \quad \partial\Omega \end{cases}$$

possesses at most one classical bounded solution, and the following proposition which can be easily deduced from the a priori estimates of Agmon-Douglis-Nirenberg, [11] :

Proposition 2. Let K be a strictly positive real number, h be a function in $L^{\infty}(\Omega)$, g a function in $L^{\infty}(R) \cap W^{2,\infty}_{loc}(R)$. Then, there exists a unique function u such that

(21) $u \in W^{1,\infty}(\Omega) \cap W^{2,p}(\Omega) \quad \forall p \in [1,\infty]$

(22) $-\Delta u + Ku = h \quad \text{in} \quad \Omega$

(23) $u_{|\partial\Omega} = g$

If w assume that $h \in C^{\gamma}_{loc}(\Omega)$, $g \in C^{2,\gamma}_{loc}(\mathbb{R})$, then the solution of (21) - (23) is in $C^{2,\gamma}_{loc}(\Omega)$; if we assume that h and g are non-negative, so is u.

II.3. Study of an associated ordinary differential equation.

We wish to obtain information considering the x-independent case of (4). Thus we shall study an ordinary differential equation (24) and the results concerning it are given by next proposition :

Proposition 3. The equation

(24) $$\begin{cases} -v'' = f(v) & \text{on } \mathbb{R}^+ \\ v(0) = \alpha \geq 0 \end{cases}$$

admits a unique bounded, non-negative, not identically zero solution of class C^2. For $\alpha = 0$, (24) admits the zero solution. Moreover, if $v \not\equiv 0$, the following estimate holds :

(25) $$v(y) \geq \alpha_o - (\alpha_o - \alpha) \exp(-y\sqrt{k\lambda})$$

where $k = \sup |f'|$.

Proof. From (24), v must be concave; by contradiction, the limit at infinity of v' must be zero. Therefore v is increasing. Still by contradiction, the limit at infinity of v cannot be smaller than α_o. Solutions for $\alpha \geq \alpha_o$ are merely constants, and $\lim_{y\to\infty} v(y) = \alpha_o$, for all $\alpha \in [0,\alpha_o]$, $v \not\equiv 0$

Multiplying (24) by v', and integrating we obtain :

(26) $$\frac{1}{2}|v'(0)|^2 + \lambda F(\alpha) = \frac{1}{2}|v'(y)|^2 + \lambda F(v(y)).$$

where F is a primitive of f.

Passing to the limit in the right hand side of (26), we obtain

(27) $$v'(0) = \left[2\lambda \int_{\alpha}^{\alpha_o} f(\beta)\, d\beta \right]^{1/2}$$

which is the unique possible initial value for v'. It is then straightforward to show that the initial conditions $v(0) = \alpha$ and $v'(0)$ given by (27) provide a bounded non-negative solution (24). By (27) this solution must be unique.

To obtain the estimate (25), set $w = \alpha_o - v$.

Then w satisfies

$w'' = \lambda f(\alpha_o - w)$.

If $\alpha \geq \alpha_o$, (35) is clear. Therefore, assume $\alpha < \alpha_o$. By Lipschitz continuity of f :

(28) $w'' = \lambda(f(\alpha_o - w) - f(\alpha_o)) \leq k\lambda w$.

Multiply (28) by w' and integrate from s to t :

$$\frac{|w'(t)|^2}{2} - \frac{|w'(s)|^2}{2} \geq k\lambda \left(\frac{w(t)^2}{2} - \frac{w(s)^2}{2}\right) \qquad \forall t \geq s.$$

Let t tend to infinity, assuming $w \not\equiv 0$; then

(29) $$w'(s) \geq -\sqrt{k\lambda}\ \ w(s) \qquad s \geq 0$$

By a standard Gronwall inequality, (29) implies (25). ∎

II.4. Proof of existence by supersolutions and subsolutions.

To prove existence we use a general algorithm which is described in Courant-Hilbert, [12] p. 369, and has been used by crowds of authors. Suppose we know a couple of functions u^o and u_o, such that :

$$u^o \geq u^o$$

$$\begin{cases} -\ \Delta u^o \geq \lambda f(u^o) \\ u^o{}_{|\Gamma} \geq g \end{cases}$$

$$\begin{cases} -\ \Delta u_o \leq \lambda f(u_o) \\ u_{o|\Gamma} \leq g \end{cases}$$

The function u^o is called a supersolution and the function u_o is called a subsolution. Then we can define two monotonous sequences u^n and u_n by

$$\begin{cases} -\ \Delta u^{n+1} + \lambda K u^{n+1} = \lambda f(u^n) + \lambda K u^n \\ u^{n+1}{}_{|\Gamma} = g \end{cases}$$

$$\begin{cases} -\ \Delta u_{n+1} + \lambda K u_{n+1} = \lambda f(u_n) + \lambda K u_n \\ u_{n|\Gamma} = g. \end{cases}$$

where $K > \sup |f'|$.

It is standard to prove that the sequence u^n decreases to a limit u^∞ , and that the sequence u_n increases to a limit u_∞ . These limits u^∞ and u_∞ are solutions of (4), they may or may not be equal. The details of the argument will be left to the reader; proposition 2 is used for this purpose.

The point is now to exhibit explicitly super- and subsolutions.

Upper supersolution :

(30) $\bar{w}_\lambda = C = \max(\alpha_o, \sup g)$

Upper subsolution :

$\underline{w}_\lambda$ is the solution of (24) for $\alpha = 0$, which is not identically zero. Denote by $\bar{u}_\lambda$ the limit of the sequence $\bar{w}_\lambda^n$ defined from $\bar{w}_\lambda$. Clearly $\bar{u}_\lambda$ is the maximum solution of (4); let us prove that :

(31) $$\lim_{y\to\infty} \bar{u}_\lambda(x,y) = \alpha_o.$$

We know from (25) that

(32) $$\bar{u}_\lambda(x,y) \geq \alpha_o(1 - \exp(-y\sqrt{k\lambda})).$$

The unique bounded solution v_λ of

(33) $$\begin{cases} -\Delta v_\lambda = \alpha_o\, k\, \lambda \exp(-y\sqrt{k\lambda}), \\ v_\lambda(x,0) = g(x) \end{cases}$$

is given by

(34) $$v_\lambda(x,y) = \alpha_o(1 - \exp(-y\sqrt{k\lambda})) + \frac{1}{\pi}\int \frac{y g(x')dx'}{|x-x'|^2+y^2}.$$

Thanks to (32)

(35) $$f(\bar{u}_\lambda) - f(\alpha_o) \leq k[\alpha_o - \max(\alpha_o, \bar{u}_\lambda)] \leq k\alpha_o \exp(-y\sqrt{k\lambda})$$

and therefore

(36) $$-\Delta\bar{u}_\lambda \leq \alpha_o\, k\, \lambda \exp(-y\sqrt{k\lambda})$$

Comparing (36) to (33), there obtains

$$\alpha_o(1 - \exp(-y\sqrt{k\lambda})) \leq \bar{u}_\lambda(x,y) \leq v_\lambda(x,y)$$

which gives (31), i.e. the asymptotic behavior of the maximum solution for all positive λ.

- Lower supersolution (for λ small enough) :

(37) $$\bar{z}_\lambda(x,y) = L\left(\frac{y+a}{x^2+(y+a)^2}\right)^{n(s)} = L(h(x,y))^{n(s)},$$

provided a, L and $n(s)$ are such that

$$a > 0, \quad L > 0, \quad n(s) \in \left(\frac{2}{s-1}, 1\right),$$

and g satisfies the condition :

(38) $$g(x) \leq L\left(\frac{a}{x^2+a^2}\right)^{n(s)} \qquad x \in \mathbb{R}.$$

Notice that (38) can always be fulfilled if g decreases faster than $\frac{1}{s^{2n(s)}}$.

Let us prove that $\bar{z}_\lambda$ is a supersolution; first notice that h is harmonic; thus

$$- \Delta \bar{z}_\lambda = - \Delta [h^{n(s)}] = - n(s)(n(s)-1)h^{n(s)-2} |\nabla h|^2 ,$$

and
$$- \Delta \bar{z}_\lambda \geq \lambda f(\bar{z}_\lambda) \quad \text{if}$$

$$\lambda \leq \frac{n(s)(1-n(s))}{L^{1-s} D} \cdot a^{n(s)(s-1)-2} ,$$

where D is such that

$$f(\alpha) \leq D \alpha^s \qquad \forall \alpha .$$

- Lower solution

(39) $\quad \underline{z}_\lambda = 0$

We can now deduce the following theorem :

Theorem 4. The problem (13) admits two branches of particular solutions :

- a branch of maximum solutions $\bar{u}_\lambda$ with the asymptotic behavior for all $\lambda > 0$

(40) $$\lim_{y \to \infty} \bar{u}_\lambda (x,y) = \alpha_o$$

- a branch of minimum solutions $\underline{u}_\lambda$ with the asymptotic behavior for λ small enough

(41) $$\lim_{|x|+y \to \infty} \underline{u}_\lambda (x,y) = 0$$

II.5. Behavior at infinity of the solutions.

This paragraph is intended to prove that the solutions of (4) can either tend to α_o or to zero as y goes to infinity, and that this last behavior is possible only for finite values of λ .

Proposition 5. Any solution of problem (13) has a limit when y goes to infinity. This limit is either zero or α_o, and is uniform in x.

Proof : Let B be the ball of center (x_o,y_o) and radius r_o, $(y_o > r_o)$, and 1_B the function equal to 1 in B and zero outside. Set

(42) $$t = \inf_{x,y \in B} f(u(x,y)) ,$$

and let w be the bounded solution of

$$(43)\qquad \begin{cases} -\Delta w = 1_B & \text{in } \mathbb{R}^2_+ , \\ w_{|\Gamma} = 0 . \end{cases}$$

From the maximum principle :

$$(44)\qquad u \geq \lambda t w \quad \text{in } \mathbb{R}^2_+ .$$

We can compute explicitly w, it is given by

$$(45)\qquad \begin{cases} w(x,y) = \dfrac{r_o^2 - (|x-x_o|^2 + |y-y_o|^2)}{4} + \dfrac{r_o^2}{4} \operatorname{Log} \dfrac{|x-x_o|^2 + |y+y_o|^2}{r_o^2} , \\ \qquad\qquad \text{if } |x-x_o|^2 + |y-y_o^2| \leq r_o^2 ; \\ w(x,y) = \dfrac{r_o^2}{4} \operatorname{Log} \dfrac{|x-x_o|^2 + |y+y_o|^2}{|x-x_o|^2 + |y-y_o|^2} \quad \text{if } |x-x_o|^2 + |y-y_o|^2 > r_o^2 . \end{cases}$$

Therefore, we can deduce from the estimate $u \leq \max(\alpha_o, \text{sug } g) = C$ and from (44) taken at the center of the ball B :

$$(46)\qquad C \geq u(x,y) \geq \lambda(\min_B f(u(x',y'))\, \frac{r_o^2}{4} (1+2 \operatorname{Log} \frac{2y}{r_o}) .$$

As we have an estimate on ∇u in L^∞ norm :

$$(47)\qquad \operatorname{Max} |\nabla u| \leq C_1(1+\lambda) ,$$

(see proposition 2)

where C_1 depends only on g and sup f, we see that for any (x,y) in B :

$$(48)\qquad f(u(x,y)) \leq (\min_B f(u(x',y')) + K\, C_1(1+\lambda)\, 2r_o .$$

Therefore from (46)

$$(49)\qquad f(u(x,y) \leq \frac{C}{\lambda \frac{r_o^2}{4}(1+2 \operatorname{Log} \frac{2y}{r_o})} + 2KC_1(1+\lambda)r_o \qquad \forall x,y \in \mathbb{R}^2_+ .$$

Now, r_o is a parameter that we can choose to our convenience, provided it is not larger than y. Taking for $r_o(y)$ some decreasing function of y such that, however, $(r_o^2(y) \operatorname{Log} \frac{2y}{r_o(y)})$ tends to infinity, like $(\operatorname{Log}(2y))^{-1/4}$ for example, we deduce that :

$$(50)\qquad \lim_{y \to \infty} f(u(x,y)) = 0 ,$$

and this limit is uniform in x. We know, however from theorem 4 that

(51) $$\overline{\lim_{y\to\infty}}\; u(x,y) \le \alpha_o ,$$

as $u(x,y) < \overline{u}_\lambda(x,y)$ and $\lim_{y\to\infty} \overline{u}_\lambda = \alpha_o$. Thus proposition 5 is proved, because 0 and α_o are the only values of u for which $f(u) = 0$ in this range. ∎

To prove that the interval where (41) holds is finite, we shall need the following result :

<u>Lemma 6</u> Let w be a function defined in a bounded open smooth set Ω_1, such that

$$\begin{cases} -\Delta w \ge \lambda f(w) , \\ \quad w \ge h \quad \text{on} \quad \partial\Omega_1 \end{cases}$$

where h is non negative and not identically zero. Then on every relatively compact open subset ω of Ω_1, w is bounded from below by a number $k(\lambda, \omega, h)$:

$$w(x,y) \ge k(\lambda,\omega) \quad \text{on} \quad \omega ,$$

and k approaches α_o, when $\lambda \to \infty$

$$\lim_{\lambda\to\infty} k(\lambda,\omega) = 0.$$

<u>Proof</u>. Let ω' be an open, smooth, relatively compact subset of Ω_1 and

$$\bar{\omega} \subset \omega' \subset \omega' \subset \Omega_1 .$$

Let us introduce $z = \alpha_o - w$, and a strictly increasing function H_δ such that

(52) $$f(\alpha) \ge H_\delta(\alpha_o - \alpha) \quad \text{if} \quad \alpha \ge \delta .$$

The hypotheses on f imply that H_δ, defined in $[0, \alpha_o - \delta]$ exist and that $H_\delta(0) = 0$. In ω', z satisfies, as a consequence, the inequality

(53) $$-\Delta z + \lambda H_\delta(z) \le 0 \quad \text{in} \quad \omega' .$$

By the maximum principle, w is positive in the interior of Ω_1, so z satisfies the boundary inequality:

(54) $$z \le \alpha_o \quad \text{on} \quad \partial\omega' .$$

In order to see that when λ is large, z is small, let us compare it to a function ζ defined on ω' by

(55) $$\zeta(s,y) = a(|x-x_o|^2 + |y-y_o|^2) + b ,$$

with $(x_o,y_o) \in \omega$. This function satisfies the following inequality

(56) $$-\Delta\zeta + \lambda H_\delta(\zeta) \ge -4a + \lambda H_\delta(b) \quad \text{on} \quad \omega' .$$

It satisfies also a boundary inequality which insolves a minimum distance, η between points of $\partial\omega'$ and points of ω :

(57) $\eta = \inf \{ (|x-x'|^2 + |y-y'|^2)^{1/2} \mid (x,y) \in \omega , (x',y') \in \omega' \}$

We have then

(58) $\zeta(x,y) \geq a\,\eta^2$ for $x,y \in \Omega\omega'$.

Comparing the pair of inequalities (53),(54) vith (56),(58), we see that, provided we choose :

(59) $a = \dfrac{\alpha_o}{\eta^2}$,

and b such that the second member of (56) be positive, i.e.

(60) $\lambda H_\delta (b) \geq 4a = \dfrac{4\,\alpha_o}{\eta^2}$.

then we will be sure that z is smaller than ζ on ω', and a fortiori on ω. Finally as (x_o,y_o) is arbitrary in ω, we can write :

(61) $z \leq b$ in ω

or equivalently :

$$w(x,y) \geq \alpha_o - b$$

Let us call b_λ the smallest value for which (71) holds :

(62) $b_\lambda = H_\delta^{-1} \left(\dfrac{4\,\alpha_o}{\eta^2}\right)$

When λ tends to infinity, b_λ tends to zero, because $H_\delta(0) = 0$. The number

(63) $k(\lambda,\omega,h) = \alpha_o - b_\lambda$

then tends to α_o .

We can now prove the following result ∎

<u>Theorem 7</u> The interval $(0, \lambda_o)$ where the minimal solutions tends to zero as $|x|$, y go to infinity is bounded.

The idea to prove this is to use recursively the lemma 7 on a sequence of overlapping domains leading progressively to infinity. We start with domains :

(64) $\begin{cases} \Omega_1^{(1)} = (a,b) \times (0,2) \\ \\ \omega^{(1)} = (a',b') \times (1/2,3/2) \end{cases}$

where (a,b) is chosen so that :

(65) $\max_{[a,b]} g(x) = g(x_o) > 0$,

and (a',b') is strictly included in (a,b) :

(66) $a < a' < b' < b$,

and $g(x) \geq \alpha_1 = \frac{1}{2}(\min(g(x_o), \alpha_o))$ on (a',b').

(67)

Let us take :

(68) $$\begin{cases} h^{(1)}(x,0) = \begin{cases} \alpha_1 & \text{on } [a',b'] \\ 0 & \text{elsewhere on } (a,b), \end{cases} \\ h^{(1)} = 0 \quad \text{elsewhere on } \partial\Omega_1^{(1)}. \end{cases}$$

From lemma 6, we know that the solution of problem (4) will satisfy the inequality :

$$u(x,y) \geq k(\lambda, \omega^{(1)}, h^{(1)}) \quad \text{on} \quad \omega^{(1)}$$

and by choosing λ large enough, say $\lambda \geq \lambda_o$, we can make sure that $k(\lambda, \omega^{(1)}, h)$ be larger than α_1. Hence $u(x,1) > \alpha_1$ on (a',b') for $\lambda \geq \lambda_o$. We can then reproduce the preceding argument taking now the new domains and the new boundary function :

$$\begin{cases} \Omega_1^{(2)} = (a,b) \times (1,3), \\ \omega^{(2)} = (a',b') \times (3/2,5/2), \\ h^{(2)}(x,1) = h^{(1)}(x,0), \\ h^{(2)} = 0 \quad \text{elsewhere on} \quad \partial\Omega_1^{(2)}, \end{cases}$$

then $k(\lambda, \omega^{(2)}, h^{(2)}) = k(\lambda, \omega^{(1)}, h^{(1)})$ and $u(x,2) \geq \alpha_1$ for $\lambda \leq \lambda_o$, $x \in (a',b')$

By reccurence :

$$\begin{cases} \Omega_1^{(n)} = (a,b) \times (n-1,n+1), \\ \omega^{(n)} = (a',b') \times (n-1/2,n+1/2), \\ h^{(n)}_{(x,n-1)} = h^{(1)}(x,0) \quad \text{on} \quad (a,b), \\ h^{(n)} = 0 \quad \text{elsewhere on} \quad \partial\Omega_1^{(n)}, \end{cases}$$

and

(69) $u(x,n) \geq \alpha_1$ for $\lambda \geq \lambda_o$, $x \in (a',b')$.

We then see that for $\lambda \geq \lambda_o$ the limit of u as y goes to infinity cannot be zero, and therefore λ_c is bounded. ∎

II.6. Bifurcation in Frechet space.

It is a classical result that $\bar{u}_\lambda$ and $\underline{u}_\lambda$ are increasing functions of λ ; to sketch the proof, for example for the upper solution, just consider the two sequences

$$w_\lambda^o = C \; ;$$

(70)
$$w \begin{cases} -\Delta w^{n+1} + \lambda K\, w^{n+1} = \lambda f(w^n) + \lambda K w^n , \\ w_\lambda^{n+1}\big|_\Gamma = g . \end{cases}$$

$$W_\mu^o = C \; ;$$

(71)
$$\begin{cases} -\Delta w_\mu^{n+1} + \lambda K\, w_\mu^{n+1} = \mu f(w^n) + \lambda K\, w_\mu^n \\ w_\mu^{n+1}\big|_\Gamma = g \end{cases}$$

where $\lambda > \mu$. Check by induction that $w_\mu^n \leq w_\lambda^n$ for all n. Then, obvionsly (70') converges to $\bar{u}_\lambda$, and as $\bar{u}_\mu$ is <u>the</u> maximum solution (71) converges to $\bar{u}_\mu$.

Let now λ_n be a sequence decreasing to zero. For any relatively compact set ω in $\mathbb{R} \times \mathbb{R}^+$ we have the estimate

$$|\bar{u}_{\lambda_n}|_{W^{2,p}(\omega)} \leq C(p, \omega) \qquad \forall n \geq n_o .$$

Therefore a subsequence can be extracted such that

$$\bar{u}_{\lambda_n} \to \bar{u}_o \quad \text{in} \quad W^{2,p}(\omega) \qquad \forall \omega .$$

Passing to the limit in (4) there obtains

$$\begin{cases} -\Delta \bar{u}_o = 0 \\ \bar{u}_o|_\Gamma = g \end{cases}$$

As $\bar{u}_o$ is bounded, $\bar{u}_o = u_o$, the unique solution of (4) for $\lambda = 0$.
But the lower branch can be easily seen to tend to u_o as goes to zero. Thus the lower branch and the upper branch meet at $\lambda = 0$ and the convergence is in $W^{2,p}_{loc}(\Omega)$ for all $p \in (1, \infty)$. Here loc means for all relatively compact ω , with $\bar{\omega} \subset \bar{\Omega} = \mathbb{R} \times \mathbb{R}^+$, because the problem behaves nicely at the boundary.

III. NUMERICAL COMPUTATIONS.

III.1. Reduction to a bounded domain.

Numerical computations were made on the following system (where a and b are parameters with "large" values)

$$(72)\quad \begin{cases} -\Delta u(x,y) = \lambda\, f(u(x,y)) \quad \text{for all} \quad \lambda \in \Omega_{ab} \\ \text{with} \quad \Omega_{ab} = \{(x,y) \mid |x| < a, \quad 0 < y < b\} \; ; \end{cases}$$

$$(73)\quad u(x,0) = g(x) \quad \text{for all} \quad |x| \leq a \; ;$$

$$(74)\quad \begin{cases} \frac{\partial u}{\partial x}(x,y) = 0 \quad \text{for} \quad |x| = a \quad \text{and} \quad 0 \leq y \leq b \,, \\ \frac{\partial u}{\partial y}(x,y) = 0 \quad \text{for} \quad |x| \leq a \quad \text{and} \quad y = b. \end{cases}$$

Thus we reduce the original problem (4) in the half plane domain Ω to the system (72), (73), (74) in the bounded domain Ω_{ab}.
The mathematical results of part II together with the physical meaning, lead us to introduce the Neumann boundary condition (74), hoping that for large values of parameters a and b, the set of solutions of (72, (73), (74) is quite a good approximation of the set of solutions of (13) restricted to Ω_{ab}.

III.2. Minimum and maximum solutions.

As Ω_{ab} is bounded the following results are classical (see Courant-Hilbert) :

(i) There exist one minimum solution $\underline{u}_\lambda \in H^1(\Omega_{ab})$ and one maximum solution $\bar{u}_\lambda \in H^1(\Omega_{ab})$ for all λ R. The functions $\lambda \to \underline{u}_\lambda$ and $\lambda \to \bar{u}_\lambda$ are increasing.

(ii) Let k be such that $k + f'(\alpha) \geq 0, \forall \alpha \in \mathbb{R}$.
Let u_n be defined by $u_o = 0$ and by

$$(75)\quad \begin{cases} -\Delta u_n + \lambda k\, u_n = \lambda(f(u_{n-1}) + k\, u_{n-1}) \\ + \text{boundary conditions } (*) \end{cases}$$

Note : * This means :

$$u_n(x,0) = g(x), \text{ for } |x| \leq a$$

$$\frac{\partial u_n}{\partial x}(x,y) = 0 \quad \text{for} \quad |x| = a; \quad a \leq y \leq b$$

$$\frac{\partial u_n}{\partial y}(x,y) = 0 \quad \text{for} \quad |x| \leq a; \quad y = b$$

We have $u_{n+1} \geq u_n$ for all n and u_n converges to $\underline{u}_\lambda$ in $H^1(\Omega)$.

(iii) Let $u_o = \alpha_o$ (see (6c) for the definition of α_o). Let u_n be defined again by (75). Then we have $u_{n+1} \leq u_n$ for all n, and u_n converges to $\bar{u}_\lambda$ in $H^1(\Omega_{ab})$.

Points (ii) and (iii) give an algorithm to compute the minimum and maximum solutions, $\underline{u}_\lambda$ and $\bar{u}_\lambda$. But the procedure is extremely slow for the minimum solution $\underline{u}_\lambda$ if k is large ang g small. We shall first give an algorithm which is faster in this case.

Staded roughly this new algorithm stems from the idea that it is not absolutely necessary to take for k in (75) a constant value. We could imagine a generalization of (75) based on the principle of Newton's method; namely, having some approximation u_n, obtain a "better" approximation by linearizing the second nomber of the equation i.e. replacing

$$- \Delta u_{n+1} = \lambda f(u_{n+1}) = \lambda f(u_{n+1} - u_n + u_n)$$

by the "approximation"

$$- \Delta u_{n+1} = \lambda (f(u_n) + (u_{n+1} - u_n) f'(u_n)) ,$$

or else :

$$- \Delta u_{n+1} - \lambda u_{n+1} f'(u_n) = \lambda f(u_n) - \lambda u_n f'(u_n) .$$

It is, of course, hoped that this algorithm gives a fast (quadratic) rate of convergence. However the operator acting on u_{n+1} would have to be changed on each step of the computation, which means that if we use a finite element method a large matrix should be inverted at each step anew, which consumes a lot of time. Hence, we could prefer to use a quasi-Newtonian method by keeping the function entering as an argument in $f'(u_n)$ for several steps. At the same time, because the rate of convergence gets smaller the larger the value of $\lambda f'(u_n)$, one should also like to keep the factor in front of $f'(u_n)$ smaller than λ wherever possible. We would then like to investigate the convergence properties of an algorithm which would give u_{n+1} from u_n by

$$(76) \qquad - \Delta u_{n+1} - \mu f'(v_n) = \lambda f(u_n) - \mu f'(v_n) u_n$$

One could also change from time to time the value of μ, and even work with a sequence λ_n converging to λ instead of keeping always the same λ value at the second member of (76).

As algorith (75), this new algorith (76) has good monotonicity and convergence properties, which we shall now explain.

Algorithm

The sequence of functions (u_n) and (v_n) are defined by induction as follows : when u_{n-1} has been defined, choose any $v_n \in H^1(\Omega_{ab})$ such that $0 \le v_n \le u_{n-1}$ and define u_n as the solution of the linear equation :

$$(77) \qquad -\Delta u_n - \mu_n f'(v_n)u_n = \lambda_n f'(u_{n-1}) - \mu_n f'(u_n)\, u_{n-1}$$

where (λ_n) is an increasing sequence of numbers converging towards λ and (μ_n) is a sequence of numbers such that $0 \le \mu_n \le \lambda_n$. We start the process with $u_o = 0$.

The convergence properties of this algorithm are studied in Appendix III of [14] It can be proved that for λ in a certain interval, $0 \le \lambda \le \lambda_{Max}$, we have for all $n \ge 1$:

$$(78) \qquad \begin{cases} u_{n+1} \ge u_n \; ; \quad u_n \le \underline{u}_{\lambda_n} \\ -\Delta u_n < \lambda f(u_n) \\ u_n \text{ converges to } \underline{u}_\lambda \text{ in } H^1(\Omega_{ab}) \end{cases}$$

The proof of these results, which are somewhat similar to those of F. Mignot and J.P. Puel [13] and of M.G. Grandall and P.H. Rabinowitz [16] , is based on a study of conditions under which $\underline{u}_\lambda$ remains small enough for $f(\underline{u}_\lambda)$ to be increasing and locally convex : see lemma III.1. of Appendix III of [14] ; the convergence is related to positivity properties of f and a lemma giving positivity properties of a function v in terms of positivity properties of $(-\Delta v - gv)$ (lemma III.2. of Appendix III of [14]); λ_{max} is also precised in Appendix III. of [14] .

Fast computation of $\underline{u}_\lambda$ then g and λ are not too large.

By g "not too large" we mean that $\beta > \max g$ where β is the constant appearing in lemma III.1. of Appendix III of [14].

By λ "not too large" we mean that λ is smaller than ν_o and ν_1 , the constants which appear in lemmas III.1. and III.2. of Appendix III. of [14] .

We are interested in the approximation of $\underline{u}_\lambda$ for a (finite) number of values of λ, say $0 < \lambda^{(1)} < \ldots < \lambda^{(m)} \ldots < \lambda^{(p)}$. The algorithm is :

1. $m = 0, \qquad n = 0, \qquad w_{-1} = 0$;

2. $u_o = w_{m-1}, \quad v_o = w_{m-1}$;

3. $n \to n+1$, take $v_n = v_{n-1}$ or $v_n = u_{n-1}$;

choose any $\mu_n \in [0, \lambda^{(m)}]$ and define u_n by

$$\begin{cases} -\Delta u_n - \mu_n f'(v_n)\, u_n = \lambda^{(m)} f(u_{n-1}) - \mu_n f'(v_n) u_{n-1} \\ + \text{boundary conditions (73) (74) on } u_n . \end{cases}$$

4. go to 3 or go to 5 ;

5. store the present value of u_n (call it w_m), change m to m+1, go to 2

The function w_m are lower approximation of $\underline{u}_{\lambda^{(m)}}$: from lemma III.3. of Appendix III. of [14] , we can infer that $w_m \leq u_{\lambda^{(m)}}$ and that we can make $|w_m - u_{\lambda^{(m)}}|_{H^1(\Omega)}$ as small as we want by iterating the step 4 → 3 a sufficient number of times.

As explained earlier, to obtain fast convergence, the best theoretical choice of μ_n in 3 is $\mu_n = \lambda^{(m)}$ and $v_n = u_{n-1}$. But a lot of computing time can be saved if $v_n = v_{n-1}$ and $\mu_n = \mu_{n-1}$. In our computations we found it interesting to keep $v_n = v_{n-1}$ about five to ten times and then to take $v_n = u_{n-1}$. Another reason of taking $\mu_n < \lambda^{(m)}$ near the turning point will be explained below in part IV.C.

III.3. Computation of intermediate solutions.

We have used a continuation method based on the hypothesis that there were turning points but no secondary bifurcations. More precisely we suppose that we have :

(79) $$\begin{cases} \dim [D_{u,\lambda}] = 1 \quad \text{for all } (u,\lambda) \in H^1(\Omega_{ab}) \times \mathbb{R}, \\ \text{solution of (81), (82), (83)} \end{cases}$$

where $D_{u,\lambda}$ is the tangent space to the curve we are looking for, namely the space of pairs (v, μ) solutions of (89), (90), (91) :

(80) $$-\Delta v = \lambda f'(u)v + \mu f(u)$$

(81) $$v(x,0) = g(x) \quad \text{for all } |x| \leq a$$

(82) $$\begin{cases} \frac{\partial v}{\partial x}(x,y) = 0 & \text{for } |x| = a \text{ and } 0 \leq y \leq b \\ \frac{\partial v}{\partial y}(x,y) = 0 & \text{for } |x| < a \text{ and } y = b \end{cases}$$

The reader will find in H.B. Keller [15] the details, and more information on the specific problem in [14]. Let us describe the idea in a few words : Once the point (u_n, λ_n) is reached, let us solve the system (80), (81), (82) by a finite elements method, which determines the pair (v_n, μ_n) up to a multiplicative constant (hypothesis (79)).

Let

$$\begin{cases} \tilde{u}_n = u_n + h_n v_n , \\ \tilde{\lambda}_n = \lambda_n + h_n \mu_n \end{cases}$$

where h_n is small enough (see below).

The pair $(\tilde{u}_n, \tilde{\lambda}_n)$ is now used as a starting point for a quasi-Newtonian method of solution for the following non-linear system, where $(\tilde{u}, \tilde{\lambda})$ is the unknown :

$$(83) \qquad \begin{cases} - \Delta\tilde{u} = \tilde{\lambda} f(\tilde{u}) , \\ \text{Boundary conditions (73, (74),} \\ \mathcal{L}_n(\tilde{u}, \tilde{\lambda}) = c_n , \end{cases}$$

where $\mathcal{L}_n$ is a linear form on $H^1(\Omega_{ab}) \times \mathbb{R}$ and c_n a real number such that

$$\mathcal{L}_n(v_n, \mu_n) \neq 0$$

and

$$\mathcal{L}_n(\tilde{u}_n, \tilde{\lambda}_n) = c_n .$$

If h_n is small enough, the quasi-Newtonian sequence built in that manner stays in a neighborhood of $(\tilde{u}_n, \tilde{\lambda}_n)$, and (83) satisfies an implicit function theorem, thanks to (79). Practically, we took either $\mathcal{L}(\tilde{u}, \tilde{\lambda}) = \tilde{\lambda}$ as long as possible, and near the turning points $\mathcal{L}(\tilde{u}, \tilde{\lambda}) = \tilde{u}(A)$, where A is a point of Ω_{ab} such that $\tilde{v}(A) \neq 0$.

Numerical results.

We made the computations in the domain $\Omega_{6,3}$ with boundary conditions (73), (74) with the following function g

$$g(x) = \begin{cases} 0.2 & \text{for} \quad |x| \leq 1.3 \\ 1.5 - |x| & \text{for} \quad 1.3 \leq |x| \leq 1.5 \\ 0 & \text{for} \quad |x| \geq 1.5 \end{cases}$$

The equation in the domain was

$$- \Delta u(x,y) = \lambda f(u(x,y)) \qquad \forall x,y \in \Omega_{6.3} ,$$

with

$$f(\alpha) = \begin{cases} \alpha^4 (1-\alpha)^2 & \text{if} \quad \alpha \in [0,1] , \\ 0 & \text{if} \quad \alpha \notin [0,1] . \end{cases}$$

We computed the branch of symmetric solutions $(u(x,y) = u(-x,y))$ starting from $\lambda = 0$, $u = u_o$, where u_o is the solution of $\Delta u_o = 0$ in $\Omega_{6.3}$ with the boundary (73), (74). The results of these computations are shown in figures (8).

In the finite domain problem, however, there exists a finite range of λ values, $0 \leq \lambda \leq \lambda_{min}$, in which only the closed solution exists, contrary to the semi-infinite domain.

The following figures show

- the S-shaped diagram of bifurcation, fig. 2.
- the level lines of the computed solutions for different values of parameter λ, on the low, intermediate and high branches, figures 3 to 7.

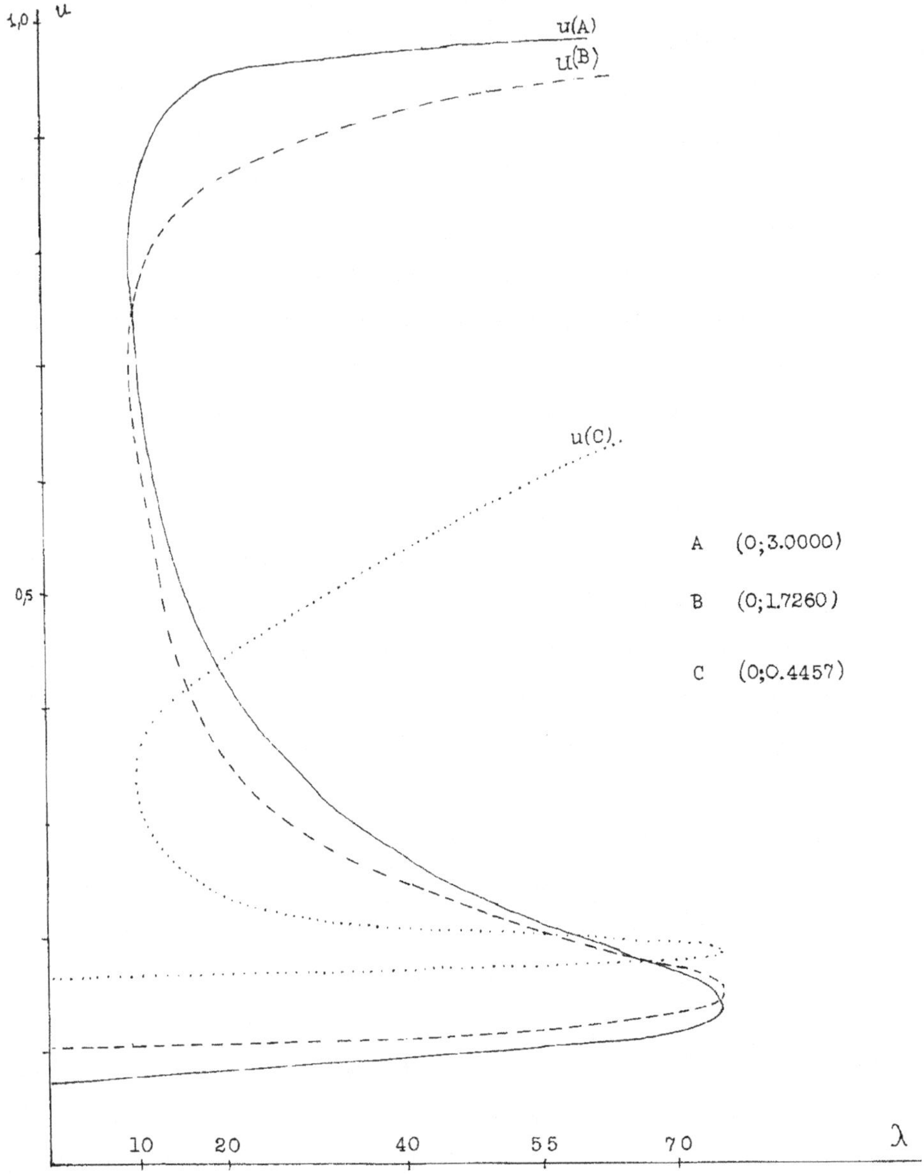

Fig. 2

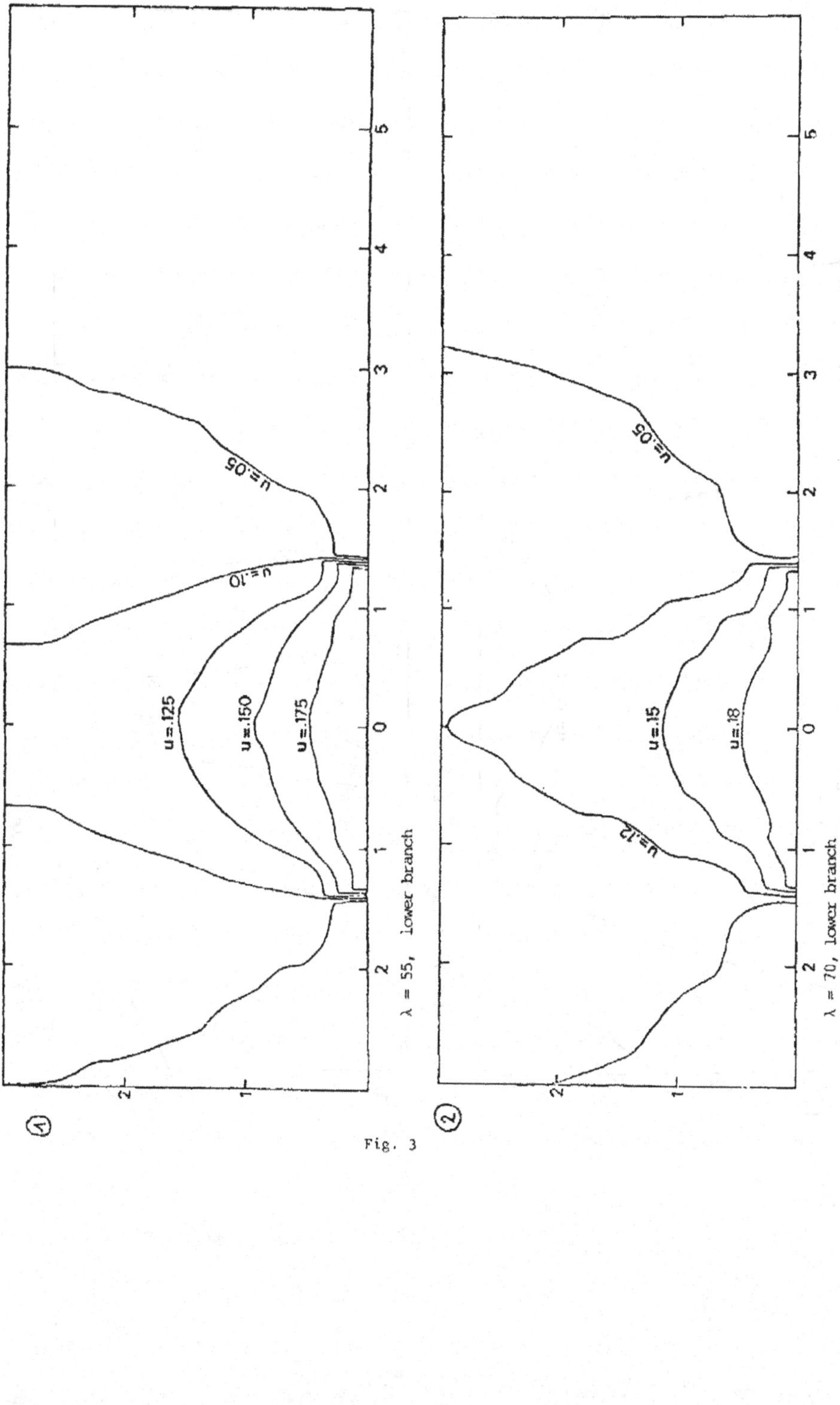
u=.05
u=.10
u=.125
u=.150
u=.175
λ = 55, lower branch
u=.05
u=.12
u=.15
u=.18
λ = 70, lower branch

Fig. 3

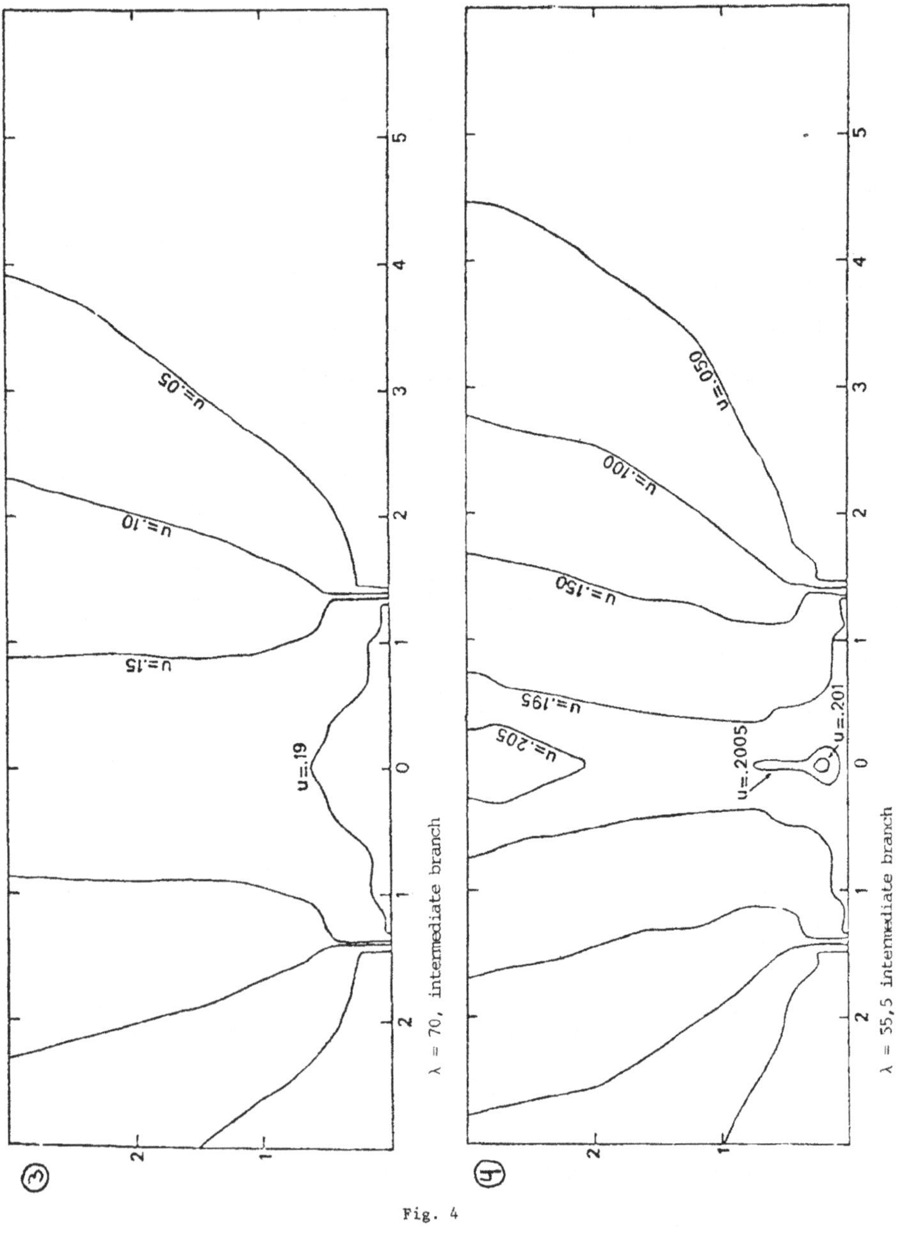
③
u=.05
u=.10
u=.15
u=.19
λ = 70, intermediate branch
④
u=.050
u=.100
u=.150
u=.195
u=.205
u=.2005
u=.201
λ = 55,5 intermediate branch

Fig. 4

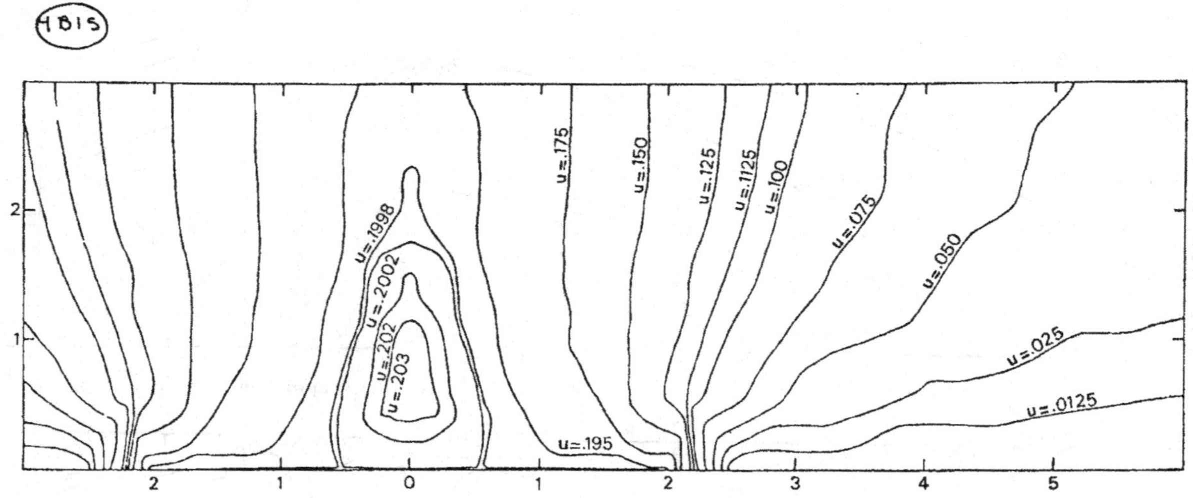

λ = 32,84, intermediate branch ; the boundary condition has be modified so as to show more clearly the " bubble ".

Fig. 5

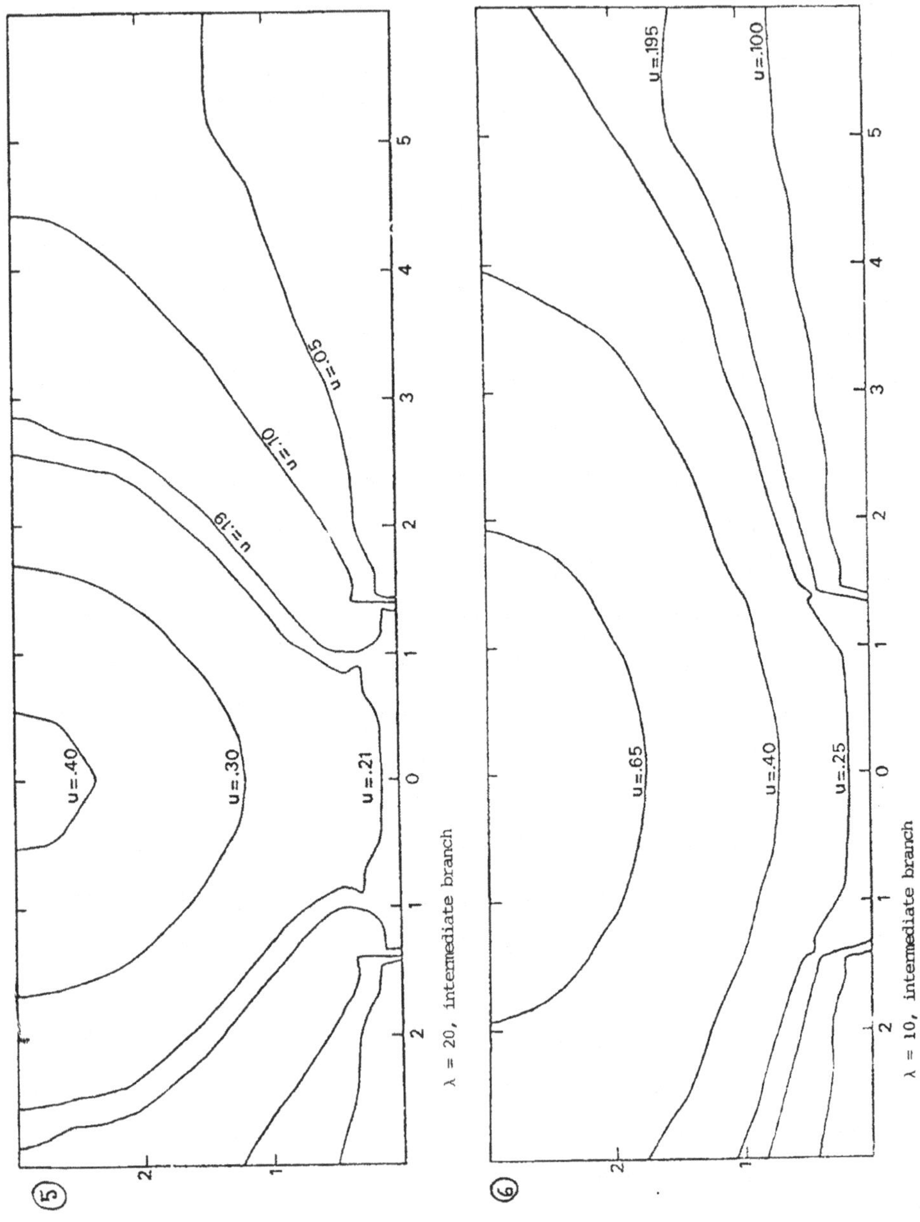

5
u=.40
u=.30
u=.21
u=.19
u=.10
u=.05
λ = 20, intermediate branch
6
u=.65
u=.40
u=.25
u=.195
u=.100
λ = 10, intermediate branch

Fig. 6'

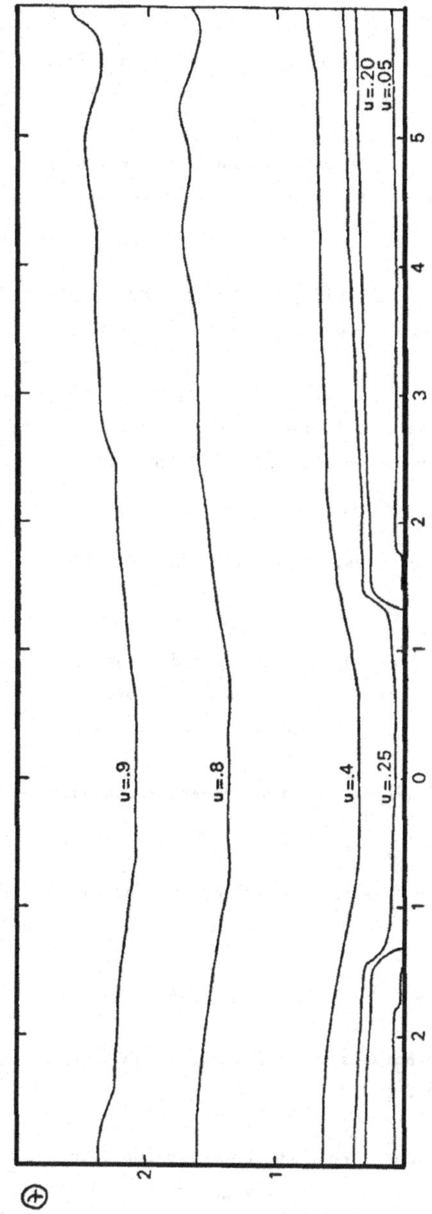

λ = 20, higher branch

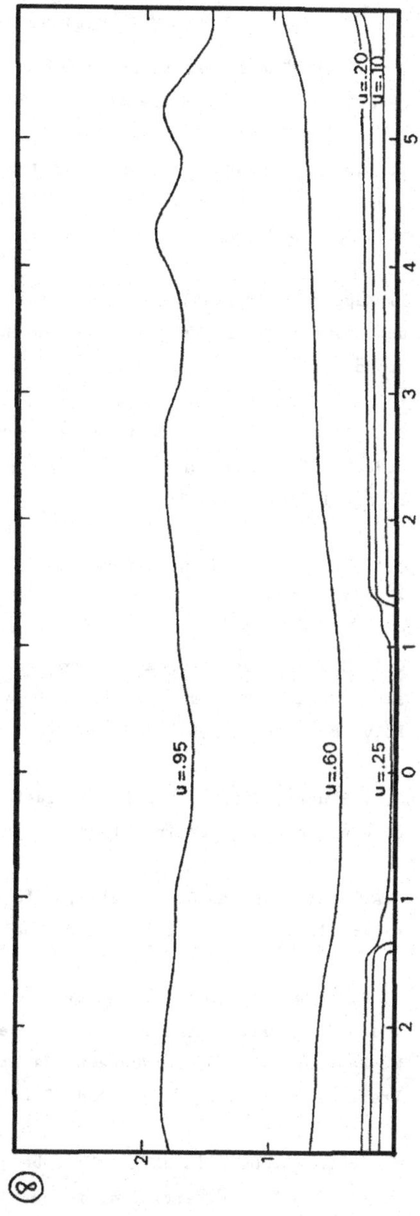

λ = 55, higher branch

Fig. 7

Bibliography

[1] T. Küpper, The lowest point of the continuous spectrum as a bifurcation point, report 78-12 of the Mathematics Department of the University of Köln, 1978.

[2] T. Küpper, Necessary and sufficient conditions for bifurcation from the continuous spectrum, report 78-23 of the Mathmatics Department of the University of Köln, 1978.

[3] T. Küpper, On minimal nonlinearities which permit bifurcation from the continuous spectrum, report 78-25 of the Mathematics Department of the University of Köln, 1978.

[4] C.A. Stuart, Global properties of components of solutions of non linear second order ordinary differential equations on the half-line, Ann. Sc. Norm. Sup. Pisa, 11 (1975) 265-286.

[5] C.A. Stuart, Des bifurcations sans valeurs propres, C.R. Acad. Sc. Paris 284 (1977), 1373-1375

[6] C.A. Stuart, Des bifurcations pour des problèmes de Dirichlet et de Neumann sans valeurs propres, rapport du Département de Mathématiques de l'Ecole Polytechnique Fédérale de Lausanne, 1979.

[7] C.A. Stuart, Bifurcation for variational problems when the linearization has no eigenvalues, preprint, 1978.

[8] C.A. Stuart and R. Chiapinelli, Bifurcation when the linearized problem has no eigenvalues, to appear in the Journal of Differential Equations.

[9] K. Kirchgässner and J. Scheurle, On the bounded Solutions of a semilinear elliptic equation in a strip, to appear in J. Diff. Eq.

[10] M. Protter and H. Weinberger, Maximum principles in differential equations, Prentice Hall, 1971, Englewood Cliffs, N.J., U.S.A.

[11] S. Agmon, A. Douglis and L. Nirenberg, Estimates near the boundary for solutions of elliptic partial differential equations satisfying general boundary conditions, I, Comm. Pure Appl. Math 12 (1959), 623-727.

[12] R. Courant and D. Hilbert, Methods of Mathematical Physics, II, Interscience, 1963, New York.

[13] F. Mignot and J.P. Puel, Sur une classe de problèmes non linéaires avec non-linéarité positive, croissante, convexe. Proceedings of the International Meeting on Recent Methods in non-linear Analysis, Pitagora Editrice, 1979, Bologna.

[14] J. Heyvaerts, J.M. Lasry, M. Schatzman and P. Witomski, Solar flares : blowing up of force free magnetic configurations, Pré-publications Mathématiques de l'Université Paris-Nord, 1979.

[15] H.B. Keller, Constructive methods for bifurcations and non-linear eigenvalue problems, 3ème colloque international sur les méthodes de calcul scientifique et technique, I.R.I.A., Rocquencourt, 1977.

[16] M.G. Crandall and P.H. Rabinowitz, Some Continuation and variational methods for positive solutions of nonlinear elliptic eigenvalue problems, Arch. Rat. Mech. Anal. 58 (1975) 207-218.

BIFURCATION OF INVARIANT TORI IN $\mathbb{R}^3$

by G. IOOSS

I . INTRODUCTION .

Let us consider a one parameter (μ) family of two-dimensional invariant tori for an autonomous family of differential equations E_μ in $\mathbb{R}^3$, such that the restriction of the trajectories on these tori admits a cross section, i.e. circles γ_μ on which the first return map (the Poincaré map) is defined. For each μ , we extend this first return map to a cross section of the flow in a neighborhood of the invariant torus containing γ_μ (see Fig. 1).

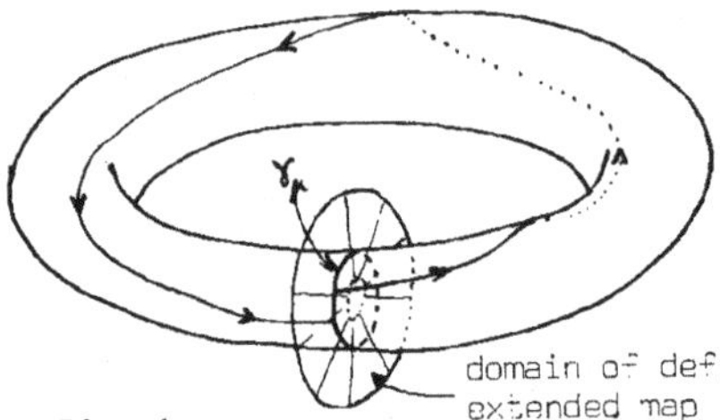

Fig. 1

These assumptions are verified in the case when we are close to a Hopf bifurcation of a closed orbit into a two dimensional invariant torus.

We give in the following, sufficient conditions implying bifurcation of the family of tori of sections γ_μ into another family of tori. Their stabilities (or attractivities) are exchanged as in the usual bifurcations into lower dimensional objects.

The work developped here is extremely simplified due to the limitation to $\mathbb{R}^3$. The general case (Banach space)is treated by A. Chenciner and the author in 2 papers [C. I. 1] , [C. I. 2] where the reader will find most of the proofs.

Our problem may be written in the following way : the map $F_\mu : T^1 \times \mathcal{V} \longrightarrow T^1 \times \mathbf{R}$ is of class C^k , $\mathcal{V}$ is a neighborhood of 0 in $\mathbf{R}$ and T^1 is the circle $\mathbf{R}/\mathbf{Z}$ and

$$(1) \qquad F_\mu(T^1 \times 0) \subseteq T^1 \times 0 \quad .$$

We shall give elements of answer to the following questions :

1/ How to study the stability of $T^1 \times 0$?

2/ Suppose that for $\mu < 0$, $T^1 \times 0$ is attractive and that for $\mu > 0$ it is repelling, what is the new attractor if there exists a bifurcated one ?

3/ What are sufficient, general enough, conditions to obtain the persistence of an invariant circle in a neighborbood of $T^1 \times 0$ for the map F_μ is we only assume its existence for $\mu = 0$ (we don't assume (1) for $\mu \neq 0$) ? (of course we are here in the critical situation for the stability of $T^1 \times 0$).

NOTATIONS . We note

(2) $F_\mu \; : \; (\theta, x) \in T^1 \times \mathcal{V} \longmapsto (f(\theta, x, \mu), \Phi(\theta, x, \mu)) \in T^1 \times \mathbb{R}$

where $f(., x, \mu)$ is a diffeomorphism of T^1 keeping the orientation and where we assume $\Phi(\theta, 0, \mu) = 0$.

II . STUDY OF THE STABILITY OF $T^1 \times 0$.

In this paragraph we suppress μ which is not useful. We consider the map

$$F \; : \; T^1 \times \mathcal{V} \longrightarrow T^1 \times \mathbb{R}$$

defined by

$$F(\theta, x) = (f(\theta, x), \Phi(\theta, x))$$

of class C^k , where $f(., x)$ is a diffeomorphism of T^1 keeping the orientation and where $\Phi(\theta, 0) = 0$. Hence we have

$$F(T^1 \times 0) = T^1 \times 0 \qquad .$$

We define the " linearized map " around $T^1 \times 0$ by

$$G \; : \; T^1 \times \mathbb{R} \longrightarrow T^1 \times \mathbb{R}$$

with

$$G(\theta, x) = (g(\theta), a(\theta)x)$$

$$g(\theta) = f(\theta, 0) \; , \quad a(\theta) = \frac{\partial \Phi}{\partial x}(\theta, 0) \qquad .$$

DEFINITION . For $\ell \leq k-1$, the ℓ - spectrograph" of F is the spectrum of the linear operator

$$\mathcal{A} \; : \; C^\ell(T^1; \mathbb{R}) \longrightarrow C^\ell(T^1; \mathbb{R})$$

defined by

$$(\mathcal{A}X)(\theta) = a[g^{-1}(\theta)] \, . \, X[g^{-1}(\theta)] \qquad ,$$

i.e. $\quad \mathrm{graph}(\mathcal{A}X) = G(\mathrm{graph}\, X) \quad$ where $X \in C^\ell(T^1; \mathbb{R})$.

We may also define the map $\mathcal{F}$ from a neighborhood $\mathcal{O}$ of 0 in $C^\ell(T^1; \mathbb{R})$, taking values in $C^{\ell}(T^1; R)$, by

$$\text{graph}(\mathcal{F} X) = F(\text{graph } X)$$

but $\mathcal{F}$ is not differentiable at 0 in $C^{\ell}(T^1; R)$. In fact the composed map

$$\mathcal{O} \subset C^{\ell}(T^1; R) \xrightarrow{\mathcal{F}} C^{\ell}(T^1; R) \hookrightarrow C^{\ell-1}(T^1; R)$$

is differentiable at 0 , of derivative $\mathcal{A} \in \mathcal{L}[C^{\ell}(T^1; R)]$! Remark that 0 is a fixed point of $\mathcal{F}$.

For the study of the asymptotic behavior of the iterated $F^n(\theta, x)$ it is natural to study the iterated of the linearized map $G^n(\theta, x) = (g^n(\theta), a_n(\theta)x)$, where

$$a_n(\theta) = a[g^{n-1}(\theta)] \ldots\ldots a[g(\theta)].a(\theta) .$$

The spectral radius of the operator $\mathcal{A}$ in $C^0(T^1; R)$, noted $s_0(F)$, satisfies

$$(3) \qquad s_0(F) = \lim_{n \to \infty} \|\mathcal{A}^n\|_0^{1/n} = \lim_{n \to \infty} (\sup_{\theta \in T^1} |\mathcal{A}_n(\theta)|)^{1/n} ,$$

and it is easy to show that if $s_0(F) < 1$ then $T^1 \times 0$ is attractive.

A much stronger result is the following.

THEOREM 1 . Assume F is C^k , $k \geq 2$ and $a(\theta) > 0$ on T^1 , and assume either g is structurally stable, or the rotation number of g is irrational, then the asymptotic stability of $T^1 \times C$ for G is equivalent to the fact $s_0(F) < 1$ and it leads to the asymptotic stability of $T^1 \times 0$ for F (the full nonlinear map).

REMARK . (i) A structurally stable g means the existence of a finite number of stable cycles of periodic points (period q) alternating with an equal number of unstable cycles. The rotation number of g is rational in this case ; $\rho(g) = m/q$ and we can show that

$$s_0(F) = [\ \ell.u.b.\ a_q(\theta)\ ]^{1/q}$$

where the $\ell.u.b.$ is taken on the set of periodic points of g . In this case the spectograph may be thick for instance it may be a thick annulus centered at 0 (see an exeample in [C.I.1.]) .

(ii) If the rotation number $\rho(g)$ of g is irrational then it may be shown that the ℓ - spectrograph of F is contained in the circle, centered at 0 of radius

$$(4) \qquad e^{\int_{T^1} \text{Log } a(\theta)\, d\nu(\theta)}$$

where ν is the unique probability measure on T^1 , invariant under g .

Moreover, if g is C^ℓ - conjugate to the rotation R_ω ($\omega = \rho(g)$) , then the ℓ - spectrograph is the entire circle of radius (4) .

We may understand some difficulties of this case even in making the assumption of C^ℓ - conjugation :

$$g = h^{-1} \circ R_\omega \circ h$$

where h is C^ℓ and $\omega \notin \mathbb{Q}$, then by a change of variable in T^1 we may assume, without loss of generality, that

$$G(\theta, x) = (\theta + \omega, a(\theta)x) ,$$

with an abuse of notation about $R_\omega \theta$. Let us look for a real eigenvalue λ of $\mathcal{A}$:

$$(5) \qquad a(\theta)\, X(\theta) = \lambda\, X(\theta + \omega) .$$

Thanks to the assumption $a(\theta) > 0$, we may take the Log of (5) and obtain

$$(6) \qquad \operatorname{Log} X(\theta + \omega) - \operatorname{Log} X(\theta) + \operatorname{Log} \lambda = \operatorname{Log} a(\theta) .$$

This equation admits a solution only if

$$\lambda = \exp \int_{T^1} \operatorname{Log} a(\theta)\, d\theta ,$$

which gives a positive eigenvalue. In fact, (6) has not necessarily a periodic solution $\operatorname{Log} X(\theta)$ if ω is too well approximated by rationals (see [H] for this type of properties). In this last case, the spectrum $s_0(F)$ is identical with the circle of radius (4) , but contains no eigenvalue.

Let us now assume the diophantine approximation assumption :

(7) $\exists\, \epsilon > 0$, $c > 0$ such that

$$|q\omega - p| > \frac{c}{|q|^{1+\epsilon}} , \qquad \forall q \in \mathbb{Z} \setminus \{0\},\ p \in \mathbb{Z} .$$

Note that almost all numbers ω satisfy (7) (in the sense of Lebesgue measure). In this case the equation (6) determines the eigenfunction X with a loss of differentiability relatively to $a(.)$ (see [H]) .

In fact the condition (7) corresponds to the existence on the invariant torus of a quasi periodic flow with two fundamental periods (Kolmogorov's theorem).

In what follows we make this assumption at criticality $\mu = 0$, which does not implies that it is realized for $\mu \neq 0$!

III . BIFURCATION AND PERSISTENCE OF INVARIANT CIRCLES UNDER F_μ

Assumptions . We assume in this paragraph that $F_0(T^1 \times 0) = T^1 \times 0$ and with the notations of I :

(i) $g(.) = f(., 0, 0)$ is C^k – conjugate to the rotation R_ω : $g = h^{-1} \circ R_\omega \circ h$, $h \in C^k$, with an irrational ω satisfying the diophantine condition (7) .

(ii) $a(\theta) = \frac{\partial \Phi}{\partial x}(\theta, 0, 0)$ is never * 0 on T^1 (i.e. $a(\theta) > 0$),

and

$$\int_{T^1} \mathrm{Log}\,[\, a(\theta)\,]\; d\nu(\theta) = 0$$

where ν is the unique probability measure on T^1 , invariant under g .

REMARKS . These assumptions mean that at $\mu = 0$ the flow is quasi periodic on the invariant torus. Note that we only assume here the existence of the torus for $\mu = 0$. Hence, there is a problem of persistence of this invariant torus for $\mu \neq 0$. The condition of C^k conjugation to R_ω is in general realized in the sens of Lebesgue measure on the rotation numbers of g (see [H]).

From the paragraph II we know that 1 is an eigenvalue of $\mathcal{A}_0$, whose spectrum is just the entire unit circle. The following change of variables in $T^1 \times R$:

$$(\theta', x') = (h(\theta), x\,[\, X(\theta)\,]^{-1}) ,$$

where X is the positive eigenfunction such that

$$X\,[\, g(\theta)\,] = a(\theta)\, X(\theta) , \quad \theta \in T^1 ,$$

leads to the new form (2) of F_μ with

$$(8) \quad \begin{cases} f(\theta, x, \mu) = \theta + \omega + \mu f_1(\theta) + x f_2(\theta) + O(|\mu| + |x|)^2 \\ \Phi(\theta, x, \mu) = [\, 1 + \mu a_1(\theta)\,]x + \mu b_1(\theta) + \mu^2 b_2(\theta) + c_2(\theta)x^2 \\ \qquad\qquad + O(|\mu| + |x|)^3 \end{cases}$$

where the new variables are still noted (θ, x) .

Now making the change of variables in $T^1 \times R$:

$$(\theta', x') = (\theta, x + \mu\, Y_1(\theta))$$

where Y_1 is solution of the equation

$$(9) \qquad Y_1(\theta + \omega) - Y_1(\theta) + b_1(\theta) = \int_{T^1} b_1(\theta)\, d\theta \overset{\text{def}}{=} b_1$$

* By construction $a(\theta)$ cannot change of sign.

we obtain a new form (8) of the map F_μ but with b_1 constant, the other coefficients being changed (new $a_1(\theta)$), and higher order coefficients.

To solve (9) is possible thanks to the diophantine condition (7), but we loose some differentiability ($\simeq 1+\epsilon$).

In the same way we may, by suitable change of variables, make constant the coefficients of μ and x in f and of μx, μ^2, x^2 in Φ. The change of variable will be of the form :

$$(\theta', x') = (\theta + \mu\,\delta_1(\theta) + x\,\delta_2(\theta),\ x + \mu\,\alpha(\theta)x + \mu^2\,\gamma_2(\theta) + x^2\,\alpha_2(\theta)) \quad ,$$

where δ_1, δ_2, α, γ_2, α_2 are solutions of equations of the form (9). We then obtain the map F_μ on the form (2) with

$$(10)\quad \begin{cases} f(\theta, x, \mu) = \theta + \omega + \beta_1\,\mu + \beta_2\,x + O\,(|\mu| + |x|)^2 \\ \Phi(\theta, x, \mu) = (1 + \mu\,a_1)x + \mu\,b_1 + \mu^2\,b_2 + c_2\,x^2 + O\,(|\mu| + |x|)^3 \end{cases}$$

where we note again (θ, x) the new variables.

Let us suppress the higher order terms, we may look for invariant circles, close to $T^1 \times 0$, for the <u>truncated</u> map $\tilde{F}_\mu$. This corresponds to look for fixed points close to 0 of the map in $\mathbb{R}$:

$$(11)\qquad x \longmapsto (1 + \mu\,a_1)x + \mu\,b_1 + \mu^2\,b_2 + c_2\,x^2 \quad .$$

We then see that if $b_1 \neq 0$, which is the <u>generic situation</u>, for μ close to 0, we have in a neighborhood of $T^1 \times 0$ two families of invariant circles for the truncated $\tilde{F}_\mu$, identical with $T^1 \times 0$ for $\mu = 0$, only for $\mu > 0$ or $\mu < 0$.

The assumption of persistence of invariant circles under the map F_μ in a whole neighborhood of $\mu = 0$ corresponds to do the assumptions (the simplest ones):

$$(12)\qquad \begin{cases} b_1 = 0 \\ \lambda_1^2 = a_1^2 - 4\,b_2\,c_2 > 0 \end{cases} \quad .$$

This leads to the existence of two real non tangent branches of fixed points of the map (11).

Now making the assumption (12) we make the new change of variables in $T^1 \times \mathbb{R}$:

$$(\theta', x') = (\theta,\ x + \mu\eta) \quad ,$$

where η satisfies $c_2\,\eta^2 - a_1\,\eta + b_2 = 0$. Then F_μ becomes on the form : (we still note x instead of x')

$$(13)\quad \begin{cases} f(\theta, x, \mu) = \theta + \omega + \omega_1 \mu + \beta_2 x + O(|\mu| + |x|)^2 \\ \Phi(\theta, x, \mu) = (1 + \mu\lambda_1)x + c_2 x^2 + O(|\mu| + |x|)^3 \end{cases} .$$

It may then be shown the following

THEOREM 2 . Under the assumptions (i), (ii) and (12) , we have in general the persistence of an invariant circle and the bifurcation of an invariant circle under the map F_μ . These both families of circles are close to $T^1 \times 0$ and are identical for $\mu = 0$. The stabilities of these circles are exchanged when μ crosses 0 .

REMARK 1 . The word " in general " means that some coefficients are assumed to be not 0 . For instance λ_1 in (12) is $\neq 0$, and if $c_2 = 0$ we have to assume that a coefficient of x^n , $n > 2$ in the Taylor series of Φ , has a non zero mean value. It may happen that the bifurcated circle only appears on one side of $\mu = 0$. In this case we shall have 2 bifurcated circles on one side of $\mu = 0$, and none on the other side (see Fig. 2) .

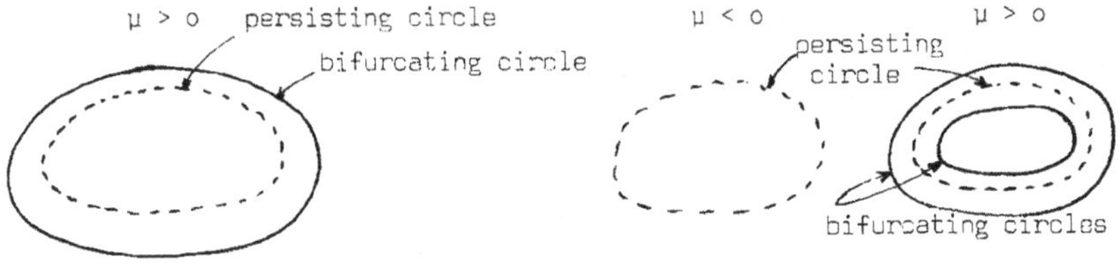

transcritical bifurcation

The fig. for $\mu > 0$ and $\mu < 0$ are deduced in exchanging the doted circle and the full circle.

supercritical bifurcation

Bifurcated circles only for $\mu > 0$ (subcritical will correspond to $\mu < 0$) .

Fig. 2

REMARK 2 . The assumption (12) corresponds to the well known fact of non-persistence in general of a fixed point of a map in R^n when 1 is an eigenvalue of the derivative at the fixed point.

We observed that the results are easy to prove on the reduced truncated map $\tilde{F}_\mu$. In fact the proof of the theorem, from the form (13) is based on technics used by RUELLE - TAKENS [R.T.] in the case of Hopf bifurcation for diffeomorphisms and consists in looking for the fixed point of a map in a functional space of closed curve of the plane.

IV . AN EXAMPLE . Let us consider the map

$$F_\mu(\theta, x) = (\theta + \omega, (1+\mu)\, a(\theta)(x - x^3))$$

where $a(\theta) > 0$ is C^∞ , and $\omega \notin \mathcal{U}$, and

$$\int_{T^1} \operatorname{Log} a(\theta)\, d\theta = 0 .$$

The ℓ - spectrograph of F_μ is the circle of center 0 and radius $1+\mu$, so, in this example, it is clear that the persisting invariant circle $T^1 \times 0$ has its stability changing when μ crosses 0 . Now, if we try to obtain a bifurcated circle in the form of the graph of $\theta \longmapsto \mu^\alpha x_0(\theta) + o(|\mu|^\alpha)$, with a smooth x_0 then we obtain

$$\operatorname{Log} x_0(\theta+\omega) - \operatorname{Log} x_0(\theta) = \operatorname{Log} a(\theta) ,$$

which may have no solution (even measurable) x_0 if ω is too well approximated by rational numbers.

In this simple example it may be easily shown that there is an attractor close to $T^1 \times 0$ for $\mu > 0$, but the explicit form of this attractor is not known (of course the assumptions of the theorem on ω are not realized).

BIBLIOGRAPHY

[C. I. 1.] A. CHENCINER, G. IOOSS Bifurcations de tores invariants, Arch. Rat. Mech. Anal., 69, 109–198 (1979)

[C. I. 2.] A. CHENCINER, G. IOOSS Persistance et Bifurcation de tores invariants, Arch. Rat. Mech. Anal. (to appear)

[H] M.R. HERMAN Mesure de Lebesgue et nombre de rotation. Lect. Notes in Maths, 597, p. 271–293, Springer Verlag, Berlin 1977

[R.T.] D. RUELLE, F. TAKENS On the nature of turbulence. Comm. Math. Phys. 20, p. 167–192 (1971)

PATTERN FORMATION AND WAVE PROPAGATION IN THE s-a SYSTEM

J.P. KERNEVEZ, G. JOLY, D. THOMAS

Université de Technologie de Compiègne
B.P. 233
60206 Compiègne (France)

and

B. BUNOW

National Institute of Health
Bethesda, Maryland 20014 (U.S.A.)

ABSTRACT

A system of two coupled reaction-diffusion equations involving substrate inhibited enzyme kinetics is studied with a view to describing and explaining stable non uniform steady state solutions and propagating wave front solutions which they admit. The pattern formation phenomenon, reminiscent of morphogenesis, is compared to the predictions of Kauffman for sequential compartment formation in Drosophila imaginal disks. A modified perturbation technique is used to obtain the emerging bifurcation branches.

Numerical analysis of pattern formation needs methods to follow branches of solutions including turning points and bifurcation points. A simple dissipative structure is given in order to test such algorithms.

INTRODUCTION

Kauffman, Shymko, and Trabert [13] proposed a model for morphogenesis of imaginal disks in Drosophila involving spontaneous development of concentration patterns from the interaction of reaction and diffusion within the disks :

$$\begin{cases} s_t - \Delta s + \gamma[F(s,a) - (s_o - s)] = 0 \\ a_t - \beta\Delta a + \gamma[F(s,a) - \alpha(a_o - a)] = 0 \\ \text{with zero-flux boundary conditions.} \end{cases}$$

We have simulated this model (the s-a system) on domains with the same geometry as the disks, using an enzyme-catalyzed reaction with realistic kinetics :

$$F(s,a) = \rho\, as/(1 + s + ks^2)$$

The results are, as Kauffman predicted, the sequential partitioning of the domain into regions of high and low concentration as the domain is enlarged.

From the mathematical point of view, the Prigogine model [25] has been the testing-bench of many methods for analysing the bifurcating solutions. For instance Meurant and Saut [21] applied results of Iooss [12] on the existence and stability of bifurcated solutions. Boa [3] applied the two timing method of Matkowsky [19] and Kogelman and J.B. Keller [18].

These methods apply without much change to our problem. It is the reason why we prefer to develop here a third method, the modified perturbation method of H.B. Keller [34], which gives easily all the useful information about the nature of the bifurcation. After a linear stability analysis of the basic state $(\tilde{s},\tilde{a})$, we apply this perturbation method to seek a parametric representation of the branch which bifurcates from the basic state at those critical points γ_0 where there is a change of stability.

The idea that chemical reaction and diffusion can give rise to a propagating front dates back to Turing [33] at least, and has been developed, among others, by Ortoleva and Ross [26], Fife [8], [10], and Othmer [27]. Murray [5] has presented such a model with substrate inhibition kinetics, very similar to our s-a system.

We present in §2 numerical results obtained in the s-a system with parameter values given to us by J.D. Murray [17]. The evolution takes place in 4 epochs, each with its characteristic time scale, and can be explained by arguments very similar to those of Fife [8].

1. PATTERN FORMATION

1.1 Kauffman's model of morphogenesis in Drosophila

The origin of biological form is a problem which has challenged embryologists since the discipline was founded. Many models have been proposed, but few were sufficiently specific as to provide critically testable predictions.

The genetic program, by itself, is insufficient to specify all of the elements of development, and some additional information is provided by the interaction of developing cells with their environment.

Turing [33] pointed out that the interaction of diffusive transport and nonlinear chemical kinetics can lead to instability of a concentration field, with the result that initially homogeneous domain develops regional non-uniformities.

The idea that chemical reaction and diffusion can give rise to spatial patterns has recently been amplified by the school around Prigogine (see review Nicolis and Prigogine [25]).

Quite recently, Kauffman, Shymko, and Trabert [13] elaborated Turing's hypothesis into a qualitative model for the detailed development of the wing imaginal disk of the fruitfly Drosophila. For details of the experimental embryology of Drosophila, the reader is referred to the references in Kauffman, et al. For our discussion, it is sufficient to recollect that an imaginal disk is a planar (approximately) amorphous block of cells, out of which the adult organs of the fly develop. There are a number of such disks, one for each organ structure in the adult. The disks differ from one another in their size and shape. As development procedes, the disks enlarge, while (more or less) retaining their original shape. Experiments tracing the cell lineage of the disks have shown that there are a series of compartmental lines which form, defined by the observation that descendents of primordial cells never cross the lines. The lines are formed sequentially, and successively subdivide the disk into progressively smaller and anatomically more specialized regions. Kauffman et al. observed that the position of the compartmental lines on the wing imaginal disk was reminiscent of the nodal lines of the eigenfunctions of a laplacian operator on an elliptical domain with Neumann boundary conditions. This apparently arcane similarity was the basis of suggesting that the compartmental lines were formed by cells responding to a concentration field which developed as result of a reaction-diffusion instability. In a linear approximation, the form of the concentration instability would resemble the eigenfunction of a laplacian, with high concentration on one side of the nodal line, and low on the other. Which eigenfunction is selected for amplification depends on parameter values. Since the imaginal disk is growing, the parameters are changing, and it is reasonable that a sequence of successive patterns should appear. The particular sequence of patterns to be observed depends upon the shape of the domain.

Kauffman et al used an elliptical domain as a crude model for the imaginal disk of the wing, whose actual shape is shown in Fig. 1. Since Laplace's equation is separable on the ellipse, the form of the eigenfunctions is well known to mathematical physicists, and results could be obtained without much calculation. On an irregular domain, however, such an analysis is not possible and numerical methods are required. In this publication, we propose to obtain the patterns which appear sequentially as a domain with the shape of each of the several imaginal disks of the fruitfly undergoes growth without deformation. The instability which leads to pattern formation under these conditions was chosen on the basis of previous research [16] on reaction-diffusion instability of an immobilized enzyme, urate oxidase. We have no illusions about the role of this enzyme in embryonic development: none whatever. However, it is an enzyme with well characterized kinetics, and the regulatory properties which lead to instability here are found widely in metabolism. For example, in the imaginal disk of flies the enzyme aldehyde oxidase appear in a clear-cut pattern (Sprey [30]).

1.2 Pattern formation by an immobilized enzyme

The enzyme urate oxidase (E.C.1.7.3.3.) has been studied upon immobilization as an artificial enzyme membrane [32], and shown to produce multiple steady states or hysteresis experimentally [24] and sustained oscillations theorically [16]. It seemed probable that spontaneous pattern formation could also be demonstrated once appropriate conditions were selected.

In an appropriate concentration regime, the rate of the reaction catalyzed by urate oxidase is :

$$J_r = V_M\, A\, S/[K_s + S + S^2/K_{ss}] \qquad (1)$$

where S is the concentration of the substrate uric acid, and A the concentration of the co-substrate oxygen, while V_M, K_M, K_s, and K_{ss} are all constants characteristics of the enzyme defined by kinetic study in free solution.

Mathematical model of an imaginal disk :

We consider a planar region bounded by a closed simple curve. No transport is permitted across the bounding curve (Neumann boundary conditions). Within the region, the enzyme is uniformly distributed.

Both species diffuse freely within the region, obeying Fick's second law with the coefficients of diffusion being D_s and D_a, respectively. The surface of the region is exposed to a reservoir containing the substrates at fixed concentration, S_o and A_o, respectively. Access of these substrates to the region is limited by mass transfer effects, according to equations (2) and (3).

$$J_s = P_s(S_o - S) \qquad (2)$$

$$J_a = P_a(A_o - A) \qquad (3)$$

The equation of continuity relates the rate of change of concentration at any point of the region to the resultant of the material flows in and out of that point. For our system, there are three flows: flow in the plane, flow from the reservoir, and flow through the chemical reaction. The resulting equations (4) and (5) :

$$\frac{\partial S}{\partial t} = D_s\, \nabla^2 S + P_s(S_o - S) - V_m AS \,/\, (K_s + S + S^2/K_{ss}) \qquad (4)$$

$$\frac{\partial A}{\partial t} = D_a \nabla^2 A + P_a(A_o - A) - V_m AS / (K_s + S + S^2/K_{ss}) \tag{5}$$

describe the evolution of the concentration field in the system from any initial distribution.

We choose K_s as the reference concentration, and define reduced concentrations :

$$s = S/K_s \tag{6}$$

$$a = A/K_s \tag{7}$$

Time is dimensionalized according to the diffusional time for the substrate S, where L is a characteristic dimension.

$$\tau = D_s\, t \,/\, L^2 \tag{8}$$

the following dimensionless parameters arise :

$$\rho = V_m/P_s \tag{9}$$

$$\beta = D_a/D_s \tag{10}$$

$$\alpha = P_a/P_s \tag{11}$$

$$\gamma = L^2 P_s/D_s \tag{12}$$

Conveniently, all of the information about the size of the system is now contained in the parameter γ.

The material balance equations take the form :

$$\frac{\partial s}{\partial t} = \nabla^2 s + \gamma[(s_o - s) - F(s,a)] \tag{13}$$

$$\frac{\partial a}{\partial t} = \beta\nabla^2 a + \gamma[\alpha(a_o - a) - F(s,a)] \tag{14}$$

with Neumann boundary conditions : s_o and a_o are the substrate concentrations in the reservoir, and γ, ρ, β, and α are dimensionless parameters.

$$F(s,a) = \rho as \,/\, (1 + s + ks^2) \tag{15}$$

Considering the derivative-free terms in equations (13) and (14), we find that there is always a concentration pair $(\bar{s},\bar{a})$ which nullifies those terms, equations (16),(17):

$$F(\bar{s},\bar{a}) - (s_o - \bar{s}) = 0 \tag{16}$$

$$F(\bar{s},\bar{a}) - \alpha(a_o - \bar{a}) = 0 \tag{17}$$

Obviously, $(\bar{s},\bar{a})$ is also a solution of the evolution equations (13),(14), but, it may not be stable to position-dependent perturbations. If we knew the eigenfunctions and eigenvalues of the linearized operator arising from equations (13) and (14) on the given domain, we could perform a linearized stability analysis and determine which eigenfunction would be amplified as the system evolved.

1.3 A numerical study of Kauffman's model for morphogenesis in Drosophila

With the parameter values $\alpha = 1.45$, $k = 0.1$, $s_o = 102.5$, $a_o = 79.2$, $\rho = 13$ and $\beta = 5$ (for which $\tilde{s} = 8$ and $\tilde{a} = 14$), equations (13), (14) were solved on two-dimensional domains by the finite element method [2]. Choosing γ, which scales the size of the domain, as a bifurcation parameter, we solved the equations successively for a sequence of closely spaced values of γ, the initial conditions for each subsequent value of γ being the equilibrium solution for the previous value.

This procedure corresponds to observation of the evolution of the solution on a growing domain, and thus is analogous to the model of Kauffman ,et al [13] for morphogenesis in the imaginal disk of the fruitfly, Drosophila. The present system constitutes a fully nonlinear dynamical model of the sort hypothesized by Kauffman but based only upon linearized analysis. A sequence of solutions typical of those which have been obtained is shown in Fig. 1 and 2. Increasing density of printing corresponds to higher concentrations. The concentration profiles for both s and a are very similar in this presentation. The s-a system acts, for each value of γ, like a spatial filter selecting only one pattern from all the possible wave vectors contained in a perturbation of the concentrations. As the solution grows away from the trivial state, other components also appear so that the final steady state resembles the unstable eigenfunction only approximately.

1.4 Bifurcation of non uniform steady states

Let Ω be a bounded region of R^p ($p = 1, 2, 3$), with outward normal ν and boundary Γ. The steady state solutions of (13),(14) are the solutions of :

$$\left| \begin{array}{ll} -\Delta s + \gamma[F(s,a) - (s_o - s)] = 0 & \\ & \text{in } \Omega \\ -\beta\Delta a + \gamma[F(s,a) - \alpha(a_o - a)] = 0 & \\ \dfrac{\partial s}{\partial \nu} = 0, \ \dfrac{\partial a}{\partial \nu} = 0 & \text{on } \Gamma \end{array} \right. \tag{18}$$

The parameters k and β have given values ($\beta > 1$), the parameters s_o, a_o, ρ, α can be chosen [16] so that the following assumptions (H1) - (H4) hold, and γ is the bifurcation parameter.

We shall first test the stability of the uniform steady state of (13),(14), $(\tilde{s},\tilde{a})$, to nonuniform disturbances. Then we shall seek parametric equations for the branch

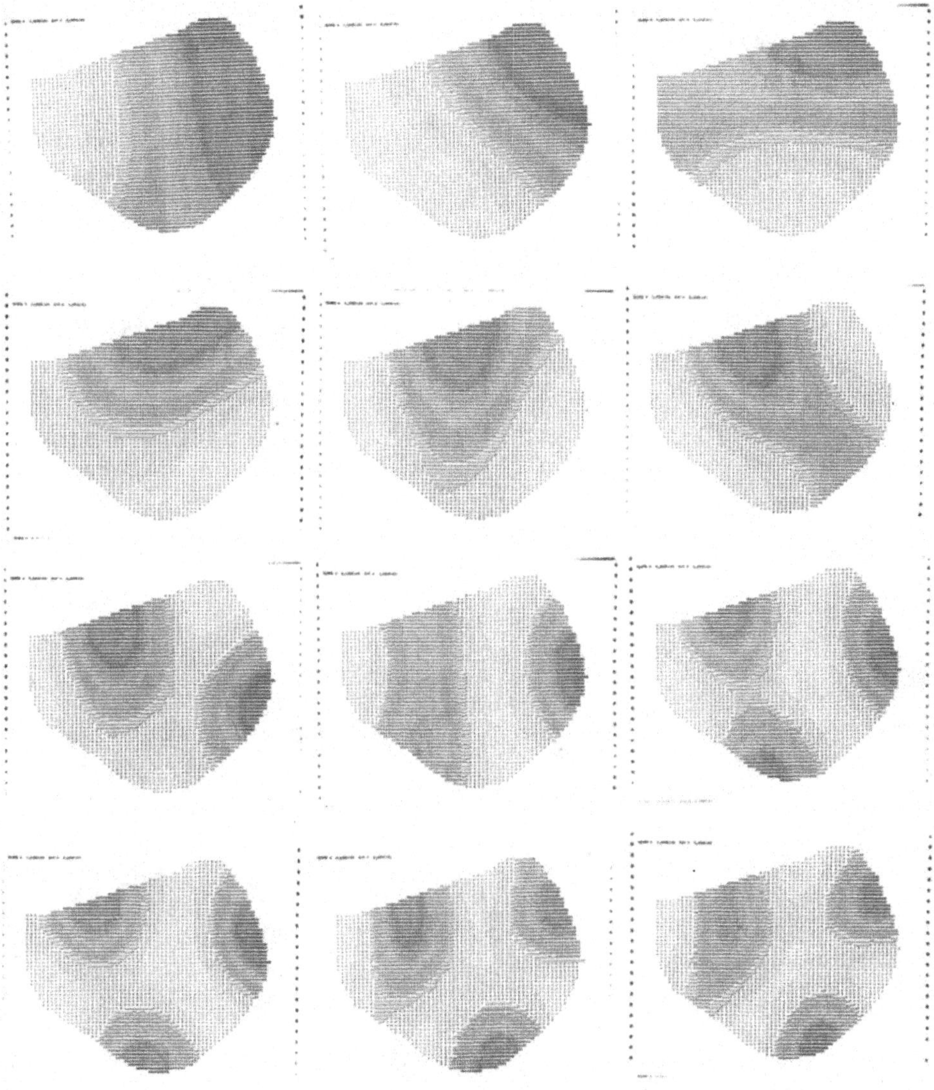

Fig. 1

SEQUENTIAL ALTERATIONS OF S CONCENTRATION PATTERNS IN A DOMAIN SIMILAR TO A WING DISK, FOR γ = 2; 8; 10; 14; 18; 26; 32; 36; 42; 46; 62; 70.

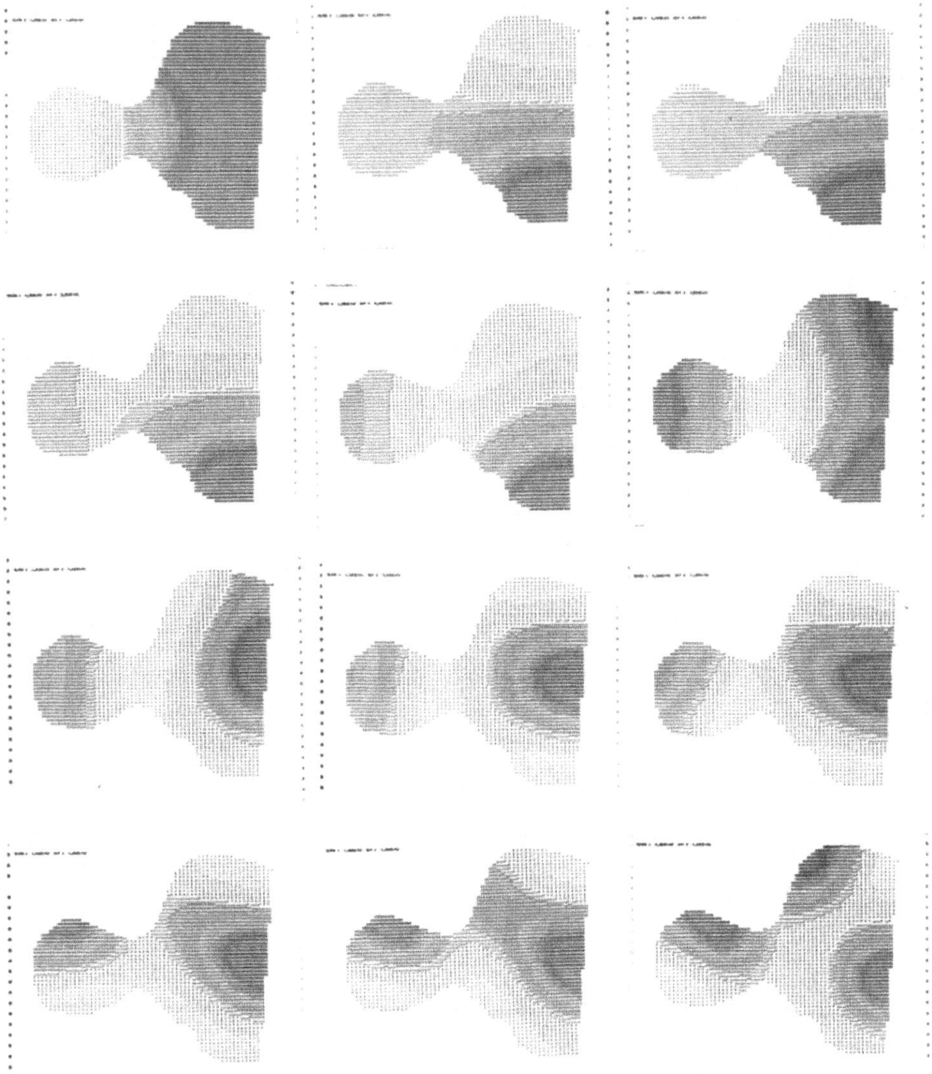

Fig. 2

SEQUENTIAL ALTERATIONS OF S CONCENTRATION PATTERNS IN A DOMAIN SIMILAR TO AN EYE-ANTENNA DISK, FOR γ = 4; 20; 28; 32; 40; 48; 76; 96; 112; 120; 148; 152.

of solutions splitting off from $(\tilde{s},\tilde{a})$ at those points γ_0 where stability is changing.

Hypotheses

(H1) The dynamical system governed by the ordinary differential equations $\frac{ds}{dt} + F(s,a) - (s_0 - s) = 0$, $\frac{da}{dt} + F(s,a) - \alpha(a_0 - a) = 0$ has exactly one equilibrium point $(\tilde{s},\tilde{a})$, which is stable, and therefore satisfies (16),(17), and

$F_s + 1 + F_a + \alpha = tr(0) > 0, \quad \alpha F_s + F_a + \alpha = det(0) > 0$

We shall denote $F_s = \frac{\partial F}{\partial s}(\tilde{s},\tilde{a})$, $F_a = \frac{\partial F}{\partial a}(\tilde{s},\tilde{a})$, $F_{ss} = \frac{\partial^2 F}{\partial s^2}(\tilde{s},\tilde{a})$, ...

(H2) $1 < \alpha < \beta$

(H3) Let $T(z) = \beta z^2 + (\beta(F_s + 1) + F_a + \alpha)z + \alpha F_s + F_a + \alpha$

We assume that $T(z)$ has 2 positive roots $0 < z' < z''$

As a consequence :

$(F_a + \alpha - \beta(F_s + 1))^2 > -4\beta F_s F_a$ and $\beta(F_s + 1) + F_a + \alpha < 0$

The latter implies :

(H4) $F_s < 0$

Stability analysis of the basic state

Stability of $(\tilde{s},\tilde{a})$ is governed by the eigenvalues of $L_\gamma = L_0 + \gamma L_1$,

$$L_0 = \begin{bmatrix} -\Delta & 0 \\ 0 & -\beta\Delta \end{bmatrix}, \quad L_1 = \begin{bmatrix} F_s+1 & F_a \\ F_s & F_a+\alpha \end{bmatrix}$$

Let (w_n,μ_n) be the eigenpairs of $-\Delta$ subject to zero-flux boundary conditions :

$$-\Delta\, w_n = \mu_n\, w_n, \quad \frac{\partial w_n}{\partial \nu} = 0, \quad \int_\Omega w_n^2\, dx = 1,\ n = 0, 1, 2, \ldots$$

For each n we have 2 eigenpairs $(\phi_n^\pm, k_n^\pm)$ of L_γ, $\phi_n^\pm$ of the form $\phi_n^\pm = \begin{bmatrix} 1 \\ M_n^\pm \end{bmatrix} w_n$, determined by :

$$\begin{bmatrix} \mu_n + \gamma(F_s + 1) - k_n^\pm & \gamma F_a \\ \gamma F_s & \beta\mu_n + \gamma(F_a + \alpha) - k_n^\pm \end{bmatrix} \begin{bmatrix} 1 \\ M_n^\pm \end{bmatrix} = \begin{bmatrix} 0 \\ 0 \end{bmatrix}$$

The $k_n^\pm$ are the (real) roots of :

$$k_n^2 - tr(n)k_n + det(n) = 0$$

$$\begin{cases} tr(n) = (\beta + 1)\mu_n + \gamma(F_s + 1 + F_a + \alpha) = (\beta + 1)\mu_n + \gamma tr(0) & (19) \\ det(n) = \gamma^2 T(\mu_n/\gamma) & (20) \end{cases}$$

The important property that $k_n^{\pm}$ are real ($k_n^- < k_n^+$) results from :

$$tr^2(n) - 4\,det(n) = [(\beta - 1)\mu_n + \gamma(F_a^{1/2} - (-F_s)^{1/2})^2 + \gamma(\alpha - 1)]$$

$$\times [(\beta - 1)\mu_n + \gamma(F_a^{1/2} + (-F_s)^{1/2})^2 + \gamma(\alpha - 1)]$$

Moreover, from (19) $k_n^- + k_n^+ = tr(n) > 0$, so that at least $k_n^+ > 0$. k_n^- will be negative if and only if $det(n) < 0$, i.e., from (20), $T(\mu_n/\gamma) < 0$, $z' < \mu_n/\gamma < z''$, $\mu_n/z'' < \gamma < \mu_n/z'$.

As a conclusion there is a family of intervals $I_n =]\mu_n/z'', \mu_n/z'[$ such that $(\tilde{s},\tilde{a})$ is unstable if γ lies within at least one of them, and $(\tilde{s},\tilde{a})$ is stable if $\gamma \in \bigcup_{n \geq 1} \overline{I_n}$. It is easily seen that I_n and I_{n+1} overlap as soon as $\mu_n/z' > \mu_{n+1}/z''$.

Bifurcation of new solutions

Our purpose now is to seek the small amplitude solutions of (18) which emerge from $(\tilde{s},\tilde{a})$ when γ crosses a critical value γ_o where there is a change of stability for $(\tilde{s},\tilde{a})$.

We can write (18) :

$$(L_o + \gamma L_1)U + \gamma M(U) = 0 \qquad (21)$$

where $u = \begin{bmatrix} u \\ v \end{bmatrix}$, $u = s - \tilde{s}$, $v = a - \tilde{a}$ and

$$M(U) = (F(\tilde{s} + u, \tilde{a} + v) - F(\tilde{s},\tilde{a}) - F_s u - F_a v) \begin{bmatrix} 1 \\ 1 \end{bmatrix}$$ is

$O(|U|_H^2)$, $H = L^2(\Omega) \times L^2(\Omega)$.

$L_\gamma = L_o + \gamma L_1$ and M possess all the properties which enable us to say that a branch of non-uniform solutions splits off from the trivial solution $U = 0$ at $\gamma = \gamma_o$, with exchange of stability between the basic state $U = 0$ and the bifurcated solutions [21]. All the useful information about the bifurcated solutions of (21) can be easily gotten by using the modified perturbation method ([34]).

We shall denote by $\phi_n^{\pm}$ (resp $\psi_n^{\pm}$) the eigenvectors of $L_o + \gamma_o L_1$ (resp. $L_o + \gamma_o L_1^T$),

$\phi_n^{\pm} = \begin{bmatrix} 1 \\ M_n^{\pm} \end{bmatrix} w_n$, $\psi_n^{\pm} = \begin{bmatrix} 1 \\ N_n^{\pm} \end{bmatrix} w_n$, $(F_a M_n^{\pm} = F_s N_n^{\pm})$, and $k_n^{\pm}$ the corresponding eigenvalues (as $\mu_{n_o}/\gamma_o = z'$ or z'', $k^-_{n_o} = 0$).

We wish to determine a one parameter family of solutions $(U(\varepsilon),\gamma(\varepsilon))$ of :

$$\begin{cases} (L_o + \gamma L_1)U(\varepsilon) + \gamma(\varepsilon)M(U(\varepsilon)) = 0 & (22)\\ (U(\varepsilon),\psi^-_{n_o}) = \varepsilon(1 + M^-_{n_o}N^-_{n_o}) & (23)\end{cases}$$

which depends differentiably on ε and which reduces to $U(0) = 0$, $\gamma(0) = \gamma_o$, at $\varepsilon = 0$. Both U and γ are assumed sufficiently differentiable with respect to ε at $\varepsilon = 0$ and we expand U and γ in Taylor series in ε about $\varepsilon = 0$.

We shall denote by $\dot U$, $\dot\gamma$, $\ddot U$, $\ddot\gamma$, etc. the derivatives of $U(\varepsilon)$ and $\gamma(\varepsilon)$ with respect to ε at $\varepsilon = 0$. To determine $\dot U$ we differentiate (22) and (23) with respect to ε and set $\varepsilon = 0$, obtaining $(L_o + \gamma_o L_1)\dot U = 0$, $(\dot U,\psi^-_{n_o}) = 1 + M^-_{n_o} N^-_{n_o}$.

Thus we have $\dot U = \bar\phi_{n_o}$

To determine $\ddot U$ and γ we differentiate (22) and (23) twice with respect to ε, set $\varepsilon = 0$, obtaining :

$$\begin{cases} (L_o + \gamma_o L_1)\ddot U + 2\dot\gamma\, L_1\, \dot U + \gamma_o M''(0)\, (\dot U,\dot U) = 0 & (24)\\ (\ddot U,\psi^-_{n_o}) = 0 & (25)\end{cases}$$

the problem of solving (24), (25) for $\ddot U$ has a unique solution only if :

$$(2\,\dot\gamma L_1\, \phi^-_{n_o} + \gamma_o M''(0)(\phi^-_{n_o}, \phi^-_{n_o}),\ \psi^-_{n_o}) = 0 \tag{26}$$

$(L_1\, \phi^-_{n_o}, \psi^-_{n_o})_H = \zeta$ and $\gamma_o(M''(0)\, \phi^{-2}_{n_o}, \psi^-_{n_o})_H = \eta$ can be calculated :

$$\zeta = -\frac{1}{\gamma_o}(L_o\, \phi^-_{n_o}, \psi^-_{n_o}) = -\frac{1}{\gamma_o}(1 + \beta M^-_{n_o} N^-_{n_o}) = \pm\mu_{n_o}\, \beta(z'' - z')/(\beta\mu_{n_o} + \gamma(F_a + \alpha))$$

which is negative or positive according to whether $\mu_{n_o}/\gamma_{n_o} = z''$ or z'.

$$\eta = \gamma_o(F_{ss} + 2F_{sa}M^-_{n_o})(1 + N^-_{n_o}) \int_\Omega w^3_{n_o}\, dx$$

which is equal to 0 if :

$$\int_\Omega w^3_{n_o}\, dx = 0 \tag{27}$$

Equation (26) is satisfied if and only if $\dot\gamma$ has the value :

$$\dot\gamma = -\eta/(2\zeta) \tag{28}$$

We can express the solution $\ddot U$ in the form :

$$\begin{cases} \ddot{U} = \sum_{k_n^{\pm} \neq k_{n_o}^-} \frac{1}{k_n^{\pm}} \frac{1}{1 + M_n^{\pm} N_n^{\pm}} (f, \psi_n^{\pm}) \phi_n^{\pm} \\ f = -2\dot{\gamma} L_1 \phi_{n_o}^- - \gamma_o M''(0)(\phi_{n_o}^-, \phi_{n_o}^-) \end{cases} \tag{29}$$

Differentiating (22) and (23) three times with respect to ε and setting $\varepsilon = 0$ yields

$$\begin{cases} (L_o + \gamma_o L_1)\dddot{U} + 3\ddot{\gamma} L_1 \dot{U} + 3\dot{\gamma} L_1 \ddot{U} + 3\dot{\gamma} M''(0)\dot{U}^2 + \gamma_o M'''(0)\dot{U}^3 + 3\gamma_o M''(0)\dot{U}\ddot{U} = 0 \\ (\dddot{U}, \psi_{n_o}^-) = 0 \end{cases} \tag{30}$$

Proceeding as before we find $\ddot{\gamma}$ and $\dddot{U}$.

We find as a result a one parameter family of solutions of (22), (23) with the expansion :

$$U(\varepsilon) = \varepsilon\phi_{n_o}^- + \frac{\varepsilon^2}{2}\ddot{U} + O(\varepsilon^3) \tag{31}$$

$$\gamma(\varepsilon) = \gamma_o + \varepsilon\dot{\gamma} + \frac{\varepsilon^2}{2}\ddot{\gamma} + O(\varepsilon^3) \tag{32}$$

If Ω is a planar region with the shape of an imaginal disk, without any axis of symmetry, there is no reason for (27) to hold, and $\dot{\gamma} \neq 0$. The bifurcation is bilateral, as shown in Figure 3.

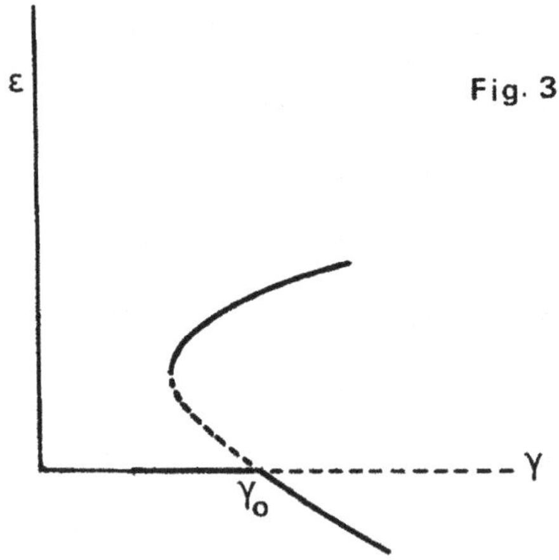

Fig. 3

If Ω is a planar region with at least one axis of symmetry Ox, the eigenfunction w_{n_o} satisfies either $w_{n_o}(x,-y) = w_{n_o}(x,y)$ or $w_{n_o}(x,-y) = -w_{n_o}(x,y)$. In the last case (27) holds, and $\dot{\gamma} = 0$. The bifurcation is supercritical if $\ddot{\gamma} > 0$ and subcritical if $\ddot{\gamma} < 0$.

It is also the case if $\Omega =]0,1[$. The eigenfunctions, $w_{n_o} = \sqrt{2}\cos \pi n_o x$, satisfy (27).

A motivation for studying this case is that it can be considered as an approximation of long ellipsoidal eggs with large axis ratio, at least for their first eigenpairs, those which create circumferential nodal lines, which segment the egg along its length. Whether the bifurcation is supercritical or subcritical depends upon the sign of $\ddot{\gamma}$.

If $\dot{\gamma} = 0$, the value of $\ddot{\gamma}$ based upon (30) is :

$$\ddot{\gamma} = -\gamma_o (M'''(0)\, \phi_{n_o}^{-\,3} + 3M''(0)(\phi_{n_o}^{-}, \ddot{U}), \psi_{n_o}^{-})$$

$$= -\gamma_o (1 + N_{n_o}^{-}) \Big[(F_{sss} + 3F_{sa} M_{n_o}^{-}) \int w_{n_o}^{4}\, dx \qquad (33)$$

$$+ 3(F_{ss} + 2F_{sa} M_{n_o}^{-}) \sum_{k_n^{\pm} \neq k_{n_o}^{-}}^{\mp\,-} \frac{1}{k_n^{\pm}} \frac{1 + N_n^{\pm}}{1 + M_n^{\pm} N_n^{\pm}} \int_{\Omega} (w_{n_o}^{2} w_n dx)^2 \,(F_{ss} + F_{sa}(M_{n_o}^{-} + M_n^{\pm})\Big]$$

If $\Omega =]0,1[$, $\int_0^1 w_{n_o}^2 w_n dx = 0, \quad (n \neq 0,\ 2n_o).$

2. WAVE PROPAGATION

Again, the system is governed by equations (13)-(14) subject to zero-flux boundary conditions, and the parameters are such that in the phase plane (s,a) the isocline curves $F(s,a) - (s_o - s) = 0$ and $F(s,a) - \alpha(a_o - a) = 0$ intersect at a point $(\tilde{s},\tilde{a})$ which is a stable steady state of the spatially homogeneous dynamical system

$$\frac{ds}{dt} + \gamma(F(s,a) - (s_o - s)) = 0, \qquad \frac{da}{dt} + \gamma(F(s,a) - \alpha(a_o - a)) = 0.$$

Moreover, $(\tilde{s},\tilde{a})$ lies to the right of the turning point D (Fig. 4a) and is a stable uniform steady state of the distributed dynamical system (13)-(14).

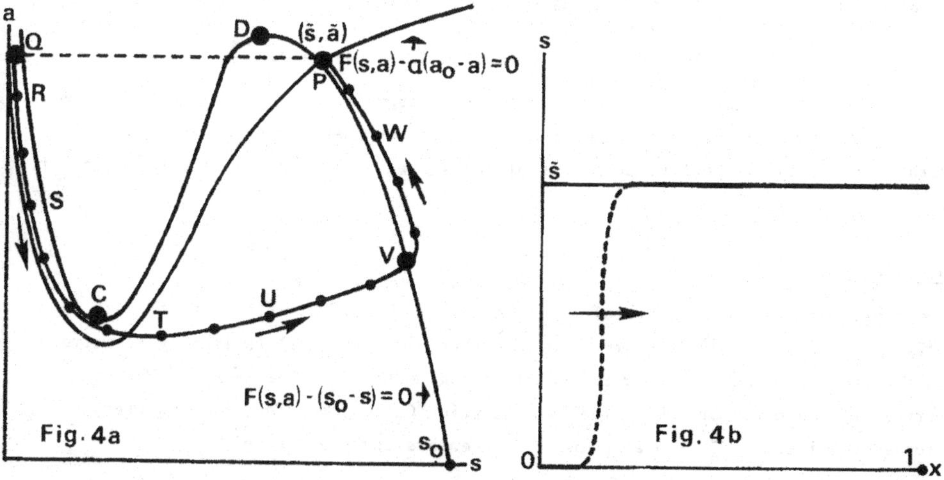

Fig. 4a

Fig. 4b

An appropriate set of parameters for this to occur is $k = 1$, $\rho = 0.9$, $s_o = 39.2$, $a_o = 784$ and $\alpha = 0.05$, for which $\tilde{s} = 22$ and $\tilde{a} = 440$ ([17]).

We describe here the results of numerical experiments done with this set of parameters, $\beta = 1$ and γ in the range $[20,100]$, the spatial domain being 1-dimensional: $\Omega =]0,1[$. The initial distribution being $s(x,0) = \tilde{s}$, $a(x,0) = \tilde{a}$, during a short time interval, the concentration of S is reduced at point $x = 0$. For example $s(0,t) = 0.01$ for $0 < t < 0.1$. This can be the result of a strong reaction of disappearance of S localized in space and time and represents a signal.

For $t > 0.1$ this excitation is suppressed and the boundary condition again is $s_x(0,t) = 0$.

The response of the system to this excitation takes place in 4 epochs (Fig. 4a):

Stage 1

At the beginning, as $s(0,t)$ is constrained to be small, the phase plane portrait is given by the dotted line PQ in Figure 4a. The corresponding profile of concentration of S (Fig. 4b) undergoes an abrupt change across a wave front which rapidly moves from $x = 0$ to $x = 1$.

At the end of this first stage, all the points $(s(x),a(x))$, $0 < x < 1$, are roughly represented by a same point R, corresponding to uniform profiles of concentration for S and A, the S profile being very low.

The subsequent evolution of the system is similar to the recovery of the stable steady state $(\tilde{s},\tilde{a})$ by the homogeneous dynamical system :

$$\frac{ds}{dt} + \gamma(F(s,a) - (s_o - s)) = 0, \quad \frac{da}{dt} + \gamma(F(s,a) - \alpha(a_o - a)) = 0,$$

the profiles of S and A remaining uniform. The larger the γ, the faster the recovery. (In the first stage too, the larger the γ, the faster the wave front propagation). With $\gamma = 20$, the respective durations of the 4 stages are about $T_1 = 0.18$, $T_2 = 1.1$, $T_3 = 0.5$, $T_4 = 2.9$.

Stage 2

The representative point moves slowly from R to T through S. $F(s,a) - (s_o - s)$ is slightly negative, very close to zero, and $F(s,a) - \alpha(a_o - a)$ is frankly positive, so that the point moves southward, nearly lying on the isocline $\frac{ds}{dt} = 0$. When it arrives in the neighborhood of the turning point C, it moves faster and faster, crosses the isocline $\frac{da}{dt} = 0$ at point T, and enters the next stage.

Stage 3

Now $\frac{ds}{dt}$ is much larger than $\frac{da}{dt}$, and the point moves rapidly from T to V through U. At point V it crosses the isocline $\frac{ds}{dt} = 0$ and the trajectory enters the region where $\frac{ds}{dt} < 0$, $\frac{da}{dt} > 0$.

Stage 4

The point moves slowly from V to P by following a path very close to the S isocline.

We compared with the propagation of the same signal in the system without reaction:

$s_t - s_{xx} + \lambda(s - \tilde{s}) = 0$

$s(0,t) = 0.01$ for $0 < t < 0.1$, $s_x(0,t) = 0$ for $t > 0.1$

$s_x(1,t) = 0$, $s(x,0) = \tilde{s}$

Fig. 5 represents the responses of both systems: diffusion alone is unable to transmit from one end to another a wave front created by an excitation limited in time.

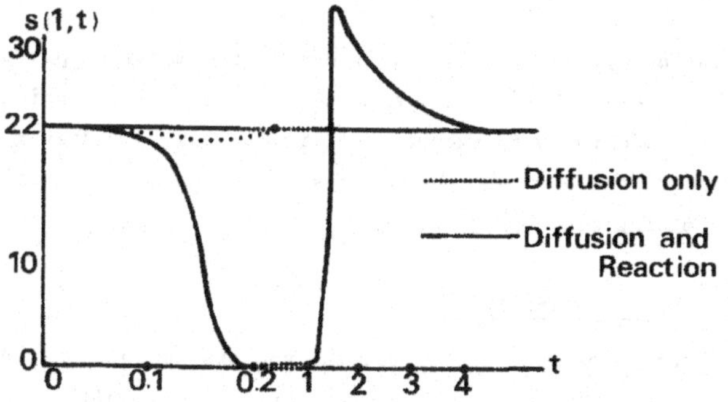

Fig. 5

Such a model could possibly help in understanding the dissemination of information in systems like slime molds.

3. SOME REMARKS ABOUT PATTERN FORMATION IN THE s-a SYSTEM

3.1 Relaxation times

Three different flows play a role in pattern formation in the s-a system: flow in the plane (longest relaxation time L^2/D_s), flow from the reservoir (relaxation times P_s^{-1} and P_a^{-1}) and flow through the chemical reaction (kinetic relaxation time V_M^{-1}). Let σ be the dimensionless parameter $\sigma = \gamma\rho = (L^2/D_s)/V_M^{-1}$ (square of the Thiele modules). Othmer has shown ([35]) that if σ is small enough then all the spatial non-uniformities decay exponentially in time. This occurs if the quantity L^2/D_s is small enough with respect to V_M^{-1}. As $\gamma = (L^2/D_s)/P_s^{-1}$, the analysis of section 1.4 yields precise conditions on L^2/D_s for the system in the spatially-uniform steady state to remain there in the face of small random concentration disturbances.

3.2 Pattern formation on the surface of a sphere as the radius increases

We shall not pursue an analysis of multiple eigenvalues here. It suffices to say that the treatment of homogeneity breakdown on the surface of a sphere increasing in size, which was proposed by Turing to account for gastrulation in blastula [33], is the same as on a planar surface.

In the stability analysis of the basic steady-state $(\tilde{s},\tilde{a})$, we only have to take :

$$\Delta = \frac{1}{\sin^2\theta}\,\frac{\partial^2}{\partial\phi^2} + \frac{1}{\sin\theta}\,\frac{\partial}{\partial\theta}\left(\sin\theta\,\frac{\partial}{\partial\theta}\right)$$

where θ and ϕ are spherical polar coordinates on the surface of the sphere, and:

$$\mu_n = n(n+1), \quad w_{n,m} = P_n^m(\cos\theta)e^{im\phi}, \quad -n \le m \le n$$

where $w_{n,m}$ are surface harmonics of degree n.

For the bifurcation analysis we can again employ the modified perturbation method as indicated by H.B. Keller [34]. This example is a good illustration of the spatial filtering effect associated to the onset of instability: whatever the irregularities giving rise to the new equilibrium, the direction of the axis of the gastrula can vary, but nothing else ([33]).

3.3 Exploration of complexity

For different choices of the parameters, the s-a system shows not only pattern formation or wave propagation, as we have seen, but also globally synchronized oscillation, asynchronous oscillation or hysteresis. If we wish to explore such a complexity, we need efficient numerical algorithms.

Up to now our study of pattern formation has consisted in obtaining stable steady states as limits of time evolutions, imposing random perturbations to those steady states to test their stability. Such a procedure is computer-time consuming and gives only the stable steady states. In order to obtain the whole family of steady states, with both the stable and unstable branches, we need algorithms able to follow curves with turning points and bifurcation points. Before using those algorithms on large systems, it is better to test them on simple models, like the two-cells system of [1] or like the following [6].

3.4 A simple dissipative structure

It is a model of 2 coupled identical cells with only one chemical species S. In each of the cells the kinetics are substrate inhibited:

$F(s) = \rho s/(1 + s + ks^2)$

Furthermore, the cells are submerged in a well-stirred bath at specified concentration λ.

The equations of change for the concentrations are:

$$\frac{ds_1}{dt} = -2s_1 + s_2 - \rho F(s_1) + \lambda$$

$$\frac{ds_2}{dt} = s_1 - 2s_2 - \rho F(s_2) + \lambda$$

This system is an example of more complex cellular networks used to model an assemblage of living cells. ([1], [6], [28], [33]). Depending on λ, multiple stable steady states can occur, solution of:

$$\begin{cases} -2s_1 + s_2 - \rho F(s_1) + \lambda = 0 \\ s_1 - 2s_2 - \rho F(s_2) + \lambda = 0 \end{cases} \tag{34}$$

Some of them are characterized by asymmetric profiles of concentration ($s_1 \neq s_2$).

Let us denote by R the total reaction rate:

$R = F(s_1) + F(s_2)$

A graph of R versus λ, for k = 1 and ρ = 100, is as shown in Figure 6:

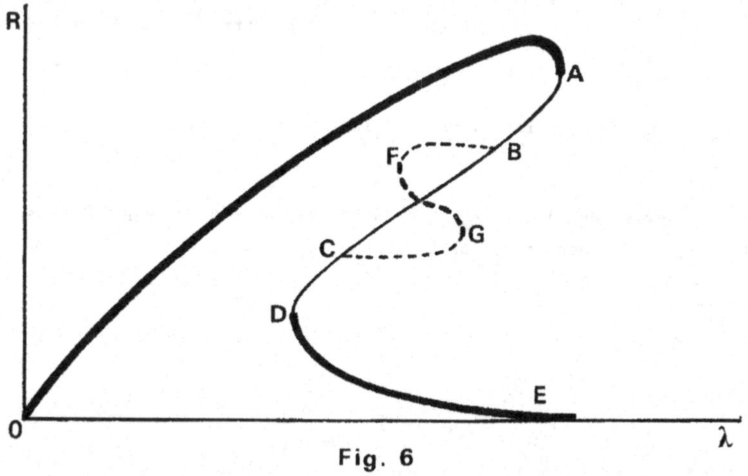

Fig. 6

There is a "trivial branch" OABCDE of "symmetric profiles" ($s_1=s_2=s$) such that:

$$s + \rho F(s) = \lambda \tag{35}$$

This branch presents 2 turning points A and D, and 2 bifurcation points B and C. The corresponding values of λ are given by (35), where s is such that $1 + \rho F'(s) = 0$ for A and D, and $3 + \rho F'(s) = 0$ for B and C. The reason is that $1 + \rho F'(s)$ and $3 + \rho F'(s)$ are the eigenvalues of the matrix of the linearization of (34) around $s_1 = s$, $s_2 = s$. Branches of "asymmetric profiles" ($s_1 \neq s_2$) bifurcate from the trivial branch at points B and C. In figure 6 two non-uniform steady states such as ($s_1 = \alpha, s_2 = \beta$) and ($s_1 = \beta$, $s_2 = \alpha$) are represented by a same point.

Solid lines denote loci of solutions which yield symmetric profiles; dashed lines denote asymmetric profiles, heavy lines denote stable solutions; light lines denote unstable solutions.

A small imperfection may appear in the right hand side of equations (34):

$$\begin{aligned} -2s_1 + s_2 - \rho F(s_1) + \lambda &= 0 \\ s_1 - 2s_2 - \rho F(s_2) + \lambda + \varepsilon &= 0 \end{aligned} \tag{36}$$

As ε departs from zero, a closed locus of solutions (Fig. 7a, dashed curve) appears which corresponds to the internal concentration difference having the opposite sign from that between the reservoirs and which is disconnected from the remaining solutions. The closed locus derives from a displacement from both the symmetric and asymmetric branches of the bifurcated solutions for $\varepsilon = 0$.

The former branch disappears from the main locus (solid curve), and on the other branch only one of two originally equivalent asymmetric mirror-image solutions remains.

Let (λ_o, s_o) be a bifurcation point on the trivial branch of (34):

$$3 + \rho F'(s_o) = 0$$

Let, in (36), ε be small and λ near λ_o. Let s be near s_o and corresponding to λ by (35). Let $\tilde{s}_1 = s_1 - s$. Then it can be shown that

$$s - s_o \sim - \frac{1}{2\beta F''(s_o)} \quad \frac{\varepsilon}{\tilde{s}_1} \qquad \text{(Fig. 7)}$$

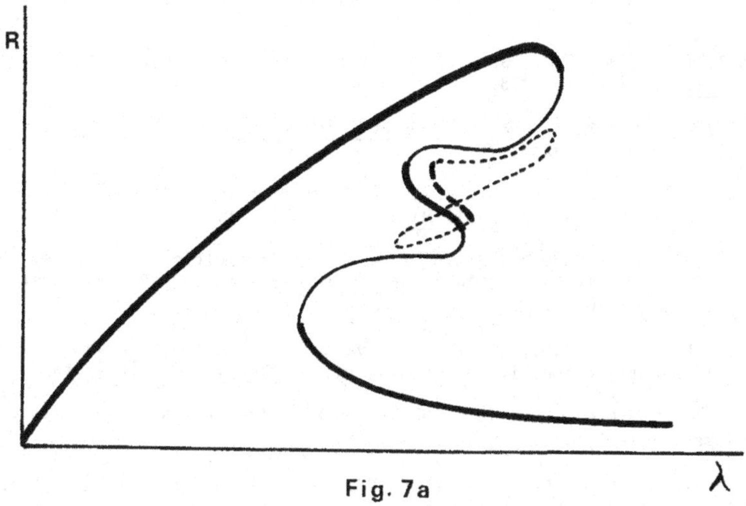

Fig. 7a

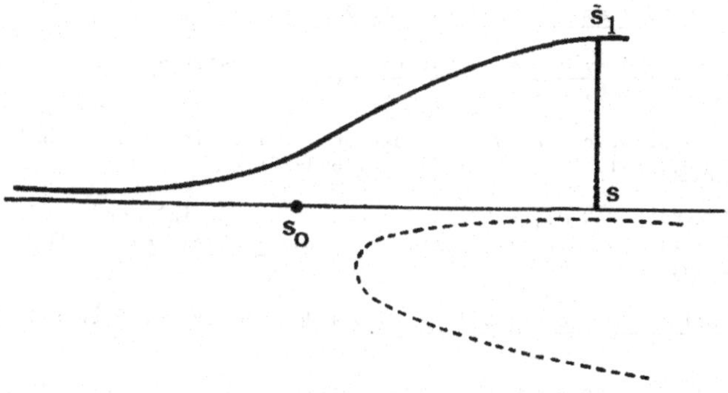

Fig. 7b

CONCLUSION

It has been shown that realistic biochemical kinetics can give rise to stable patterns of concentration or propagating wave fronts. We certainly do not propose that urate oxidase is specifically responsible for morphogenesis in the fruitfly or signal transmission. Rather, we suggest that the phenomena of diffusion limitation combined with the regulatory properties inherent in enzymes are universal characteristics of living systems. Thus, it seems reasonable to suppose that phenomena such as those which we have demonstrated here theoretically might well play a role in many biological processes.

REFERENCES

[1] ASHKENAZI M., OTHMER H.G., Spatial patterns in coupled biochemical oscillators, J. Math. Biol., Vol. 5, 1978, pp. 305-350.

[2] BATHE K.J., WILSON E.L., Numerical methods in finite element analysis, Englewood Cliffs, Prentice Hall, 1976.

[3] BOA J.A., COHEN D.S., Bifurcation of localized disturbances in a model biochemical reaction, Siam. J. Appl. Math., Vol. 30, N°1, 1976.

[4] BRAUNER C.M., NICOLAENKO B., Singular perturbation, multiple solutions, and hysteresis in a nonlinear problem, Lect. Notes in Math., N°594, Springer Verlag, 1977, pp. 50-76.

[5] BRITTON N.F., MURRAY J.D., Threshold wave and cell-cell avalanche behavior in a class of substrate inhibition oscillators, J. Theor. Biol.(in press).

[6] BUNOW B., COLTON C.K., Substrate inhibition kinetics in assemblages of cells, Biosystems , 7,1975,160-171.

[7] FIFE P.C., Pattern formation in reacting and diffusing systems, J. Chem. Phys. Vol. 64, 1976, pp. 554-564.

[8] FIFE P.C., Singular perturbations and wave front techniques in reaction-diffusion problems, Siam-AMS Proceedings, Vol. 10, 1976, pp. 23-50.

[9] FIFE P.C., Stationary patterns for reaction-diffusion equations, MRC Technical Summary Report, 1976, pp. 1-50.

[10] FIFE P.C., Asymptotic states for equations of reaction and diffusion, Bull. Am. Math. Soc., Vol. 84, N°5, 1978, pp. 693-726.

[11] HERSCHKOWITZ-KAUFFMAN M., NICOLIS G., Localized spatial structures and nonlinear chemical waves in dissipative systems, J. Chem. Phys., Vol. 56, 1972, pp. 1890-1895.

[12] IOOSS G., Bifurcation et stabilité, Cours de 3ème Cycle, Université de Paris XI, 1972.

[13] KAUFFMAN S.A., SHYMKO R.M., TRABERT K., Control of sequential compartment formation in Drosophila, Science, Vol. 199, 1978, pp. 259-270.

[14] KERNEVEZ J.P., THOMAS D., Numerical analysis and control of some biochemical systems, Appl. Math. and Opt., Vol. 1, N°3, 1975.

[15] KERNEVEZ J.P., DUBAN M.C., JOLY G., THOMAS D., Hysteresis, oscillations and morphogenesis in immobilized enzyme systems in "The significance of nonlinearity in the natural sciences" , Ed. Perlmutter and Scott, Plenum, 1977, pp.327-353.

[16] KERNEVEZ J.P., JOLY G., DUBAN M.C., BUNOW B., THOMAS D., Hysteresis, oscillations and pattern formation in realistic immobilized enzyme systems, J. Math. Biol., 7,1979,41-56.

[17] KERNEVEZ J.P., MURRAY J.D., JOLY G., DUBAN M.C., THOMAS D., Propagation d'onde dans un système à enzyme immobilisée, C.R.A.S., 287-A,1978,961-964.

[18] KOGELMAN S., KELLER J.B., Transient behavior of unstable nonlinear systems with applications to the Benard and Taylor problems, Siam. J. Appl. Math., Vol. 20, N°4, 1971, pp. 619-637.

[19] MATKOWSKY B.J., A simple nonlinear dynamic stability problem, Bull. Amer. Math. Soc., Vol. 76, 1970, pp. 620-625.

[20] MARSDEN J.E., MAC CRACKEN M., The Hopf bifurcation and its applications, Applied Mathematical Sciences 19, Springer Verlag, New York, 1976.

[21] MEURANT G., SAUT J.C., Bifurcation and stability in a chemical system, J. Math. Anal. and Appl., Vol. 59, 1, 1977, pp. 69-92.

[22] MIMURA M., MURRAY J.D., Spatial structures in a model substrate-inhibition reaction diffusion system,Z. für Natürfosch,33C,1978,580-586.

[23] MURRAY J.D., Nonlinear differential equation models in biology, Clarendon, Oxford,1977.

[24] NAPARSTEK A., ROMETTE J.L., KERNEVEZ J.P., THOMAS D., Memory in enzyme membranes, Nature , Vol. 249, 1974, p. 490.

[25] NICOLIS G., PRIGOGINE I., Self-organization in nonequilibrium systems, Wiley Interscience, 1977.

[26] ORTOVELA P., ROSS J., Theory of propagation of discontinuities in kinetic systems with multiple time scales: front, front multiplicity, and pulses, J. Chem. Phys., Vol. 63, N°8, 1975, pp. 3398-3408.

[27] OTHMER H.G., Nonlinear wave propagation in reacting systems, J. Math. Biol., Vol. 2, 1975, pp. 133-163.

[28] OTHMER H.G., SCRIVEN L.E., Instability and dynamic pattern in cellular networks J. Theor. Biol., Vol. 32, 1971, pp. 507-537.

[29] SATTINGER D.H., Topics in stability and bifurcation theory, Lect. Notes in Math., N° 309, Springer Verlag, 1973.

[30] SPREY T.H., Aldehyde oxidase distribution in the imaginal disks of some diptera, Wilhelm Roux's Archives 183, 1-15 (1977).

[31] THOMAS D., BARBOTIN J.N., DAVID A., HERVAGAULT J.F., ROMETTE J.L., Experimental evidence for a kinetic and electrochemical memory in enzyme membranes, Proc. Natl. Sci. USA, Vol. 74, N°12, 1977, pp. 5314-5317.

[32] THOMAS D., BROUN G., Artificial enzyme membranes, Methods in Enzymology, Vol. 44, 1976, pp. 901-929.

[33] TURING A.M., The chemical basis of morphogenesis, Phil. Trans. Roy. Soc., Vol. B237, 1952, pp. 37-72.

[34] KELLER H.B., Perturbation theory, Notes on a series of six lectures presented at the Department of Mathematics, Michigan State University, East Lansing, Michigan, 1968, pp. 1-64.

[35] OTHMER H.G., Current problems in pattern formation, S.A. Levin, ed., Lectures on Mathematics in the Life Sciences, Vol. 9 : Some Mathematical Questions in Biology VIII (1976), pp. 57-85.

[36] IOOSS, personal communication.

VARIATION D'UN POINT DE RETOURNEMENT
PAR RAPPORT AU DOMAINE

-=-=-=-=-=-=-

F. MIGNOT*, F. MURAT**, J.P. PUEL***

ABSTRACT.-

Let Ω be a bounded and regular domain of $\mathbb{R}^N$, and Γ be its boundary.

For positive λ we consider the problem

$$(0.1)_\lambda \qquad \begin{cases} -\Delta u = \lambda\, e^u & \text{in } \Omega \\ u = 0 & \text{on } \Gamma \end{cases} .$$

There exists a maximum value λ^* of the parameter λ with $0 < \lambda^* < +\infty$, such that $(0.1)_\lambda$ has at least one solution u in $H_0^1(\Omega) \cap L^\infty(\Omega)$, for $\lambda \in [0,\lambda^*[$. Moreover, if the dimension N is less than 10, there exists a unique solution $u^* \in H_0^1(\Omega) \cap L^\infty(\Omega)$ of problem $(0.1)_{\lambda^*}$, and the point (λ^*,u^*) is then a turning point.

In this paper, we study the variation of this turning point with respect to the open set Ω, and more precisely we give an expression of the derivative of the turning point with respect to Ω (in a sense which is correctly defined in def. 2.1).

In problem $(0.1)_\lambda$ we could have considered more general 2nd order elliptic operators and other types of positive increasing and convex nonlinearities, but for simplicity's sake we shall restrict ourselves to the particular problem stated above.

* Université de Lille I
** CNRS et Université Paris VI
*** Université de Nancy II

0 - INTRODUCTION.-

Etant donné un ouvert Ω borné, régulier, connexe de $\mathbb{R}^N$, l'ensemble des λ réels tels que le problème

$$(0.1)_\lambda \qquad \begin{cases} -\Delta u = \lambda e^u & , \quad \text{dans } \Omega , \\ u = 0 & , \quad \text{sur } \Gamma = \partial\Omega , \end{cases}$$

ait au moins une solution u dans $H_0^1(\Omega) \cap L^\infty(\Omega)$ admet une borne supérieure λ^* finie.

A ce λ^* est associé, si $N < 10$, un unique $u^* \in H_0^1(\Omega) \cap L^\infty(\Omega)$ solution de $(0,1)_{\lambda^*}$. L'ensemble des solutions (λ,u) de $(0.1)_\lambda$ présente en (λ^*,u^*) un point de retournement (cf. figure 1).

Notre but est ici d'étudier le comportement de ce point de retournement en fonction de Ω, et, plus précisément, de donner une expression de sa dérivée -dans un sens que nous définirons- par rapport à Ω.

Plus généralement nous pouvons considérer des problèmes du type

$$(0.2)_\lambda \qquad \begin{cases} Au = \lambda f(u) + g & , \quad \text{dans } \Omega , \\ u = 0 & , \quad \text{sur } \Gamma , \end{cases}$$

avec un opérateur A linéaire elliptique du deuxième ordre, symétrique, à coefficients réguliers, une fonction g positive appartenant à $L^\infty(\Omega)$ et une non linéarité f qui est en gros positive, de classe C^2 et strictement convexe. Par exemple on peut prendre $f(u) = u^m$ ($u \in \mathbb{R}^+$, $m > 1$) ou $f(u) = \frac{1}{(1-u)^k}$ ($u \in [0,1[$, $k > 0$) . (Les hypothèses précises sur A , f et g sont données dans MIGNOT-PUEL [1]).

Il existe alors un λ^* fini tel que le problème $(0.2)_\lambda$ ait au moins une solution pour $\lambda \in [0,\lambda^*[$. Moyennant une restriction sur la dimension -dépendant de la non-linéarité f- Le problème $(0.2)_{\lambda^*}$ présente un point de retournement.

Les démonstrations que nous allons faire concernant la variation par rapport au domaine, du point de retournement pourraient être effectuées dans ce cadre général ; cependant, pour simplifier l'exposé, nous nous limiterons au cas où $f(u) = e^u$.

Le plan de l'article est le suivant :

1. Rappels sur les points de retournement.
2. Variations du domaine.
3. Enoncé des résultats.
4. Démonstrations.

1 - RAPPELS SUR LES POINTS DE RETOURNEMENT.-

Dans ce paragraphe, nous rappelons les principaux résultats relatifs aux points de retournement que nous utiliserons.

Ce type de problèmes a été étudié par de nombreux auteurs, et en particulier sous l'angle qui nous intéresse ici par CRANDALL-RABINOWITZ [1] et BANDLE [1] . On en trouvera un exposé simple par exemple dans MIGNOT-PUEL [1] .

Soit Ω un ouvert de $\mathbb{R}^N$. Nous supposerons que :

(1.1) $\left\{\begin{array}{l}\Omega \text{ est un ouvert borné connexe, dont la frontière } \partial\Omega \text{ est} \\ \text{une variété de classe } C^2 \text{ , } \Omega \text{ étant localement d'un seul} \\ \text{côté de } \partial\Omega \text{ .}\end{array}\right.$

Nous désignerons par n la normale extérieure à Ω .

Soit p un réel fini assez grand $(p > \frac{N}{2})$ tel que (inclusion de Sobolev) :

$$W^{2,p}(\Omega) \subset L^\infty(\Omega) \quad .$$

Considérons le problème : Trouver les couples (λ,u) solutions de :

$$(1.2)_\lambda \qquad \begin{array}{l} \lambda \in \mathbb{R} \ , \ \lambda \geqslant 0 \ , \ u \in H_0^1(\Omega) \\ -\Delta u = \lambda\, e^u \quad \text{dans } \Omega \ . \end{array}$$

Notons que si u est solution de $(1.2)_\lambda$, il est équivalent de dire que $u \in H_0^1(\Omega) \cap L^\infty(\Omega)$ ou $u \in H_0^1(\Omega) \cap W^{2,p}(\Omega)$.

On démontre les résultats suivants :

Théorème 1.1.-

Il existe $\lambda^* > 0$, fini, tel que :

(i) Si $0 \leqslant \lambda < \lambda^*$, le problème $(1.2)_\lambda$ admet au moins une solution $u \in H_0^1(\Omega) \cap W^{2,p}(\Omega)$. Il admet de plus une solution minimum notée $u_1(\lambda)$. L'application $\lambda \to u_1(\lambda)$ est croissante et régulière.

(ii) Si $\lambda > \lambda^*$, le problème $(1.2)_\lambda$ n'admet pas de solution $u \in H_0^1(\Omega)$ vérifiant $e^u \in L^\infty(\Omega)$.

Théorème 1.2.- (Etude du cas $\lambda = \lambda^*$) .

(i) Si λ tend vers λ^* par valeurs inférieures, $u_1(\lambda)$ tend vers u^* dans $H_0^1(\Omega) \cap W^{2,q}(\Omega)$, $\forall\, q$, $1 \leqslant q < 5$ et u^* vérifie

$$(1.3) \qquad \begin{cases} u^* \in H_0^1(\Omega)\ , \ e^{u^*} \in L^q(\Omega) \cap H^{-1}(\Omega)\ , \ \forall\, q\ , \ 1 \leqslant q < 5\ , \\ -\Delta u^* = \lambda^* e^{u^*} \quad \text{dans } \Omega\ . \end{cases}$$

(ii) Le problème $(1.2)_{\lambda^*}$ admet au plus une solution $u^* \in H_0^1(\Omega) \cap L^\infty(\Omega)$.

Comme $W^{2,5}(\Omega)$ est inclus dans $L^\infty(\Omega)$ si $N < 10$, nous pouvons préciser le théorème 1.2 par le :

Théorème 1.3.-

On suppose $N < 10$. Alors :

(i) La limite u^* de $u_1(\lambda)$ appartient à $W^{2,p}(\Omega) \subset L^\infty(\Omega)$.

(ii) La première valeur propre du problème

$$\begin{cases} -\Delta v = \mu\, e^{u^*} v & , \\ v \in H_0^1(\Omega) & , \end{cases}$$

est λ^* , c'est-à-dire qu'il existe ϕ^* unique tel que :

(1.4) $$\begin{cases} \phi^* \in H_0^1(\Omega) \quad , \quad \phi^* \geq 0 \quad \text{dans } \Omega \; , \\ -\Delta\phi^* = \lambda^* e^{u^*} \phi^* \quad , \text{ dans } \Omega \; , \\ \int_\Omega e^{u^*}(\phi^*)^2 \, dx = 1 \quad . \end{cases}$$

iii) La valeur λ^* est <u>caractérisée</u> par l'existence de $(u^*,\phi^*) \in (H_0^1(\Omega) \cap L^\infty(\Omega)) \times H_0^1(\Omega)$ solution de (1.3) et (1.4).

iv) L'ensemble des solutions de $(1.2)_\lambda$ peut-être paramétré dans un voisinage de (λ^*,u^*) dans $\mathbb{R} \times (H_0^1(\Omega) \cap W^{2,p}(\Omega))$ par :

$$\lambda(t) = \lambda^* - \gamma\, t^2 + o(t^2) \qquad (\gamma > 0)$$
$$u(t) = u^* + t\, \phi^* + o(t)$$
$$t \in \left]-\varepsilon,+\varepsilon\right[$$

la branche $u_1(\lambda)$ correspond à $t \leq 0$.

La figure ci-dessous donne l'allure de la courbe (λ,u) au voisinage de (λ^*,u^*)

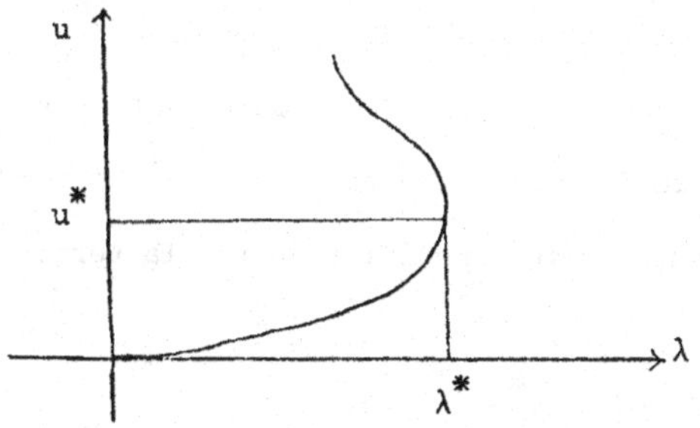

figure 1

Notons que comme $u^* \in L^\infty(\Omega)$, la solution ϕ^* de (1.4) appartient à $W^{2,p}(\Omega)$.

Remarque 1.1.-

La restriction sur la dimension, $N < 10$, est nécessaire, pour avoir $u^* \in L^\infty(\Omega)$ comme le montre un contre exemple de JOSEPH-LUNDGREN [1] .

Ces résultats nous permettent maintenant de donner la définition suivante qu'ils justifient :

Définition 1.1.-

Soit Ω un ouvert de $\mathbb{R}^N$ $(N < 10)$ vérifiant (1.1) . On appelle point de retournement le triplet (λ^*, u^*, ϕ^*) unique solution du système :

(1.5) $\lambda^* \in \mathbb{R}$, $0 < \lambda^* < +\infty$,

(1.6) $u^* \in H_0^1(\Omega) \cap W^{2,p}(\Omega)$,

(1.7) $\phi^* \in H_0^1(\Omega) \cap W^{2,p}(\Omega)$, $\phi \geq 0$ dans Ω ,

(1.8) $-\Delta u^* = \lambda^* e^{u^*}$, dans Ω ,

(1.9) $-\Delta \phi^* = \lambda^* e^{u^*} \phi^*$, dans Ω ,

(1.10) $\int_\Omega e^{u^*} (\phi^*)^2 \, dx = 1$.

Remarque 1.2.-

On désigne généralement par point de retournement le seul couple (λ^*, u^*) . Pour la commodité de notre exposé où ϕ^* intervient de façon essentielle, nous avons préféré modifier un peu la terminologie.

2 - VARIATION DU DOMAINE.-

Nous nous intéressons à la variation du point de retournement $(\lambda^*_\Omega, u^*_\Omega, \phi^*_\Omega)$ par rapport au domaine Ω sur lequel le problème est posé.

Il s'agit d'un problème du type "optimum design" ou "contrôle par un domaine géométrique". Une des principales difficultés de ce type de problèmes est de trouver un cadre convenable pour exprimer l'idée qu'un domaine est proche d'un autre. Cette difficulté peut être levée de plusieurs façons : on peut utiliser pour définir des domaines voisins de Ω :

(i) des accroissements portés sur la normale : cette méthode a été introduite par HADAMARD [1] .

(ii) une représentation par un difféomorphisme proche de l'identité : c'est la méthode des variations intérieures due à GARABEDIAN-SCHIFFER [1] .

(iii) des variations de la fonction caractéristique de l'ouvert, mesurées en norme $L^1(\mathbb{R}^N)$: cette méthode a été introduite par CEA-GIOAN-MICHEL [1] .

En utilisant la méthode d'Hadamard, BANDLE [1] a étudié la continuité par rapport au domaine de la solution minimum $u_1(\lambda)$ du problème $(1.2)_\lambda$ pour λ fixé $(\lambda < \lambda^*)$ (voir théorème 1.1), ainsi que la dérivabilité de l'énergie associée à ce problème. Nous utiliserons ici la méthode des variations intérieures. Pour une présentation des problèmes d'optimum design dans ce cadre, on pourra par exemple consulter MURAT-SIMON [1] (version simple) et [2] (version détaillée), ainsi que SIMON [1] .

Nous souhaitons donc étudier la continuité et la dérivabilité de "l'application"

$$\Omega \to (\lambda^*_\Omega, u^*_\Omega, \phi^*_\Omega)$$

dont ni la source ni le but ne sont des variétés.

Pour remédier à cette difficulté considérons un ouvert Ω

vérifiant (1.1) et définissons pour $\Theta \in C^{2,b}(\mathbb{R}^N,\mathbb{R}^N)$ (1)

$$(I + \Theta)\Omega = \{x \in \mathbb{R}^N \mid \exists\, y \in \Omega \quad x = y + \Theta(y)\} \quad .$$

Si Θ est assez petit (dans $C^{2,b}(\mathbb{R}^N,\mathbb{R}^N)$), $(I + \Theta)$ est un difféomorphisme de $\mathbb{R}^N$, d'inverse appartenant également à $C^{2,b}(\mathbb{R}^N,\mathbb{R}^N)$, et l'ouvert $(I + \Theta)\Omega$ vérifie encore (1.1). Le point de retournement $(\lambda^*,u^*,\phi^*)_{(I + \Theta)\Omega}$ est donc parfaitement défini. Nous considérerons $(I + \Theta)\Omega$ comme une variation de Ω et nous nous limiterons à de telles variations de Ω. C'est là l'essence de la méthode des variations intérieures.

Par cette méthode, nous avons remplacé le paramètre Ω par le paramètre Θ qui varie dans un ouvert $\mathcal{O}$ de l'espace de Banach $C^{2,b}(\mathbb{R}^N,\mathbb{R}^N)$. Il en résulte que l'application

$$\Theta \to \lambda^*_{(I + \Theta)\Omega}$$

va de $\mathcal{O}$ dans $\mathbb{R}$, et que l'on peut parler de sa continuité et de sa dérivabilité.

Il n'en est pas de même pour "l'application"

$$\Theta \to u^*_{(I + \Theta)\Omega}$$

car $u^*_{(I + \Theta)\Omega}$ appartient à $H^1_0((I + \Theta)\Omega)$, espace qui varie avec Θ. Nous lèverons cette difficulté de trois manières :

---

(1) On désigne par $C^{2,b}(\mathbb{R}^N,\mathbb{R}^N)$ l'espace des fonctions T de $\mathbb{R}^N$ dans $\mathbb{R}^N$, de classe C^2, uniformément continues et bornées ainsi que leurs dérivées d'ordre 1 et 2, que l'on munit de la norme

$$|||T||| = \sum_{\alpha = 0,1,2} \ \sup_{x \in \mathbb{R}^N} ||D^\alpha T(x)|| \quad .$$

<u>1ère méthode</u> : Soit $\widetilde{u^*_{(I+\Theta)\Omega}}$ la fonction de $H^1_0(\mathbb{R}^N)$ obtenue en prolongeant $u^*_{(I+\Theta)\Omega}$ par 0 en dehors de $(I+\Theta)\Omega$. On peut alors étudier la continuité et la dérivabilité de l'application

$$\Theta \to \widetilde{u^*_{(I+\Theta)\Omega}}$$

qui va de $\mathcal{O}$ dans $H^1_0(\mathbb{R}^N)$.

Remarquons que l'emploi de cette méthode est facile car nous avons un prolongement naturel à cause des conditions de Dirichlet. Dans le cas du problème de Neumann, on ne pourrait l'utiliser aussi simplement.

<u>2ème méthode</u> : Soit ω un ouvert tel que $\bar{\omega} \subset \Omega$. Pour Θ assez petit on a :

$$\omega \subset (I+\Theta)\Omega$$

et la restriction de $u^*_{(I+\Theta)\Omega}$ à ω , notée $u^*_{(I+\Theta)\Omega}\big|_\omega$, est bien définie. On peut alors étudier la continuité et la dérivabilité de l'application

$$\Theta \to u_{(I+\Theta)\Omega}\big|_\omega$$

qui va de $\mathcal{O}$ dans $W^{2,p}(\omega)$.

<u>3ème méthode</u> : Si nous effectuons le changement de variable $(I+\Theta)$ la fonction

$$v(\Theta) = (u^*_{(I+\Theta)\Omega}) \circ (I+\Theta) \tag{2.1}$$

appartient pour chaque Θ à $H^1_0(\Omega) \cap W^{2,p}(\Omega)$. On peut donc étudier la continuité et la dérivabilité de l'application

$$\Theta \to v(\Theta)$$

qui va de $\mathcal{O}$ dans $H^1_0(\Omega) \cap W^{2,p}(\Omega)$.

Les rapports entre ces 3 méthodes sont indiqués par la :

Proposition 2.1.-

Si l'application $\Theta \to v(\Theta)$ est dérivable en $\Theta = 0$ de $C^{2,b}(\mathbb{R}^N,\mathbb{R}^N)$ à valeurs dans $H_0^1(\Omega) \cap W^{2,p}(\Omega)$, alors :

(i) L'application $\Theta \to \widetilde{u^*_{(I+\Theta)\Omega}}$ est dérivable en $\Theta = 0$ de $C^{2,b}(\mathbb{R}^N,\mathbb{R}^N)$ à valeurs dans $L^2(\mathbb{R}^N)$ et sa dérivée est donnée par[2] :

(2.2) $$\frac{\partial}{\partial\Theta}\left(\widetilde{u^*_{(I+\Theta)\Omega}}\right)(0)\cdot\tau = \left\{\frac{\partial v(\Theta)}{\partial\Theta}(0)\cdot\tau - \operatorname{grad} u^*_\Omega\cdot\tau\right\}^\sim , \quad \forall\, \tau \in C^{2,b}(\mathbb{R}^N,\mathbb{R}^N) .$$

(ii) L'application $\Theta \to u^*_{(I+\Theta)\Omega}\big|_\omega$ est dérivable en $\Theta = 0$ de $C^{2,b}(\mathbb{R}^N,\mathbb{R}^N)$ à valeurs dans $W^{1,p}(\omega)$ et sa dérivée est donnée par[2] :

(2.3) $$\frac{\partial}{\partial\Theta}\left(u^*_{(I+\Theta)\Omega}\big|_\omega\right)(0)\cdot\tau = \left\{\frac{\partial v(\Theta)}{\partial\Theta}(0)\cdot\tau - \operatorname{grad} u^*_\Omega\cdot\tau\right\}\Big|_\omega , \quad \forall\, \tau \in C^{2,b}(\mathbb{R}^N,\mathbb{R}^N) .$$

Cette proposition sera démontrée au paragraphe 4. Notons que $\frac{\partial v(\Theta)}{\partial\Theta}(0)\cdot\tau \in H_0^1(\Omega) \cap W^{2,p}(\Omega)$ mais que $\operatorname{grad} u^*_\Omega\cdot\tau$ n'appartient qu'à $W^{1,p}(\Omega)$ et n'est (en général) pas nul sur $\partial\Omega$, et que donc $\frac{\partial}{\partial\Theta}\left(\widetilde{u^*_{(I+\Theta)\Omega}}\right)(0)\cdot\tau$ n'appartient pas à $H^1(\mathbb{R}^N)$.

La proposition 2.1 justifie la :

Définition 2.1.-

Si l'application $\Theta \to v(\Theta)$ définie par (2.1) est dérivable

---

(2) de façon générale, on désigne par $\tilde{\Phi}$ le prolongement de Φ par zéro en dehors de son domaine de définition, et par $\Phi|_\omega$ la restriction de Φ à un ouvert ω inclus dans le domaine de définition de Φ.

en $\Theta = 0$ de $C^{2,b}(\mathbb{R}^N,\mathbb{R}^N)$ à valeurs dans $H_0^1(\Omega) \cap W^{2,p}(\Omega)$, nous dirons que u_Ω^* est dérivable par rapport à Ω et nous définirons sa dérivée par

(2.4)
$$\left\{ \begin{array}{l} \frac{\partial u_\Omega^*}{\partial \Omega}(\Omega)\cdot\tau = \frac{\partial v(\Theta)}{\partial \Theta}(0)\cdot\tau - \text{grad } u_\Omega^*\cdot\tau \quad , \\ \qquad\qquad\qquad\qquad \forall\, \tau \in C^{2,b}(\mathbb{R}^N,\mathbb{R}^N) \quad . \end{array} \right.$$

Tout ce qui vient d'être dit s'applique évidemment à la fonction propre ϕ_Ω^* . Nous noterons de même :

(2.5)
$$\psi(\Theta) = (\phi^*_{(I+\Theta)\Omega}) \circ (I+\Theta) \quad .$$

Par analogie, nous noterons également

(2.6)
$$\mu(\Theta) = \lambda^*_{(I+\Theta)\Omega}$$

et aussi :

(2.7)
$$\frac{\partial \lambda_\Omega^*}{\partial \Omega}(\Omega)\cdot\tau = \frac{\partial \mu(\Theta)}{\partial \Theta}(0)\cdot\tau \quad , \quad \forall\, \tau \in C^{2,b}(\mathbb{R}^N,\mathbb{R}^N) \quad .$$

Remarque 2.1.-

Dans le début de ce paragraphe, nous nous sommes surtout attachés à donner un sens à la notion de dérivée par rapport à un domaine. Mais la méthode des variations intérieures peut aussi être utilisée pour étudier la continuité.

Considérons en effet l'ensemble des domaines définis à partir d'un ouvert Q vérifiant (1.1) par :

$$\mathcal{D}_Q^2 = \{\Omega \mid \exists\, T \in \mathcal{T}^2 \quad , \quad \Omega = T(Q)\}$$

où $\mathcal{T}^2$ est l'ensemble des bijections de $\mathbb{R}^N$ dans lui-même telles que $(T - I)$ et $(T^{-1} - I)$ appartiennent à $C^{2,b}(\mathbb{R}^N,\mathbb{R}^N)$.

Il est naturel d'introduire sur $\mathcal{D}_Q^2$ la topologie dont une base de voisinages de Ω est définie par :

$$\mathcal{V}_\varepsilon(\Omega) = \{\Omega' \in \mathcal{D}_Q^2 \mid d(\Omega',\Omega) < \varepsilon\}$$

où $d(\Omega',\Omega)$ est définie par :

$$d(\Omega',\Omega) = \inf_{\substack{T \in \mathcal{T}^2 \\ T(\Omega') = \Omega}} \{|||T - I||| + |||T^{-1} - I|||\} .$$

Ceci revient à dire que $\mathcal{D}_Q^2$ n'est autre que le quotient de $\mathcal{T}^2$ par le sous-groupe

$$\mathcal{T}_Q^2 = \{T \in \mathcal{T}^2 \mid T(Q) = Q\}$$

si l'on a muni $\mathcal{T}^2$ de la topologie définie par

$$\begin{cases} \delta(T,I) = |||T - I||| + |||T^{-1} - I||| \\ \delta(T_1,T_2) = \delta(T_1 \circ T_2^{-1},I) \end{cases} .$$

Malheureusement les applications d et δ ne sont pas des distances (car elles ne vérifient pas l'inégalité triangulaire mais seulement une variante de celle-ci). Pour munir $\mathcal{D}_Q^2$ d'une structure d'espace métrique, il faut lever cette difficulté : deux solutions ont été proposées par MICHELETTI [1] et par MURAT-SIMON [2] .

Remarque 2.2.-

Pour donner un sens à la notion de dérivée, nous avons utilisé des variations de Ω écrites sous la forme $(I + \Theta)\Omega$. Cette façon de prendre les choses est en accord avec la topologie que nous venons de définir sur $\mathcal{D}_Q^2$, car il existe deux constantes η (petite) et C telles que :

$$\begin{cases} \forall\, \Theta \in C^{2,b}(R^N,R^N) \quad , \quad |||\Theta||| \leq \eta \quad , \\ |||(I+\Theta)^{-1} - I||| \leq c\; |||\Theta||| \quad . \end{cases}$$

Les ensembles

$$\mathcal{V}'_\varepsilon = \{\Omega' \mid \exists\; \Theta \in C^{2,b}(R^N,\mathbb{R}^N)\; ,\; |||\Theta||| \leq \varepsilon\; ,\; \Omega' = (I+\Theta)\Omega\}$$

forment donc une base de voisinages de Ω dans la topologie de $\mathcal{D}^2_Q$.

Mais l'application $\Theta \to (I+\Theta)\Omega$ n'est pas injective, puisque par exemple si $\Theta = 0$ sur $\partial\Omega$, on a $(I+\Theta)\Omega = \Omega$. Ceci entraîne la propriété générale suivante (cf. MURAT-SIMON [2]) .

Proposition 2.2.-

Soit J une application définie sur $\mathcal{D}^2_Q$, à valeurs dans un espace de Banach E .

Si l'application (composée)

$$\Theta \to (I+\Theta)\Omega \to J((I+\Theta)\Omega)$$

est dérivable (au sens usuel) en $\Theta = 0$ de $C^{2,b}(\mathbb{R}^N,\mathbb{R}^N)$ à valeurs dans E , alors on a :

$$\begin{cases} \dfrac{\partial J((I+\Theta)\Omega)}{\partial\Theta}(0)\cdot\tau = 0 \\ \forall\, \tau \in C^{2,b}(\mathbb{R}^N,R^N) \quad \text{tel que} \quad n\cdot\tau = 0 \quad \text{sur} \quad \partial\Omega \quad . \end{cases}$$

C'est dire que la dérivée de J ne dépend de τ que par la composante normale $n\cdot\tau$ de τ sur $\partial\Omega$, ou encore que si $n\cdot\tau = 0$ sur $\partial\Omega$, $(I+\tau)\Omega$ est égal à Ω "au 2ème ordre près".

3 - ENONCE DES RESULTATS.-

Dans ce paragraphe, nous nous réferons aux résultats et

notations des paragraphes 1 et 2, et notemment à la définition 1.1, aux notations (2.1) , (2.5) et (2.6), et aux notes de bas de pages (1) et (2). Nous supposerons toujours $N < 10$.

Théorème 3.1.-

L'application $\Theta \to (\mu(\Theta), v(\Theta), \psi(\Theta))$ est continue et dérivable en $\Theta = 0$ de $C^{2,b}(\mathbb{R}^N, \mathbb{R}^N)$ à valeurs dans $\mathbb{R} \times (H_0^1(\Omega) \cap W^{2,p}(\Omega))^2$.

Nous n'avons pas explicité dans cet énoncé les équations qui définissent $\frac{\partial v}{\partial \Theta}(0)$ et $\frac{\partial \psi}{\partial \Theta}(0)$. Nous le ferons au cours de la démonstration (paragraphe 4). La dérivée $\frac{\partial \mu}{\partial \Theta}(0)$ est explicitée au Corollaire 3.1. Nous allons compléter ce théorème en donnant l'expression des dérivées "par rapport à Ω " :

Théorème 3.2.-

Le point de retournement $(\lambda_\Omega^*, u_\Omega^*, \phi_\Omega^*)$ est dérivable par rapport à Ω au sens de la définition 2.1 (voir aussi (2.6) et (2.7)) et sa dérivée est définie par :

Pour tout $\tau \in C^{2,b}(R^N, R^N)$, $(\frac{\partial \lambda_\Omega^*}{\partial \Omega}\cdot\tau \, , \frac{\partial u_\Omega^*}{\partial \Omega}\cdot\tau \, , \frac{\partial \phi^*}{\partial \Omega}\cdot\tau)$ est l'unique solution $(\Lambda(\tau)\, , U(\tau)\, , \Phi(\tau))$ du système (3.1) à (3.6) :

(3.1) $\quad (\Lambda(\tau)\, , U(\tau)\, , \Phi(\tau)) \in \mathbb{R} \times (W^{1,p}(\Omega))^2$

(3.2) $\quad -\Delta\, U(\tau) = \Lambda(\tau)\; e^{u^*} + \lambda^*\, e^{u^*}\, U(\tau)$ dans Ω

(3.3) $\quad U(\tau) = -n\cdot\tau\, \frac{\partial u^*}{\partial n}$ sur $\partial\Omega$

(3.4) $\quad -\Delta\, \Phi(\tau) = \Lambda(\tau)\, e^{u^*}\, \phi^* + \lambda^*\, e^{u^*}\, U(\tau)\, \phi^* + \lambda^*\, e^{u^*}\, \Phi(\tau)$ dans Ω

(3.5) $\quad \Phi(\tau) = -n\cdot\tau\, \frac{\partial \phi^*}{\partial n}$ sur $\partial\Omega$

(3.6) $\quad \int_\Omega e^{u^*}\, U(\tau)(\phi^*)^2\, dx + 2\int_\Omega e^{u^*}\, \phi^*\, \Phi(\tau)\, dx = 0$.

Nous verrons dans les démonstrations (paragraphe 4) que l'existence et l'unicité d'une solution du système (3.1) à (3.6) est conséquence de l'alternative de Fredholm. En appliquant celle-ci à (3.2) , (3.3) (ce qui revient à multiplier (3.2) par ϕ^* et (1.9) par $U(\tau)$) on obtient (compte tenu de la notation (2.7)), le :

Corollaire 3.1.-

On a :

$$\text{(3.7)}\quad \begin{cases} \dfrac{\partial\lambda^*_\Omega}{\partial\Omega}\cdot\tau = \Lambda(\tau) = -\dfrac{\displaystyle\int_{\partial\Omega} n\cdot\tau\,\frac{\partial u^*}{\partial n}\,\frac{\partial\phi^*}{\partial n}\,ds}{\displaystyle\int_\Omega e^{u^*}\,\phi^*\,dx} \\ \forall\,\tau\in C^{2,b}(\mathbb{R}^N,\mathbb{R}^N)\ . \end{cases}$$

On déduit immédiatement des théorèmes 3.1 et 3.2, de la proposition 2.1 et de la définition 2.1 le :

Corollaire 3.2.-

Les applications $\Theta\to(\widetilde{u^*_{(I+\Theta)\Omega}}\ ,\ \widetilde{\phi^*_{(I+\Theta)\Omega}})$ et $\Theta\to(u^*_{(I+\Theta)\Omega}\big|_\omega\ ,\ \phi^*_{(I+\Theta)\Omega}\big|_\omega)$ (prolongements de $u^*_{(I+\Theta)\Omega}$ et $\phi^*_{(I+\Theta)\Omega}$ par zéro en dehors de $(I+\Theta)\Omega$, et restrictions de ces fonctions à un ouvert ω tel que $\bar\omega\subset\Omega$) , sont dérivables en $\Theta=0$ de $C^{2,b}(\mathbb{R}^N,\mathbb{R}^N)$ à valeurs dans $(L^2(\mathbb{R}^N))^2$ et dans $(W^{1,p}(\omega))^2$. Leurs dérivées sont données par :

$$\forall\,\tau\in C^{2,b}(\mathbb{R}^N,\mathbb{R}^N)\ ,$$

$$\begin{cases} \dfrac{\partial}{\partial\Theta}\,(\widetilde{u^*_{(I+\Theta)\Omega}})\,(0)\cdot\tau = (\dfrac{\partial u^*_\Omega}{\partial\Omega}\cdot\tau)^\sim = \widetilde{U(\tau)} \\ \dfrac{\partial}{\partial\Theta}\,(\widetilde{\phi^*_{(I+\Theta)\Omega}})\,(0)\cdot\tau = (\dfrac{\partial\phi^*_\Omega}{\partial\Omega}\cdot\tau)^\sim = \widetilde{\Phi(\tau)} \end{cases}$$

$$\begin{cases} \frac{\partial}{\partial\Theta}\,(u^*_{(I+\Theta)\Omega}\big|_\omega)\,(0)\cdot\tau = (\frac{\partial u^*_\Omega}{\partial\Omega}\cdot\tau)\big|_\omega = U(\tau)\big|_\omega \\ \\ \frac{\partial}{\partial\Theta}\,(\phi^*_{(I+\Theta)\Omega}\big|_\omega)\,(0)\cdot\tau = (\frac{\partial \phi^*_\Omega}{\partial\Omega}\cdot\tau)\big|_\omega = \Phi(\tau)\big|_\omega \end{cases}$$

où $(U(\tau)\ ,\ \Phi(\tau))$ est la solution du système (3.1) à (3.6) .

Remarque 3.1. -

L'expression (3.7) de $\frac{\partial\lambda^*_\Omega}{\partial\Omega}\cdot\tau$ est bien définie : en effet le dénominateur est non nul, puisque $u^* \in W^{2,p}(\Omega) \subset L^\infty(\Omega)$, et puisque ϕ^* est positive et non nulle ; d'autre part le numérateur a un sens puisque u^* et ϕ^* sont dans $W^{2,p}(\Omega)$ et que $\tau \in C^{2,b}(\mathbb{R}^N,\mathbb{R}^N)$.

Notons que la dérivée $\frac{\partial\lambda^*_\Omega}{\partial\Omega}\cdot\tau$ ne dépend que de la valeur de $n\cdot\tau$ sur $\partial\Omega$. Il s'agit là d'un résultat général (cf. proposition 2.2).

La formule (3.7) montre que si Ω croit, λ^*_Ω diminue, puisque $\frac{\partial u^*}{\partial n} \geq 0$, $\frac{\partial\phi^*}{\partial n} \geq 0$, $\int_\Omega e^{u^*}\,\phi\,dx > 0$, et puisque dans ce cas $n\cdot\tau \geq 0$. Bien plus, étant donné un ouvert Ω , cette formule indique quelle région de $\partial\Omega$ on doit faire varier pour obtenir la variation maximum de λ^*_Ω . Cette formule permet également d'envisager l'emploi d'une méthode de gradient dans des problèmes de contrôle optimal ou d'identification de λ^*_Ω .

Remarque 3.2. -

Les dérivées calculées dans le théorème 3.2 sont celles que l'on obtient formellement par dérivation des équations du système (1.5) à (1.10).

En effet, en ce qui concerne les équations dans Ω , on obtient (3.2) et (3.4) en dérivant formellement les équations correspondantes (1.8) et (1.9). Quant aux conditions aux limites (3.3) et (3.5) et à l'équation (3.6) elles s'obtiennent en utilisant les règles de dérivation suivantes (formelles mais justifiables) :

$$(3.8) \qquad \frac{\partial}{\partial\Omega}\left(\int_\Omega f_\Omega \, dx\right)\cdot\tau = \int_\Omega \frac{\partial f_\Omega}{\partial\Omega}\cdot\tau \, dx + \int_{\partial\Omega} n\cdot\tau \, f \, ds$$

$$(3.9) \qquad \frac{\partial}{\partial\Omega}\left(\int_{\partial\Omega} g_\Omega \, ds\right)\cdot\tau = \int_{\partial\Omega} \frac{\partial g_\Omega}{\partial\Omega}\cdot\tau \, ds + \int_{\partial\Omega} n\cdot\tau \left\{\frac{\partial g_\Omega}{\partial n} + H g_\Omega\right\} ds \, .$$

(Dans ces règles de dérivation, f_Ω et g_Ω désignent des fonctions dépendant du paramètre Ω ; H est la courbure moyenne de la variété $\partial\Omega$) .

L'équation (3.6) résulte alors immédiatement de la dérivation formelle de (1.10) et du fait que $e^{u^*}(\phi^*)^2 = 0$ sur $\partial\Omega$ car $\phi^* \in H_0^1(\Omega)$. Pour obtenir (3.3) on écrit

$$u_\Omega^* = 0 \quad \text{sur} \quad \partial\Omega \iff \int_{\partial\Omega} u_\Omega^* \, \alpha \, ds = 0 \, , \ \forall\, \alpha \in \mathcal{D}(\mathbb{R}^N) \, .$$

En appliquant (3.9) à cette dernière égalité on obtient :

$$\int_{\partial\Omega} \frac{\partial u_\Omega^*}{\partial\Omega}\cdot\tau \, \alpha \, ds + \int_{\partial\Omega} n\cdot\tau \left\{\frac{\partial}{\partial n}(u_\Omega^* \alpha) + H \, u_\Omega^* \alpha\right\} ds = 0$$

ce qui donne (puisque $u_\Omega^* = 0$ sur $\partial\Omega$) :

$$\int_{\partial\Omega} \left\{\frac{\partial u_\Omega^*}{\partial\Omega}\cdot\tau + n\cdot\tau \, \frac{\partial u_\Omega^*}{\partial n}\right\} \alpha \, ds = 0 \quad , \quad \forall\, \alpha \in \mathcal{D}(\mathbb{R}^N) \quad ,$$

et ce résultat est équivalent à (3.3) . Le même raisonnement permet de déduire formellement (3.5) de la condition aux limites $\phi_\Omega^* = 0$ sur $\partial\Omega$.

Remarque 3.3.-

Le corollaire 3.2 donne des résultats de dérivabilité ; cela implique bien sûr la continuité en $\Theta = 0$ des applications

$$(3.10)\quad \begin{cases} \Theta \to (\widetilde{u^*_{(I+\Theta)\Omega}} \ , \ \widetilde{\phi^*_{(I+\Theta)\Omega}}) \\ \Theta \to (u^*_{(I+\Theta)\Omega} \ , \ \phi^*_{(I+\Theta)\Omega}) \ . \end{cases}$$

On aurait pu étudier ces problèmes de continuité sous un autre angle, en utilisant la topologie de $\mathcal{D}^2_Q$ introduite à la remarque 2.1. Compte tenu de la façon dont cette topologie est définie (par l'intermédiaire des applications d et δ) , la continuité (en $\Theta = 0$) des applications définies par (3.10) entraîne la continuité (en $\Omega = \Omega_0$) des applications

$$\begin{cases} \Omega \to (\widetilde{u^*_\Omega} \ , \ \widetilde{\phi^*_\Omega}) \\ \Omega \to (u^*_\Omega|_\omega \ , \ \phi^*_\Omega|_\omega) \ , \quad \bar{\omega} \subset \Omega \ , \end{cases}$$

où Ω varie dans $\mathcal{D}^2_Q$.

4 - DEMONSTRATIONS.-

DEMONSTRATION DE LA PROPOSITION 2.1.-

Elle repose sur le lemme suivant (pour la démonstration, voir par exemple MURAT-SIMON [2]) :

Lemme 4.1.-

Soit ϕ une application définie sur $C^{2,b}(\mathbb{R}^N,\mathbb{R}^N)$ à valeurs dans $W^{m,p}(\mathbb{R}^N)$ ($m = 1$ ou 2 , $1 \leq p < +\infty$) . On suppose que Φ est dérivable en $\Theta = 0$ de $C^{2,b}(\mathbb{R}^N,\mathbb{R}^N)$ à valeurs dans $W^{m,p}(\mathbb{R}^N)$.

Alors l'application $\Theta \to \Phi(\Theta) \circ (I + \Theta)^{-1}$ qui va de $C^{2,b}(\mathbb{R}^N,\mathbb{R}^N)$ dans $W^{m,p}(\mathbb{R}^N)$ est dérivable en $\Theta = 0$ de $C^{2,b}(\mathbb{R}^N,\mathbb{R}^N)$ à valeurs dans $W^{m-1,p}(\mathbb{R}^N)$ et sa dérivée est donnée par :

$$(4.1)\quad \begin{cases} \dfrac{\partial}{\partial\Theta}(\Phi(\Theta)\circ(I+\Theta)^{-1})(0)\cdot\tau = \dfrac{\partial\Phi(\Theta)}{\partial\Theta}(0)\cdot\tau - \operatorname{grad}\Phi(0)\cdot\tau , \\ \forall\,\tau\in C^{2,b}(\mathbb{R}^N,\mathbb{R}^N) . \end{cases}$$

Démonstration de (i).-

Soit Φ_1 l'application de $C^{2,b}(R^N,R^N)$ à valeurs dans $H^1(R^N)$

$$\Phi_1 : \Theta \to \widetilde{v(\Theta)} \quad ;$$

Φ_1 est la composée de l'application $\Theta \to v(\Theta)$ (dérivable par hypothèse) et du prolongement par 0 en dehors de Ω qui est une application linéaire continue de $H_0^1(\Omega)$ dans $H^1(R^N)$.

D'après le lemme 4.1, l'application $\Theta \to \Phi_1(\Theta)\circ(I+\Theta)^{-1}$ est donc dérivable à valeurs dans $L^2(R^N)$ (ici $m = 1$, $p = 2$) . Mais l'on a :

$$\Phi_1(\Theta)\circ(I+\Theta)^{-1} = \widetilde{v(\Theta)}\circ(I+\Theta)^{-1} = \widetilde{u^*_{(I+\Theta)\Omega}}$$

et d'autre part :

$$\frac{\partial\Phi_1(\Theta)}{\partial\Theta}(0)\cdot\tau = \frac{\partial\widetilde{v(\Theta)}}{\partial\Theta}(0)\cdot\tau = \left(\frac{\partial v(\Theta)}{\partial\Theta}(0)\cdot\tau\right)^{\sim}$$

(commutation de la dérivation avec une application linéaire continue) et

$$\operatorname{grad}\Phi_1(0)\cdot\tau = \operatorname{grad}\widetilde{v(0)}\cdot\tau = (\operatorname{grad} u^*_\Omega\cdot\tau)^{\sim} .$$

De (4.1) et de ces 3 formules on déduit (2.2).

Démonstration de (ii).-

Soit A un ouvert tel que

$$\bar{\omega}\subset A \quad , \quad \bar{A}\subset\Omega$$

et soit α une fonction de $\mathcal{D}(A)$ telle que $\alpha = 1$ sur ω . Considérons l'application Φ_2 de $C^{2,b}(\mathbb{R}^N,\mathbb{R}^N)$ à valeurs dans $W_0^{2,p}(\mathbb{R}^N)$:

$$\Phi_2(\Theta) = (\alpha\ v(\Theta))^{\sim}$$

Φ_2 est la composée de l'application $\Theta \to v(\Theta)$ (dérivable à valeurs dans $W^{2,p}(\Omega)$ par hypothèse) par deux applications linéaires continues : la multiplication par α qui appartient à $L(W^{2,p}(\Omega)\ ;\ W_0^{2,p}(\Omega))$ et le prolongement qui appartient à $L(W_0^{2,p}(\Omega)\ ;\ W^{2,p}(R^N))$. D'après le lemme 4.1, l'application $\Theta \to \Phi_2(\Theta) \circ (I + \Theta)^{-1}$ est donc dérivable à valeurs dans $W^{2,p}(R^N)$. Si on compose avec la restriction à ω , qui appartient à $L(W^{2,p}(R^N)\ ;\ W^{2,p}(\omega))$, on en déduit que $\Theta \to \Phi_2(\Theta) \circ (I + \Theta)^{-1}|_\omega$ est dérivable à valeurs dans $W^{2,p}(\omega)$.

Mais on a, pour Θ assez petit, d'après le choix de α :

$$\Phi_2(\Theta) \circ (I + \Theta)^{-1}|_\omega = (\alpha\ v(\Theta))^{\sim} \circ (I + \Theta)^{-1}|_\omega$$
$$= u^*_{(I + \Theta)\Omega}|_\omega$$

et de (4.1) on déduit simplement (2.3).

DEMONSTRATION DU THEOREME 3.1.-

1) *Changement de variable.*-

Nous noterons grad Φ comme un vecteur colonne ; $(I + \Theta)'$ désignera la dérivée de $(I + \Theta)$, qui en tout point $x \in \mathbb{R}^N$ est une matrice $N \times N$ de coefficients $(\delta_{ij} + \frac{\partial \Theta_i}{\partial x_j})$; la transposée de la matrice inverse sera noté ${}^t(I + \Theta)'^{-1}$. On a alors (pour Θ assez petit) :

$$(\text{grad}\ \Phi) \circ (I + \Theta) = {}^t(I + \Theta)'^{-1}\ \text{grad}(\Phi \circ (I + \Theta)) .$$

D'autre part, quand $\tilde{w}$ décrit $H_0^1((I + \Theta)\Omega)$, $w = \tilde{w} \circ (I + \Theta)$ décrit $H_0^1(\Omega)$.

Compte tenu de ces remarques, on effectue dans la formulation variationnelle du système (1.5) à (1.10) qui définit le point de retournement $(\lambda^*, u^*, \phi^*)_{(I+\Theta)\Omega}$ le changement de variable $(I + \Theta)$. En notant pour $\Theta \in C^{2,b}(\mathbb{R}^N, \mathbb{R}^N)$ (assez petit) :

$$\begin{cases} \mu(\Theta) = \lambda^*_{(I+\Theta)\Omega} & , \\ v(\Theta) = u^*_{(I+\Theta)\Omega} \circ (I+\Theta) & , \\ \psi(\Theta) = \phi^*_{(I+\Theta)\Omega} \circ (I+\Theta) & , \end{cases}$$

on obtient le système :

(4.2) $$\mu(\Theta) \in \mathbb{R} \quad , \quad 0 < \mu(\Theta) < +\infty \quad ,$$

(4.3) $$v(\Theta) \in H^1_0(\Omega) \cap W^{2,p}(\Omega) \quad ,$$

(4.4) $$\psi(\Theta) \in H^1_0(\Omega) \cap W^{2,p}(\Omega) \quad , \quad \psi(\Theta) \geq 0 \text{ dans } \Omega \; ,$$

(4.5) $$\begin{cases} \int_\Omega {}^t(I+\Theta)'^{-1} \operatorname{grad} v(\Theta) \cdot {}^t(I+\Theta)'^{-1} \operatorname{grad} w \, |\det(I+\Theta)'| dx \\ = \int_\Omega \mu(\Theta) \, e^{v(\Theta)} \, w \, |\det(I+\Theta)'| dx \quad , \quad \forall \, w \in H^1_0(\Omega) \; , \end{cases}$$

(4.6) $$\begin{cases} \int_\Omega {}^t(I+\Theta)'^{-1} \operatorname{grad} \psi(\Theta) \cdot {}^t(I+\Theta)'^{-1} \operatorname{grad} w \, |\det(I+\Theta)'| dx \\ = \int_\Omega \mu(\Theta) \, e^{v(\Theta)} \, \psi(\Theta) \, w |\det(I+\Theta)'| dx \quad , \quad \forall \, w \in H^1_0(\Omega) \; , \end{cases}$$

(4.7) $$\int_\Omega e^{v(\Theta)} \, (\psi(\Theta))^2 \, |\det(I+\Theta)'| dx = 1 \quad .$$

2) *Continuité*.-

Le triplet $(\mu(\Theta) , v(\Theta) , \psi(\Theta))$ étant l'unique solution du système (4.2) à (4.7) , la continuité en $\Theta = 0$ de l'application $\Theta \to (\mu(\Theta) , v(\Theta) , \psi(\Theta))$ qui va de $C^{2,b}(\mathbb{R}^N, \mathbb{R}^N)$ dans

$\mathbb{R} \times (H_0^1(\Omega) \cap W^{2,p}(\Omega))$ se démontre de façon classique : on observe d'abord que si Θ est assez petit, les différentes inconnues sont bornées ; on extrait des sous-suites qui convergent faiblement, on identifie leurs limites grâce à l'unicité de la solution du système (4.2) à (4.7) ; on démontre enfin la convergence forte.

3) *Utilisation du théorème de dérivation de la solution d'une équation implicite.* -

Pour démontrer la dérivabilité de $\Theta \to (\mu(\Theta)\ ,\ v(\Theta)\ ,\ \psi(\Theta))$, nous allons utiliser une variante du théorème des fonctions implicites (voir par exemple SCHWARTZ [1]) :

Lemme 4.2. -

Soient E,F,G trois espaces de Banach et soient $\mathcal{U}(0)$ (resp. $\mathcal{V}(\eta_0)$) un voisinage de 0 (resp. η_0) dans E (resp. F).

On considère 2 applications f et g telles que :

(4.8)
$$\begin{cases} f : \mathcal{U}(0) \times \mathcal{V}(\eta_0) \to G \quad , \quad g : \mathcal{U}(0) \to \mathcal{V}(\eta_0) \quad , \\ g(0) = \eta_0 \quad , \quad f(\zeta, g(\zeta)) = 0 \quad , \quad \forall\, \zeta \in \mathcal{U}(0) \quad . \end{cases}$$

(4.9) f est dérivable (Fréchet) au point $(0,\eta_0)$,

(4.10) $\begin{cases} \text{La dérivée partielle } \dfrac{\partial f}{\partial \eta}(0,\eta_0) \text{ est un} \\ \text{isomorphisme de } F \text{ sur } G \quad , \end{cases}$

(4.11) g est continue en $\zeta = 0$.

Alors g est dérivable (Fréchet) en $\zeta = 0$ et sa dérivée est définie par :

$$(4.12)\quad \begin{cases} \dfrac{\partial f}{\partial \zeta}(0,\eta_0)\cdot\tau + \dfrac{\partial f}{\partial \eta}(0,\eta_0)\cdot\dfrac{\partial g}{\partial \zeta}(0)\cdot\tau = 0 \ , \\ \forall\, \tau \in E \ . \end{cases}$$

Notons que cette variante du théorème des fonctions implicites ne suppose pas f dérivable en dehors du point $(0,\eta_0)$. Par contre elle suppose connue a priori l'existence d'une fonction implicite g .

Nous prendrons ici $E = C^{2,b}(\mathbb{R}^N,\mathbb{R}^N)$, $F = \mathbb{R} \times (H_0^1(\Omega) \cap W^{2,p}(\Omega))^2$ et $G = (L^p(\Omega))^2 \times \mathbb{R}$; $\mathcal{U}(0)$ sera un ouvert $\mathcal{O}$ de $C^{2,b}(\mathbb{R}^N,\mathbb{R}^N)$ de Θ assez petits pour que $(I + \Theta)$ soit une bijection ; nous prendrons $\eta_0 = (\mu(0),v(0),\psi(0)) = (\lambda_\Omega^*,u_\Omega^*,\phi_\Omega^*)$ et $\mathcal{V}(\eta_0) = F$.

L'application $f = (f_1,f_2,f_3)$ sera définie par :

$\forall\, \Theta \in \mathcal{O}$, $\forall (\mu,v,\psi) \in \mathbb{R} \times (H_0^1(\Omega) \cap W^{2,p}(\Omega))^2$,

$$(4.13)\quad f_1(\Theta,\mu,v,\psi) = -\mathrm{div}\{|\det(I + \Theta)'|\,(I + \Theta)'^{-1}\,{}^t(I + \Theta)'^{-1}\,\mathrm{grad}\, v\} - \mu\, e^v\, |\det(I + \Theta)'| \ ,$$

$$(4.14)\quad f_2(\Theta,\mu,v,\psi) = -\mathrm{div}\{|\det(I + \Theta)'|\,(I + \Theta)'^{-1}\,{}^t(I + \Theta)'^{-1}\,\mathrm{grad}\, \psi\} - \mu\, e^v\, \psi\, |\det(I + \Theta)'| \ ,$$

$$(4.15)\quad f_3(\Theta,\mu,v,\psi) = \int_\Omega e^v\, \psi^2\, |\det(I + \Theta)'|\, dx - 1 \ .$$

L'application g sera définie par

$$g : \Theta \to (\mu(\Theta) \ , \ v(\Theta) \ , \ \psi(\Theta))$$

où ce triplet est l'unique solution du système (4.2) à (4.7) .

Dans ces conditions l'hypothèse (4.8) est vérifiée puisque (4.13) , (4.14) et (4.15) correspondent à l'interprétation des équations

(4.5) , (4.6) et (4.7) . D'autre part l'hypothèse (4.11) n'est autre que la propriété de continuité que nous avons étudiée au point 2).

Si nous vérifions les hypothèses (4.9) et (4.10), le lemme 4.2 nous donnera la dérivabilité de l'application $\Theta \to (\mu(\Theta), v(\Theta), \psi(\Theta))$ et (4.12) définira la dérivée ; le théorème 3.1 sera alors démontré.

4) *Vérification des hypothèses (4.9) et (4.10).*-

La vérification de l'hypothèse (4.9) est une conséquence du lemme suivant (pour la démonstration voir par exemple MURAT-SIMON [2]) :

Lemme 4.3.-

Les applications $\Theta \to |\det(I + \Theta)'|$ et $\Theta \to (I + \Theta)'^{-1}$ sont dérivables en $\Theta = 0$ de $C^{2,b}(\mathbb{R}^N, \mathbb{R}^N)$ à valeurs dans $C^{1,b}(\mathbb{R}^N, \mathbb{R})$ et $C^{1,b}(\mathbb{R}^N, \mathbb{R}^{N^2})$ et leurs dérivées sont données par :

$$(4.16) \qquad \begin{cases} \forall \tau \in C^{2,b}(\mathbb{R}^N, \mathbb{R}^N) , \\ \dfrac{\partial}{\partial\Theta} (|\det(I + \Theta)'|)(0)\cdot\tau = \operatorname{div} \tau , \\ \dfrac{\partial}{\partial\Theta} ((I + \Theta)'^{-1})(0)\cdot\tau = -\tau' . \end{cases}$$

En effet, une fois connu ce lemme, il suffit de remarquer que f_1, f_2 et f_3 sont définies par des applications multilinéaires continues de $|\det(I + \Theta)'|$, $(I + \Theta)'^{-1}$, μ , v , ψ et e^v (qui est lui-même une fonction dérivable de v) .

Vérification de l'hypothèse (4.10). L'application $\frac{\partial f}{\partial \eta}(0,\eta_0)$ étant continue, il suffit, d'après le théorème de Banach de montrer qu'elle est bijective pour démontrer qu'elle est un isomorphisme. Compte tenu des expressions de la dérivée de f au point $(0,\eta_0) = (0,(\lambda^*_\Omega,u^*_\Omega,\phi^*_\Omega))$, ceci revient à montrer le :

Lemme 4.4.-

Etant donné $h = (h_1,h_2,h_3) \in (L^p(\Omega))^2 \times \mathbb{R}$, il existe une unique solution du système

(4.17) $\bar{\mu},\bar{v},\bar{\psi} \in \mathbb{R} \times (H^1_0(\Omega) \cap W^{2,p}(\Omega))^2$,

(4.18) $-\Delta\bar{v} - \bar{\mu}\, e^{u^*} - \lambda^*\, e^{u^*}\, \bar{v} = h_1$ dans Ω ,

(4.19) $-\Delta\bar{\psi} - \bar{\mu}\, e^{u^*}\, \phi^* - \lambda^*\, e^{u^*}\, \bar{v}\, \phi^* - \lambda^*\, e^{u^*}\, \bar{\psi} = h_2$ dans Ω ,

(4.20) $\int_\Omega e^{u^*}\, \bar{v}(\phi^*)^2\, dx + 2\int_\Omega e^{u^*}\, \phi^*\, \bar{\psi}\, dx = h_3$.

Démonstration du lemme 4.4.-

Elle repose sur l'alternative de Fredholm appliquée à l'opérateur

$$\begin{cases} -\Delta\Phi - \lambda^*\, e^{u^*}\, \Phi \ , \\ \Phi \in H^1_0(\Omega) \ . \end{cases}$$

En effet (cf. théorème 1.3 (ii)) l'équation (4.18) a (au moins) une solution $\bar{v} \in H^1_0(\Omega)$ si et seulement si :

(4.21) $\bar{\mu}\int_\Omega e^{u^*}\, \phi^*\, dx + \int_\Omega h_1\, \phi^*\, dx = 0$

ce qui détermine $\bar{\mu}$ puisque $\int_\Omega e^{u^*}\, \phi^* \neq 0$ (voir remarque 3.1) ; on a alors

(4.22) $$\bar{v} = \bar{v}_0 + s\,\phi^*$$

où $\bar{v}_0$ est déterminée de façon unique et où s est un paramètre réel.

Mais l'équation (4.19) a (au moins) une solution $\bar{\psi} \in H_0^1(\Omega)$ si et seulement si :

$$\bar{\mu} \int_\Omega e^{u^*} (\phi^*)^2 \, dx + \lambda^* \int_\Omega e^{u^*} \bar{v} (\phi^*)^2 \, dx + \int_\Omega h^2 \phi^* \, dx = 0$$

$\bar{v}$ étant donné par (4.22), cette équation détermine s de façon unique puisque $\lambda^* \int_\Omega e^{u^*} (\phi^*)^3 \, dx \neq 0$; on a alors

$$\bar{\psi} = \bar{\psi}_0 + t \, \phi^*$$

où $\bar{\psi}_0$ est déterminée de façon unique et où t est un paramètre réel.

Le paramètre t est alors déterminé de façon unique par l'équation (4.20) puisque $2 \int_\Omega e^{u^*} (\phi^*)^2 \, dx \neq 0$.

Il existe donc une solution unique $(\bar{\mu},\bar{v},\bar{\psi}) \in \mathbb{R} \times (H_0^1(\Omega))^2$ du système (4.18) à (4.20) . En utilisant les résultats de régularité $W^{2,p}(\Omega)$, qui s'appliquent ici puisque $\partial\Omega$ est de classe C^2 , que $(h_1,h_2) \in (L^p(\Omega))^2$ et que $(u^*,\psi^*) \in (L^\infty(\Omega))^2$ on en déduit que cette solution appartient en fait à $\mathbb{R} \times (H_0^1(\Omega) \cap W^{2,p}(\Omega))^2$. C'est ce qu'il fallait démontrer.

5) *Expression des dérivées* $\frac{\partial\mu(\Theta)}{\partial\Theta}(0)$, $\frac{\partial v(\Theta)}{\partial\Theta}(0)$, $\frac{\partial\psi(\Theta)}{\partial\Theta}(0)$.-

Compte tenu des expressions des dérivées de f , et en raison du lemme 4.4, ces dérivées sont définies par (4.12) de la façon suivante :

Pour tout $\tau \in C^{2,b}(\mathbb{R}^N,\mathbb{R}^N)$, $\left(\frac{\partial\mu(\Theta)}{\partial\Theta}(0)\cdot\tau \right.$, $\frac{\partial v(\Theta)}{\partial\Theta}(0)\cdot\tau$, $\left.\frac{\partial\psi(\Theta)}{\partial\Theta}(0)\cdot\tau\right)$ est l'unique triplet $(M(\tau)$, $V(\tau)$, $\Psi(\tau))$ solution du système (4.23) à (4.26)

(4.23) $$(M(\tau)\ ,\ V(\tau)\ ,\ \Psi(\tau)) \in \mathbb{R} \times (H_0^1(\Omega) \cap W^{2,p}(\Omega))^2$$

(4.24) $$\begin{cases} -\Delta V(\tau) - M(\tau)\, e^{u^*} - \lambda^* e^{u^*} V(\tau) \\ \quad - \operatorname{div}\{(\operatorname{div}\tau - \tau' - {}^t\tau')\operatorname{grad} u^*\} - \lambda^* e^{u^*} \operatorname{div}\tau = 0 \quad \text{dans } \Omega\ , \end{cases}$$

(4.25) $$\begin{cases} -\Delta\Psi(\tau) - M(\tau)\, e^{u^*} \phi^* - \lambda^* e^{u^*} V(\tau)\, \phi^* - \lambda^* e^{u^*} \Psi(\tau) \\ \quad - \operatorname{div}\{(\operatorname{div}\tau - \tau' - {}^t\tau')\operatorname{grad} \phi^*\} - \lambda^* e^{u^*} \phi^* \operatorname{div}\tau = 0, \text{ dans } \Omega, \end{cases}$$

(4.26) $$\int_\Omega e^{u^*} V(\tau)(\phi^*)^2\, dx + 2\int_\Omega e^{u^*} \phi^* \Psi(\tau)\, dx + \int_\Omega e^{u^*} (\phi^*)^2 \operatorname{div}\tau = 0\ .$$

Démonstration du corollaire 3.1.-

Elle repose sur l'identité suivante (qui se démontre en régularisant u, en développant chaque membre et en identifiant terme à terme) :

Lemme 4.5.-

Si $u \in H^2(\Omega)$, si $\tau \in C^{2,b}(\mathbb{R}^N,\mathbb{R}^N)$, on a :

(4.27) $$\begin{cases} - \operatorname{div}\{(\operatorname{div}\tau - \tau' - {}^t\tau')\operatorname{grad} u\} + \Delta u \operatorname{div}\tau = \\ \quad = \Delta(\operatorname{grad} u\cdot\tau) - \operatorname{grad}(\Delta u)\cdot\tau\ . \end{cases}$$

D'après le lemme 4.4, le système (4.23) à (4.26) a une solution unique, et $M(\tau)$ est donné par (cf. (4.21)) :

(4.28) $$\begin{cases} M(\tau)\displaystyle\int_\Omega e^{u^*}\phi^*\, dx = -\int_\Omega \operatorname{div}\{(\operatorname{div}\tau - \tau' - {}^t\tau')\operatorname{grad} u^*\}\, \phi^*\, dx \\ \qquad - \lambda^* \displaystyle\int_\Omega e^{u^*} \operatorname{div}\tau\, \phi^*\, dx\ . \end{cases}$$

Grâce au lemme 4.5 et au fait que $-\Delta u^* = \lambda^* e^{u^*}$, le second membre de (4.28), que nous noterons Π, s'écrit :

$$\Pi = + < \Delta(\text{grad } u^* \cdot \tau) , \phi^* > - < \text{grad}(\Delta u^*) , \tau \phi^* >$$

où $<\ ,\ >$ désigne la dualité $H^{-1}(\Omega)$, $H_0^1(\Omega)$.

Mais on a :

$$- < \text{grad}(\Delta u^*) , \tau \phi^* > = < \text{grad}(\lambda^* e^{u^*}) , \tau \phi^* >$$

$$= \int_\Omega \lambda^* e^{u^*} \text{grad } u^* \cdot \tau \phi^* dx$$

$$= -\int_\Omega \Delta \phi^* \text{grad } u^* \cdot \tau dx$$

et donc, grâce à la formule de Green :

$$\Pi = < \Delta(\text{grad } u^* \cdot \tau) , \phi^* > - \int_\Omega \Delta \phi^* \text{grad } u^* \cdot \tau dx$$

$$= << \frac{\partial}{\partial n} (\text{grad } u^* \cdot \tau) , \phi^* >> - \int_{\partial\Omega} \frac{\partial \phi^*}{\partial n} \text{grad } u^* \cdot \tau ds$$

où $<<\ ,\ >>$ désigne la dualité $H^{-1/2}(\partial\Omega)$, $H^{1/2}(\partial\Omega)$. Comme $\phi^* = 0$ sur $\partial\Omega$, et comme $\text{grad } u^* = n \frac{\partial u^*}{\partial n}$ sur $\partial\Omega$ puisque $u^* = 0$ sur $\partial\Omega$, on a finalement :

$$\Pi = - \int_{\partial\Omega} n \cdot \tau \frac{\partial u^*}{\partial n} \frac{\partial \phi^*}{\partial n} ds \quad .$$

Si l'on se rappelle la notation (2.7)

$$\frac{\partial \lambda_\Omega^*}{\partial \Omega} \cdot \tau = \frac{\partial \mu(\Theta)}{\partial \Theta} (0) \cdot \tau = M(\tau) \quad ,$$

on voit qu'on a démontré le corollaire 3.1.

Démonstration du théorème 3.2.-

Le fait que le point de retournement $(\lambda_\Omega^*, u_\Omega^*, \phi_\Omega^*)$ soit dérivable par rapport à Ω au sens de la définition 2.1 résulte de cette définition et du théorème 3.1.

Par définition, les dérivées sont données par

$$\begin{cases} \forall\, \tau \in C^{2,b}(\mathbb{R}^N, \mathbb{R}^N) \quad , \\ \dfrac{\partial \lambda_\Omega^*}{\partial \Omega} \cdot \tau = \dfrac{\partial \mu(\Theta)}{\partial \Theta}(0) \cdot \tau = M(\tau) \quad , \\ \dfrac{\partial u_\Omega^*}{\partial \Omega} \cdot \tau = \dfrac{\partial v(\Theta)}{\partial \Theta}(0) \cdot \tau - \text{grad}\, u^* \cdot \tau = V(\tau) - \text{grad}\, u^* \cdot \tau \quad , \\ \dfrac{\partial \phi_\Omega^*}{\partial \Omega} \cdot \tau = \dfrac{\partial \psi(\Theta)}{\partial \Theta}(0) \cdot \tau - \text{grad}\, \phi^* \cdot \tau = \Psi(\tau) - \text{grad}\, \phi^* \cdot \tau \quad , \end{cases}$$

que nous conviendrons (pour abréger les notations) de désigner par $\Lambda(\tau)$, $U(\tau)$, $\Phi(\tau)$.

Il est évident que $\Lambda(\tau) \in \mathbb{R}$, et que $U(\tau)$, $\Phi(\tau) \in (W^{1,p}(\Omega))^2$ et vérifient les conditions aux limites (3.3) et (3.5) : en effet $\dfrac{\partial v(\Theta)}{\partial \Theta}(0) \cdot \tau \in H_0^1(\Omega) \cap W^{2,p}(\Omega)$ et sur $\partial\Omega$, $\text{grad}\, u^* = n \dfrac{\partial u^*}{\partial n}$.

Montrons maintenant que $\Lambda(\tau)$, $U(\tau)$ vérifient (3.2) (les démonstrations de (3.4) et (3.6) sont analogues). D'après les définitions de $\Lambda(\tau)$ et $U(\tau)$ et d'après (4.24), on a :

$$\begin{cases} -\Delta\, U(\tau) - \Lambda(\tau)\, e^{u^*} - \lambda^*\, e^{u^*}\, U(\tau) = \\ \quad = \text{div}\{(\text{div}\, \tau - \tau' - {}^t\tau')\text{grad}\, u^*\} + \lambda^*\, e^{u^*}\, \text{div}\, \tau \\ \quad + \Delta(\text{grad}\, u^* \cdot \tau) + \lambda^*\, e^{u^*}\, (\text{grad}\, u^* \cdot \tau) \text{ , dans } \Omega \ . \end{cases}$$

D'après le lemme 4.5 et le fait que $-\Delta u^{*} = \lambda^{*} e^{u^{*}}$, le second membre de cette équation est nul, ce qui démontre (3.2) .

Le triplet $(\frac{\partial \lambda^{*}_{\Omega}}{\partial \Omega} \cdot \tau \, , \, \frac{\partial u^{*}_{\Omega}}{\partial \Omega} \cdot \tau \, , \, \frac{\partial \phi^{*}_{\Omega}}{\partial \Omega} \cdot \tau)$ est donc solution du système (3.1) à (3.6) . Ce système a une solution unique, puisque il se déduit par le changement d'inconnues

$$\Lambda(\tau) = M(\tau) \quad ,$$

$$U(\tau) = V(\tau) - \text{grad } u^{*} \cdot \tau \quad ,$$

$$\Phi(\tau) = \Psi(\tau) - \text{grad } \phi^{*} \cdot \tau \quad ,$$

du système (4.23) à (4.27) , qui a lui même une solution unique d'après le lemme 4.4 (alternative de Fredholm). On pourrait d'ailleurs démontrer directement que le système (3.1) à (3.6) a une solution unique, en lui appliquant une alternative de Fredholm convenablement adaptée.

B I B L I O G R A P H I E

C. BANDLE

[1] Existence theorems, qualitative results and a priori bounds for a class of non linear Dirichlet problems. Arch. Rat. Mech. Anal. Vol. 49 (1973), pp. 241-269.

J. CEA - A. GIOAN - J. MICHEL

[1] Quelques résultats sur l'identification de domaine. Calcolo Vol. 10, 3-4, (1973), pp. 208-232.

M.G. CRANDALL - P.H. RABINOWITZ

[1] Some continuation and variational methods for positive solutions of non linear elliptic eigenvalue problems. Arch. Rat. Mech. Anal. Vol. 58 (1975), pp. 207-218.

P.R. GARABEDIAN - M. SCHIFFER

[1] Convexity of domain functionals. J. d'Analyse Math., 2, (1952-53), pp. 281-368.

J. HADAMARD

[1] Mémoire sur le problème d'analyse relatif à l'équilibre des plaques élastiques encastrées, (1907). Oeuvres de J. HADAMARD, Vol. 2, Ed. du CNRS, Paris (1968).

D.D. JOSEPH and T.S. LUNDGREN

[1] Quasilinear Dirichlet problems driven by positive sources. Arch. Rational Mech. Anal. Vol. 49 (1973), pp. 241-269.

A.M. MICHELETTI

[1] Metrica per famiglie di domini limitati e proprieta generiche degli autovalori. Ann. Sc. Norm. Sup. Pisa 26 (3) (1972), pp. 683-694.

F. MIGNOT - J.P. PUEL

[1] Sur une classe de problèmes non linéaires avec non linéarité positive croissante convexe. Comptes-rendus du Congrés d'Analyse non linéaire, Rome, (mai 1978). A paraître.

F. MURAT - J. SIMON

[1] Etude de problèmes d'optimal design. Proceedings of the 7 th IFIP Conférence, Nice, Sept 75, Part 2. Lecture Notes in Computer Sciences n° 41, Springer Verlag (1976), pp. 54-62.

[2] Sur le contrôle par un domaine géométrique. Publication n° 76 015 du Laboratoire d'Analyse Numérique de l'Université Paris VI (déc. 76).

L. SCHWARTZ

[1] Analyse mathématique. Cours professé à l'Ecole Polytechnique de Paris - Hermann, Paris (1967).

J. SIMON

[1] Dérivation par rapport à un domaine d'équations posées sur le domaine ou sur son bord. A paraître.

Dynamic Pade' Approximant and Behavior Singularities in Nonlinear Physico-Chemical Systems

Peter Ortoleva
Department of Chemistry
Indiana University
Bloomington, Indiana 47401

Abstract

Spatio-temporal phenomena in nonlinear systems have been found to be of great variety including periodic and chaotic structures.[1] We shall find here that the method of Pade' approximants may be extended to describe some of these phenomena including the chemical center wave. Catastrophe (or more generally singularity) theory is also shown here to be of great utility in obtaining classification theorems for systems with multiple space or time scales. The idea of symmetry broken singularities is introduced. Finally, unlike in the case of ordinary differential systems, it is shown that in some cases the phenomena must be understood in terms of the geometry of function space via "behavior functionals". These ideas shall be introduced through a discussion of various physical problems including crystal growth and reaction diffusion systems.

I. Discontinuity and Periodicity in Crystal Growth

The periodic zoning phenomena in Plageoclase feldspar crystals has long alluded satisfactory explanation[2]. Plageoclase feldspars are solid solutions of two types of feldspars Anorthite ($CaAl_2Si_2O_8$) and Albite ($NaAlSi_3O_8$) (denoted An and Ab respectively) which crystallize from a melt. In Nature one observes that the crystals are typically zoned, i.e. the fraction f of An ($f = An/(An + Ab)$) varies from the crystal center to the rim. What is most surprising is that in many samples this zoning is periodic. Traditionally, many geologists have believed that this phenomena was a consequence of periodic changes in the conditions of crystallization, i.e. the conditions of the magma.

It appears, however, that these patterns may be explained on the basis of the nonlinear dynamics of crystal growth.[3] Let $\underline{\psi}(\vec{r},t)$ be the column vector of concentrations of the chemical species in the melt ($\vec{r} \in \mathcal{M}$) at time t. Assuming Fickian diffusion (with matrix $\underline{\underline{D}}$) and neglecting chemical reactions in $\mathcal{M}$ we have the continuity equation

$$\partial\underline{\psi}/\partial t = \underline{\underline{D}}\nabla^2\underline{\psi}\ . \qquad (I.1)$$

The boundary condition at the crystal rim $\partial\mathcal{M}$ reflects the balance between the incorporation rate $\underline{G}$ into the crystal and the material swept out upon advancement $\vec{n}\cdot\vec{v}\,\underline{\psi}$ ($\vec{n}$ is the unit normal to $\partial\mathcal{M}$ pointing into $\mathcal{M}$) and the diffusive influx $\underline{\underline{D}}\,\vec{n}\cdot\vec{\nabla}\underline{\psi}$, i.e.

$$\underline{G} = \vec{n}\cdot\vec{v}\,\underline{\psi} + \underline{\underline{D}}\,\vec{n}\cdot\vec{\nabla}\underline{\psi},\quad \vec{r}\in\partial\mathcal{M}, \qquad (I.2)$$

where $\vec{v}$ is the velocity of advancement of the crystal rim. An equation for the form of the crystal rim (surface) $\partial\mathcal{M}$ is denoted by $S(\vec{r},t) = 0$. Kinematic considerations yield[1b]

$$\partial S/\partial t + \vec{v}\cdot\vec{\nabla}S = 0,\quad \vec{r}\in\partial\mathcal{M}. \qquad (I.3)$$

For one dimensional growth along the direction r we may take $S(r,t) = r - R(t)$ where $R(t)$ is the location of the crystal rim. From (I.3) we obtain

$$dR/dt = v\ .$$

The phenomenological law is needed which relates v to the concentrations in the melt at the rim $\underline{\psi}(r \to R(t) + 0^+, t)$ and the composition of the solid at the rim $f(r \to R(t) - 0^+, t)$,

$$v = V\left(\underline{\psi}(R(t) + 0^+, t), f(R(t) - 0^+, t)\right)\ . \qquad (I.4)$$

Finally to complete the theory we need an equation for f at $R(t) - 0^+$ which we write

$$\partial f/\partial t = 1/\varepsilon\ \mathcal{F}(\underline{\psi}, f),\quad r\in\partial\mathcal{M}. \qquad (I.5)$$

The factor $1/\varepsilon$ in front of $\mathcal{F}$ emphasizes that it takes a very short time ($\varepsilon \to 0$) for one layer of solid to be deposited after which $f(R(t) - 0^+, t)$ is given by

$$\mathcal{F} = 0 \Rightarrow f = g_{An}/(g_{An} + g_{Ab}) \tag{I.6}$$

where g_{An} and g_{Ab} are the rates of building An and Ab units of the crystal . Finally chemical kinetic considerations yield the dependence of $\underline{G}$, g_{An}, g_{Ab} and V on $\underline{\psi}$ (R(t) + 0,t) and $f(R(t) - 0^+,t)$. It is clear that the growth of crystals of this type constitutes a highly nonlinear Stephan type dynamics.

An important aspect of this phenomena appears to be the presence of a strong feedback effect in that An tends to grow faster on an An rich solid surface and more generally g_{An}, g_{Ab}, V and $\underline{G}$ depend strongly on $f(R(t) - 0^+,t)$. Because of this the "rim composition behavior surface" (I.6) on which the slow evolution of the system (on a time scale of order $\epsilon^0 = 1$) must lie, becomes multiple valued over certain ranges of $\underline{\psi}$ when expressed in the form $f = \tilde{f}(\underline{\psi})$. Thus from (I.5) it is clear that f is either smoothly varying or f makes rapid transitions on a time scale vanishing with ϵ.

One aspect of classifying possible growth phenomena of this type will thus depend on the delineation of the geometry of the surfaces $f = \tilde{f}(\underline{\psi})$ in s + 1 dimensional space ($\underline{\psi} = \{\psi_1, \psi_2, \ldots \psi_s\}$). This then reduces to determining the nature of the so called Cuspoids (s = 1 fold; s = 2 cusp, ...) as set forth by Thom[4]. This is typical of one of the classification type approaches we will emphasize.

In Reference 3 the details of the Plageoclase system are given. It is found that the only crucial dependence of the behavior function $\mathcal{F}$ is on two melt species concentrations (Al^{3+} and SiO_4^{4-}). Thus the behavior surface is with s = 2, i.e. at most a cusp. This leads to oscillations of rock composition (shown in Fig. 1) that involve a periodic sequence of up and down discontinuous (as $\epsilon \to 0$) jumps (a) for one range of parameters and a finite sequence in another case (b).

II. Multiscale Reaction Diffusion Structures

Consider a reaction diffusion system of the type

$$\partial\underline{\psi}/\partial t = \epsilon^{\underline{\underline{H}}_1}\underline{\underline{D}}\nabla^2\psi + \epsilon^{-\underline{\underline{H}}_2}\underline{F} \tag{II.1}$$

where the matrices $\underline{\underline{H}}_1$ and $\underline{\underline{H}}_2$ are constant and we assume that $\underline{\underline{D}}$ and $\underline{F}$ are finite as $\epsilon \to 0$. The presence of a smallness parameter ϵ (or a variety of such paramters) is a natural consequence of the frequent occurrence in reaction diffusion systems of widely separated length and time scales implied in the variation of rate and transport coefficients over several orders of magnitude. What one might expect is that as $\epsilon \to 0$ the spatio-temporal distribution $\underline{\psi}(\vec{r},t)$ will either be very rapid or alternatively will lie on "local behavior" surfaces of the form

$$F_i(\underline{\psi}) = 0 \tag{II.2}$$

in concentration space. In general only a subset f of the d species will be fast (i.e. $F_i = 0$, $i = 1,2,\ldots f \le d$). The intersection of these "behavior surfaces" comprises the "slow manifold" for the system. This observation has led to a variety of developments in the theory of static structures, temporal oscillations and propagating waves (see Ref. 5 and citations). This field has proceeded along two complementary directions. Catastrophe (and more generally singularity[6]) theory may be used to characterize the topology of the slow manifold. The theory is made quantitative by application of multiple scale techniques matching the region of smooth variations and short scale jumps. Thus the theory provides both a tool of quantitative analysis and a classification scheme based on the topological features of the slow manifold.

From the above remarks one might get the false impression that all the interesting qualitative features of multiple scale systems can be understood in terms of the topology of the surfaces $F_i = 0$. In the remainder of this section we demonstrate that this is indeed not the case, but that diffusion may play a nontrivial role in determining the topology of the slow manifold. We shall speak of the topological features of these surfaces as "singularities"[7] rather than catastrophes[4] since the latter is limited to the case of systems with functions derivable from a potential, a situation seldom met in chemical kinetics except for the case f = 1.

A. Two Box Model

Let us start with a discussion of symmetry breaking singularities that arise when we allow for inhomogeneity by considering a two box system. Descriptive variables X and Y are allowed to evolve in time with values $X(\alpha,t)$ and $Y(\alpha,t)$ in box α (= 1,2) according to the equations

$$\frac{dX(1,t)}{dt} = \frac{1}{\varepsilon}\{D(X(2,t) - X(1,t)) + R(X(1,t),Y(1,t))\} \tag{II.3}$$

$$\frac{dY(1,t)}{dt} = E(Y(2,t) - Y(1,t)) + S(X(1,t),Y(1,t)) \tag{II.4}$$

and similarly for box 2. In the multiscale limit $\varepsilon \to 0$ the homogeneous evolution $(X(\alpha,t) = X(t),\ Y(\alpha,t) = Y(t))$ will reside on the behavior surface

$$R(X,Y) = 0\ . \tag{II.5}$$

The qualitative dynamics of this multiscale evolution can be understood in terms of the geometry of the surfaces $X = \overline{X}(Y)$ such that $F(\overline{X}(Y),Y) = 0$. For one variable X and Y this geometry consists at most of a sequence of folds. For multiple "fast (behavior)" variables $X = \{X_1, \ldots X_f\}$ and "slow (control)" variables $Y = \{Y_1, \ldots Y_s\}$ more complex singularities specified by the f surfaces $R_i(X,Y) = 0$, $i = 1, \ldots f$ may arise (i.e. for $f = 1$, $s = 2$ we may have cusps). We now pose the question as to what new topological features - "symmetry breaking singularites" can arise in a given system (i.e. D,E,R,S) when inhomogeneous evolution is considered; furthermore to what extent will the symmetry of the equations with respect to the permutation of box labels limit the realm of the possible?

For the two box problem as $\varepsilon \to 0$ we have the slow manifold given by (in an obvious notation)

$$D[X(2,t) - X(1,t)] + R(1) = 0 \tag{II.6}$$

$$D[X(1,t) - X(2,t)] + R(2) = 0\ . \tag{II.7}$$

Insight into the inhomogeneous slow manifold can be obtained by combining (II.6,7) to eliminate X of one of the boxes (say box 2) to obtain a closed equation for $X(1,t)$ in terms of the $Y(\alpha,t)$. We find

$$R(X(1,t),Y(1,t) + R(X(1,t) - D^{-1}R(X(1,t),Y(1,t),Y(2,t)) = 0\ . \tag{II.8}$$

From this we see that the slow manifold can potentially take on higher order singularities than the homogeneous slow manifold since the dimensionality of the control space (the space of the Y variables) has doubled.

Let us demonstrate the possibility of symmetry breaking singularities arising in the trivial example $R = Y - X^2$ for the case of a single fast variable and a single slow variable. Letting $y = Y(1,t)/D^2$, $z = Y(2,t)/D^2$ and $x = X(1,t)/D$ we find the inhomogeneous manifold to be defined by

$$x^4 + 2x^3 - 2(y-1)x^2 - 2yx + y^2 - y - z = 0\ . \tag{II.9}$$

Our present considerations then reduce to the question of whether this surface in x,y,z space contains any topological features other than the homogeneous subslow manifold given by the intersection of (III.9) and the symmetry constraint plane $y = z$, i.e. the fold $y = x^2$. The symmetry breaking leads to the unfolding of a cusp into the inhomogeneous slow manifold from a homogeneous, cusp free (folded), slow manifold. A sketch of the inhomogeneous slow manifold (II.9) is shown in Fig. 2 and it is seen that symmetry breaking introduces a cusp.

The next step in a full two box theory of symmetry breaking singularities would be to investigate in more general terms what symmetry breaking singularities will arise for given homogeneous slow manifolds in systems of various dimensions ($s + f > 2$). A further important problem is to generalize the results to systems of $n(>2)$ boxes. These questions are presently being investigated.[7]

B. Continuous Systems

In the previous section we saw that new features for multiscale dynamics arise due to symmetry breaking singularities. We might expect new features to arise in the continuous case when the control space is continuously infinite dimensional, i.e. a function space. Consider the model problem

$$\frac{\partial X}{\partial t} = \frac{1}{\epsilon}[D\nabla^2 X + R(X,Y)] \qquad (II.10)$$

$$\frac{\partial Y}{\partial t} = E\nabla^2 Y + S(X,Y). \qquad (II.11)$$

As $\epsilon \to 0$ the X evolution is rapid or is constrained to the slow manifold

$$D\nabla^2 X + R(X,Y) = 0. \qquad (II.12)$$

This equation is the continuous analogue of (II.6,7). For any distribution $Y(\vec{r},t)$ we can solve this equation to obtain a real bounded solution(s) $X(\vec{r},t) = \overline{X}(X_o,\vec{r}|Y]$. These solution(s) depend on the value of X at one or more points in space (denoted X_o), position $\vec{r}$ and are functionals of the instantaneous distribution $Y(\vec{r},t)$.

Let us specialize our considerations to one dimension denoted r on $-\infty < r < \infty$. If Y is constant then there are homogeneous solutions for X such that

$$R(X,Y) = 0. \qquad (II.13)$$

Furthermore if there are multiple such homogeneous solutions for a given Y then there may exist coexistence structures consisting of smooth variations taking X to one steady state at $-\infty$ and another at $+\infty$ if Y takes on isolated values[7].

For inhomogeneous variations $Y(r,t)$ great variety can occur. To see this in a manner analogous to the analysis of the two box problem let us integrate (II.12) from $-\infty$ to ∞ and assuming that the spatial derivative of X vanished at infinity we obtain

$$\mathcal{R}(X_o|Y) \equiv \int_{-\infty}^{\infty} dr R(\overline{X},Y) = 0. \qquad (II.14)$$

The solutions to this equation yield X_o as a functional of $Y(r,t)$. It is clear that there can exist regions of function space (for Y) such that the Y distribution passes locally (in r) through values so that (II.13) has multiple solutions. Thus in many cases X_o will have multiple solutions. On the borders (in Y function space) these multiple solutions will terminate as for folds, cusps, umbilics, etc. as for the case of finite dimensional control (Y) spaces. Finally we expect that the solutions of (II.14) will, in some cases, form a continuum (as for the coexistence structures where X_o can take all values between the two homogeneous steady states.

Clearly the delineation of the topological features (singularities) of the surfaces $X_o[Y]$ in the direct product of the discrete (X_o) and function (Y) spaces presents itself as a challenging mathematical problem. It is tempting to conjecture that a great reduction of the freedom in the Y space can lead to quite tractible problems. For example limiting Y to monotonic functions might well be very analogous to the two box problem or functions with a single maximum might be closely related to the three box problems.

III. Center Waves

In a two dimensional reacting continuum it has been found experimentally[8] and in computer simulation[9] that spiral and circular patterns of concentration may be obtained. Theoretical work on these phenomena has led to only incomplete results as a result of divergences arising in distorted oscillator or wave approaches, bifurcation theory and non-truncatable expansion schemes about the center of the wave[10]. Furthermore no results on stability of these structures existed. Making the observation that these patterns are typically plane wave like far from the center, a dynamic Pade' approximant scheme has been introduced to solve these problems[10].

A. A Model Problem

To illustrate the dynamic Pade' approach let us consider a simple model system often used in the theory of chemical waves[11] and furthermore shall limit our considerations to circular waves in a two dimensional medium. We consider two "species" X and Y to evolve according to

$$\frac{\partial}{\partial t}\begin{bmatrix} X \\ Y \end{bmatrix} = \{D\nabla^2 + \begin{bmatrix} B & -A \\ A & B \end{bmatrix}\}\begin{bmatrix} X \\ Y \end{bmatrix} \tag{III.1}$$

where D is a diffusion constant and A and B are functions of a radial variable R, $R^2 = X^2 + Y^2$. It is convenient to change to a set of phase plane coordinates R, Φ such that

$$X = R\cos\Phi,\ Y = R\sin\Phi\ . \tag{III.2}$$

With this (III.1) becomes

$$\partial R/\partial t = RB(R) + D[\nabla^2 R - R|\vec{\nabla}\Phi|^2]$$

$$\partial\Phi/\partial t = A(R) + D[\nabla^2\Phi + 2\vec{\nabla}R\cdot\vec{\nabla}\Phi/R]\ . \tag{III.3}$$

Our problem shall be to find solutions of (III.3) with circular (concentric ring) geometry.

B. Plane Waves

Inspection of (III.3) shows the existence of plane waves with

$$\Phi^\infty = \vec{k}\cdot\vec{r} + wt$$

$$B(R^\infty) = k^2 D$$

$$w = A(R^\infty)\ . \tag{III.4}$$

As mentioned earlier the center waves of interest here are plane wave like and hence these solutions will serve to characterize these phenomena far from the center (core) of the wave.

C. Core Expansion

In the core of a center wave one finds large deviations from the plane wave function R^∞, Φ^∞. Indeed, as mentioned earlier, the phase of the plane wave function is found to diverge as one enters the core in a weak distortion of plane waves (or homogeneous cycles) scheme.[10,11c] Thus we turn to an expansion of the concentrations around the center of the wave to obtain some exact information there. We assume the existence of an expansion of R and Φ in the radial distance (r) from the wave center:

$$R = \sum_{n=0}^{\infty} R_n r^n$$

$$\Phi = \sum_{n=0}^{\infty} \Phi_n r^n\ . \tag{III.5}$$

For circular waves the coefficients R_n and Φ_n will depend on time while for spiral waves they will depend on an angular variable as well. To determine the core coefficients R_n, Φ_n we insert the core expansions into (III.3) and collect like powers of r. The first few terms in this hierarchy are

$$dR_0/dt = R_0 B(R_0) + 4DR_2$$

$$dR_2/dt = (RB(R))'_0\, R_2 - 4DR_0\Phi_2^2 + 16DR_4$$

$$d\Phi_0/dt = A(R_0) + 4D\Phi_2$$

$$d\Phi_2/dt = A'_0 R_2 + 8DR_2\Phi_2/R_0 + 16D\Phi_4\ . \tag{III.6}$$

Higher order equations take a similar form. For circular waves the odd coefficients R_1, Φ_1 etc. vanish identically.

Note that the R_n equation contains R_{n+2} (and similarly for Φ_n) and hence the core expansion has no natural truncation scheme.

D. Dynamic Pade' Matching Scheme

The core expansion does not truncate because of lack of information away from the center. The plane wave approach suffers near the center. We now construct a Pade' approximant which incorporates information obtained from the two extreme domains to form a uniformally valid center wave function. First we write R(r,t) in the form of a generalized Pade' approximant $R^P(r,t)$,

$$R^P(r,t) \sim \frac{a_o + a_1 r + \ldots + a_L r^L R^\infty}{b_o + b_1 r + \ldots + b_L r^L} \quad . \tag{III.7}$$

The coefficient functions a_i and b_i depend on time for circular waves (and on an angular variable for spiral waves). Since we want the proper outer behavior we put $b_L = a_L$. Thus clearly

$$R^P \underset{r\to\infty}{\sim} R^\infty \quad . \tag{III.8}$$

Next we use the core expansion to fix the a and b coefficients by expanding R^P in small r and finally using R^∞ to obtain an expression for R_{n+2} when a truncation to R_n is desired.

E. Phase Branch Point

Establishing a Pade' approximant is not quite as straightforward for Φ as for R. From the core expansion we see that near $r = 0$ that Φ is an even function of r. Yet far from the origin $\Phi \sim \Phi^\infty$ which we see from (III.4) implies that Φ is odd ($\sim$ kr) as $r \to \infty$. Thus it is clear that a transition on analyticity must occur - i.e. there must be a branch point in Φ as a function of r. Such a situation can be found from a simple formula like

$$\Phi^P(r,t) = wt + \sqrt{\alpha_o^2 + (kr)^2} \tag{III.9}$$

which is seen to be even near $r = 0$ and linear in r as $r \to \infty$. Clearly other forms with fractional powers of appropriate ratios of polynomials in r would also suffice and generate higher order Pade' approximants.

F. Truncation and Closure

The theory can now be completed by using the Pade' approximants to close the truncated core expansion. In a lowest order theory we have used (III.9) and

$$R^P(r,t) = \frac{a_o + a_2 r^2 R^\infty}{1 + a_2 r^2}$$

to generate three coupled equations for $\alpha_o(t)$, $a_o(t)$ and $a_2(t)$. Thus the dynamic Pade' scheme reduces the problem of two dimensional partial differential equations. The scheme not only provides a method of studying the stationary profiles (periodic or chaotic) but also provides an approximate stability analysis. Finally the dynamics near the core, i.e. $R_n(t)$, $\Phi_n(t)$ for small n, may have an inherent frequency different from the periodicity in R^∞, Φ^∞. Thus we can forsee all the possibilities (multimodes, mode locking, chaos, frequency quenching) found in the theory of forced nonlinear oscillation[12].

IV. Remarks

The multiple scale/singularity and the dynamic Pade' approximant approaches have several common features. They are both inherently nonlinear methods which are not just expansions in the small as is done in bifurcation theory. Furthermore they provide at least a partial (if not rigorous) stability analysis. Perhaps most interesting is that they suggest methods of classifying phenomena which potentially may (and have) led to the prediction of phenomena.

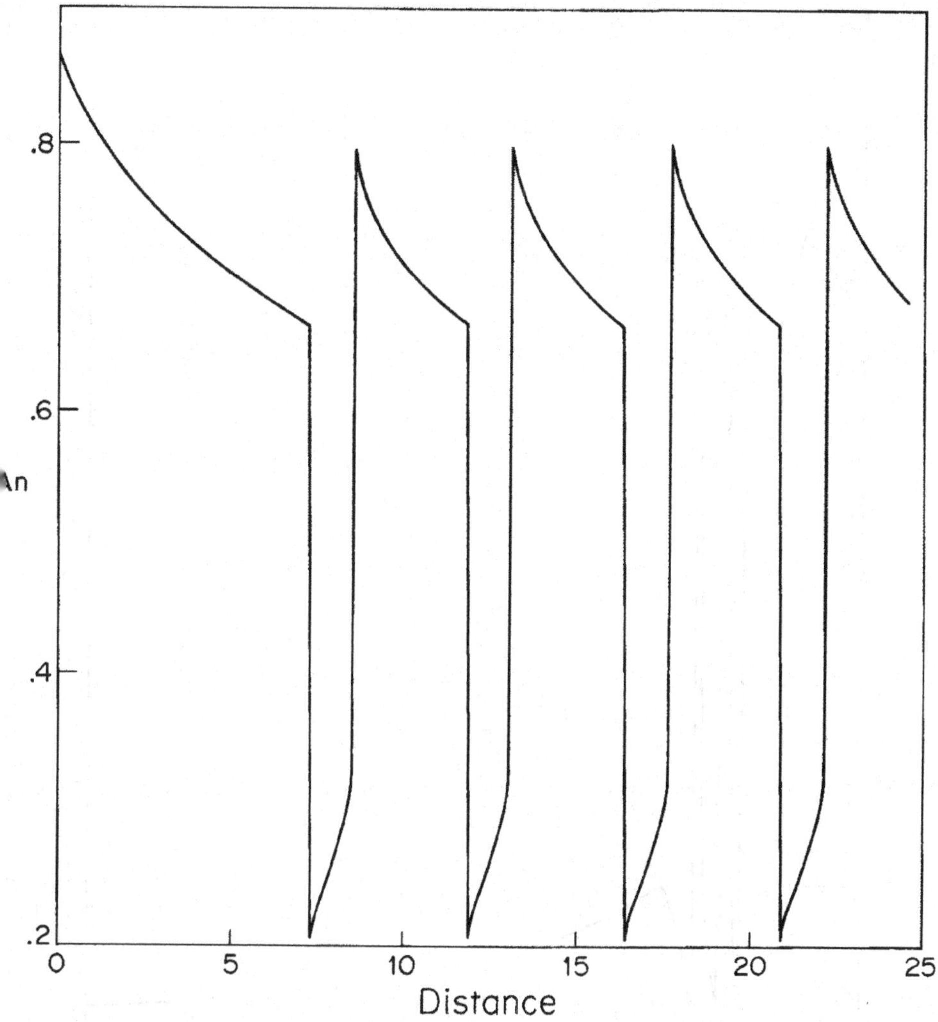

Fig. 1a

Figure 1: Periodic zoning phenomena shown resulting from crystal growth theory (Ref. 3) with a fold (a) and a cusp (b) surface mole fraction f behavior surface.

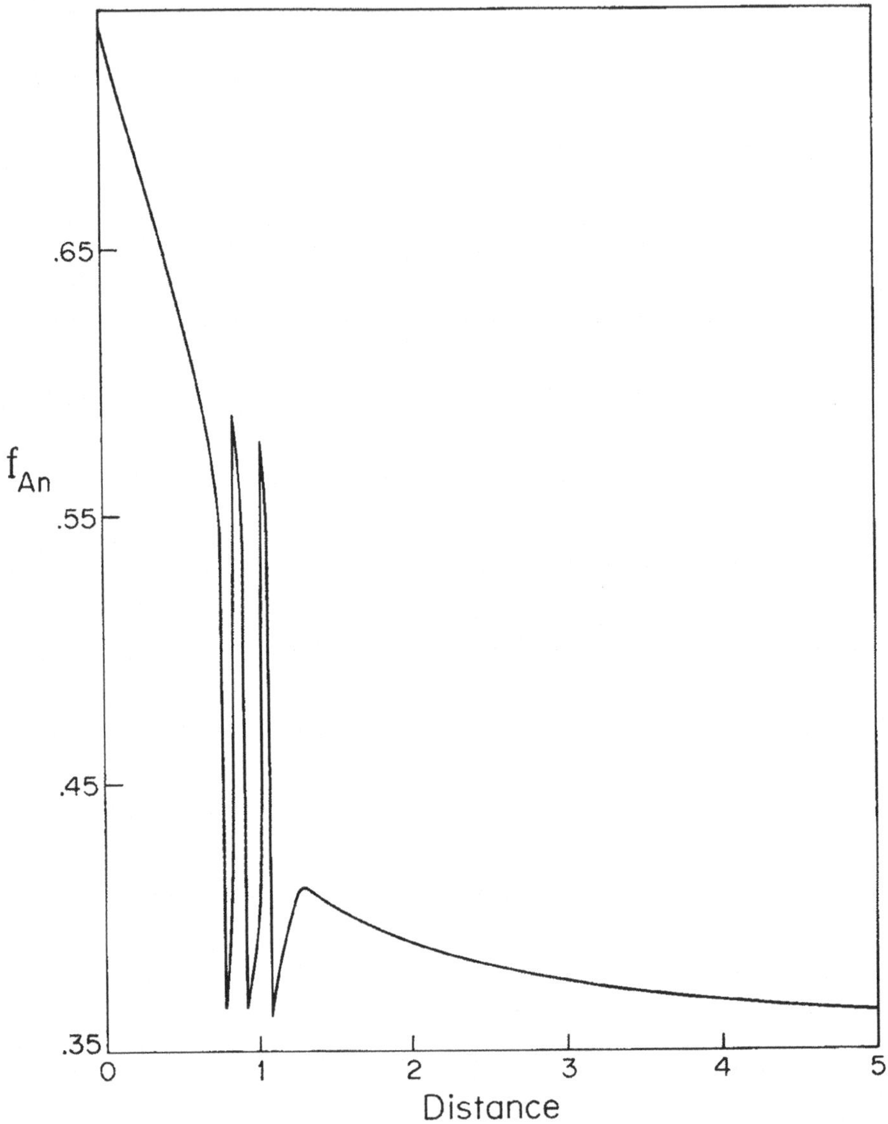

Fig. 1b

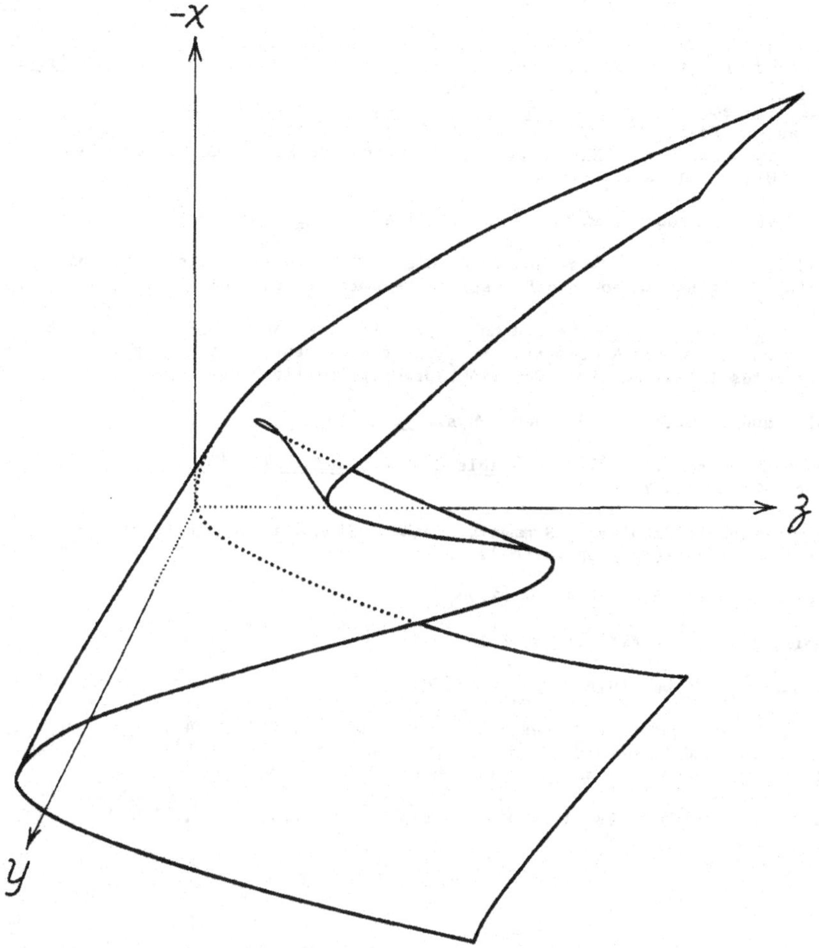

Figure 2: Behavior surface for two box problem discussed in Sect. 2A showing symmetry breaking of fold singularity.

REFERENCES

1. a. P.Glansdorff and I. Prigogine, Thermodynamic Theory of Structure Stability, and Fluctuations (Wiley, N.Y., 1971).
   b. H.Eyring ed., Periodicies in Chemistry and Biology, Theoret. Chem. 4, (Academic Press, 1978).
   c. Adv. in Chem. Phys. XXXVIII ed. I. Prigogine and S.A. Rice (Interscience, John Wiley and Sons, 1978).

2. Y. Bottinga, A. Kudo and P. Weill, Amer. Mineral., 51, 792 (1966).

3. J. Chadam, D. Feinn, S. Hasse and P. Ortoleva, "Chemical Kinetic Theory of Periodic Zoning in Plageoclose Feldspars" (submitted for publication).

4. R. Thom, Stability, Structure and Morphogenesis (Benjamin, N.Y., 1972); A.E. Woodcock and T. Poston A Geometric Study of the Elementary Catastrophes, Lecture Notes in Mathematics, No. 373 (Springer, Berlin, 1974).

5. D. Feinn and P. Ortoleva, J. Chem. Phys. 67, 5 (1977).

6. M. Colubitsky and V. Guillemin, Stable Mappings and Their Singularities (Springer-Verlag, N.Y., 1973).

7. S. Schmidt and P. Ortoleva, "Symmetry Breaking Singularities in Reaction-Diffusion Systems" (in preparation).

8. A.T. Winfree, Sci. Amer. 230, 82 (1974).

9. See Refs. 3 and 4 in citation 10 below.

10. P. Ortoleva, J. Chem. Phys. 69, 300 (1978).

11. a. J. Dreitlein and M.-L. Smoes, J. Theor. Biol. 46, 559 (1974);
    b. N. Kopell and L. Howard, Stud. in Appl. Math, 52, 291 (1973);
    c. P. Ortoleva and J. Ross, J. Chem. Phys. 60, 5090 (1974).

12. N. Minorsky, Nonlinear Oscillations, (Kreiger, Huntington, N.Y., 1974).

REMARKS ON A NON LINEAR EQUATION
ARISING IN POPULATION GENETICS.

J.C. SAUT

U.E.R. de Mathématiques
Université Paris VII
75221 PARIS CEDEX 05

I - INTRODUCTION AND NOTATIONS.

This conference is based on a joint work with B. Scheurer [13].

Let Ω be a bounded open set of class C^2 of $\mathbb{R}^m$, $\partial\Omega = \Gamma$. In this paper, we look for non trivial solutions of the equation :

$$(1.1) \quad \begin{cases} \Delta u + \lambda F(x,u) = 0 \text{ in } \Omega , \\ \dfrac{\partial u}{\partial n}\Big|_\Gamma = 0. \end{cases}$$

Here, n is the unit exterior normal on Γ, λ is a non negative real parameter, and $F(.,0) \equiv F(.,1) = 0$. The function F does not verify any sign hypothesis and is not regular in x.

Problems of this kind arise in population genetics (Cf. for instance Slatkin [10]) : u is then the frequence of some population, and we shall only focus our attention on solutions satisfying $0 \leq u(x) \leq 1$.

Problem (1.1) has been recently studied under very interesting aspects by W.H. Fleming [6], F. Hoppensteadt [12], and Peletier-Fife [5]. These authors give existence and stability results for (1.1) using mainly, for existence, variational [6], or sub and super solutions methods [5]. In [6], Fleming uses stability arguments to get further bifurcation from the trivial branches $u \equiv 0$, $u \equiv 1$. Using contraction mapping theorem, Peletier-Fife [5] give an example of secondary bifurcation occuring from the branch $\lambda = 0$. Several proofs are specific of the space dimension 1, and stated for rather special F's.

Our aim in this paper is to explicit completely, for a fairly large classe of non-linearities and in any space dimension, the structure of the solution set of (1.1), in the neighborhood of $\lambda = 0$.

More precisely, we investigate secondary bifurcation from the vertical primary bifurcated branch at $\lambda = 0$. For this purpose, we use Lyapunov-Schmidt method. To study the bifurcation equation, we apply differential methods, in particular, the Morse lemma and the k-determination theorem of J. Mather (Cf. [10], [11] and the Appendix).

This allow us to count the number of branches of non trivial solutions, under integral conditions on the non-linearity F, and so to complete and generalize some results of [5], [6].

The plan is as follows. First we give a criterion of non existence of non trivial solutions of (1.1), lying essentially on the simple eigenvalue bifurcation theorem of Crandall and Rabinowitz [3].

Then we give three results on the structure of the solution set of (1.1) in a neighborhood of $\lambda = 0$, corresponding to increasing degeneracy levels of the bifurcation equation.

Lastly we prove an estimate which seems to be new, on the first positive eigenvalue λ_1 of the linearized problem

$$(1.2)\qquad \begin{cases} \Delta u + \lambda r(x) u = 0 \text{ in } \Omega \\ \\ \dfrac{\partial u}{\partial n}\Big|_{\Gamma} = 0 \end{cases}$$

The function r belongs only to $L^\infty(\Omega)$, without any sign hypothesis. The result leads of course to an estimate on the first value λ of bifurcation with respect to the branch $u \equiv 0$, or the branch $u \equiv 1$, for problem (1.1).

For the reader's convenience, the result of Mather used in this work is recalled in an appendix.

We shall denote by $W^{m,p}(\Omega)$ the Sobolev space of $L^p(\Omega)$ functions which derivatives up to the order m are in $L^p(\Omega)$, and $W^{m,2}(\Omega) = H^m(\Omega)$.

For $0 < \alpha < 1$, $C^{k+\alpha}(\Omega)$ is the space of functions having bounded derivatives up the order k, the kth derivatives being Hölder continuous with exponent α.

We then set :

$$S = \{(\lambda,u) \in \mathbb{R}_+ \times H^2(\Omega) \ ; \ (\lambda,u) \text{ satisfies (1.1) and } 0 \le u(x) \le 1\}.$$

We make on F the following hypothesis :

(1.3) There exists $K > 0$ such that $\|F(.,\xi)\|_{L^\infty(\Omega)} \le K, \ \xi \in [0,1]$.

(1.4) For almost every $x \in \Omega$, $F(x,.)$ is C^k, $k \ge 2$.

(1.5) For every $\xi \in [0,1]$, $F_u^{(i)}(.,\xi) \in L^2(\Omega)$, $i = 0,1,2$.

(1.6) $F(.,0) \equiv F(.,1) \equiv 0$.

In application to population genetics, one encounters in particular :

Example 1 : $F(x,u) = g(x)f(u)$, $f(0) = f(1) = 0$ (cf [6])

Example 2 : $F(x,u) = f(u)(u-a(x))$, $f(0) = f(1) = 0$ (cf. [5]).

Let us remark that in these examples g and a are not necessarily continuous (cf. Slatkin [10]).

By (1.3) and the classical regularity results of linear elliptic equations (Agmon, Douglis, Nirenberg [1]), if $(\lambda,u) \in S$, then $u \in W^{2,p}(\Omega)$ for every p, $1 < p < \infty$, and therefore $u \in C^{1+\alpha}(\Omega)$ for some α, $0 < \alpha < 1$, by Sobolev imbedding theorem.

II - We first give a result on non existence of non trivial solutions of (1.1) in the neighborhood of $\lambda = 0$.

Theorem 1

Let us suppose that F satisfies (1.3), (1.4), (1.5), (1.6) and the two following conditions :

(2.1) $$\int_\Omega \frac{\partial F}{\partial u}(x,\varepsilon)\, dx \neq 0, \qquad \varepsilon = 0,1$$

(2.2) $$\int_\Omega F(x,c)\, dx \neq 0, \qquad c \in]0,1[.$$

Then there exists $\lambda_0 > 0$ such that $(\lambda,u) \in S$ with $0 < \lambda < \lambda_0$ implies $u \equiv 0$ or $u \equiv 1$.

Remark 1

In example 1, condition (2.1), (2.2) are equivalent to $\int_\Omega g(x)\, dx \neq 0$ and $\frac{\partial f(\varepsilon)}{\partial u} \neq 0$, $\varepsilon = 0,1$.

In example 2, where $f(u) = u(1-u)$ (cf. Peletier-Fife [5]), conditions (2.1) and (2.2) are equivalent to $\int_\Omega (c-a(x))\, dx \neq 0$, $c \in [0,1]$.

Proof of Theorem 1 :

The hypothesis (2.1) insures that the transversality condition in the simple eigenvalue bifurcation theorem of Crandall, Rabinowitz ([3] p. 325) is satisfied at $\lambda = 0$. Consequently, S consists only in the branches $(\lambda = 0, u \equiv 0)$, $(\lambda = 0, u \equiv 1)$

in the neighborhood of (0,0) and (0,1).

(i) We show now that for every $c \in]0,1[$, there exists in $\mathbb{R}_+ \times H^2(\Omega)$ a ball $B = B(0,c) ; \varepsilon_c)$, $\varepsilon_c > 0$, such that $S \cap B = B \cap [\{0\} \times [0,1]]$.

If it were not the case, there will exist $c \in]0,1[$ and a sequence $(\lambda_n, u_n) \in S$ such that $\lambda_n \neq 0$ for every n, and $\lambda_n \to 0$, $u_n \to c$ in $H^2(\Omega)$. So there will exist a subsequence $u_n \to c$ a.e. But equation (1.1) (and $\lambda_n \neq 0$) imply $\int_\Omega F(x,u_n)\,dx = 0$, and by (1.3) and Lebesgue theorem, $\int_\Omega F(x,c)\,dx = 0$, which violates condition (2.2).

(ii) Let us suppose now the conclusion of the theorem to be false. Then, there will exist a sequence $(\lambda_n, u_n) \in S$, $\lambda_n \neq 0$, $\lambda_n \to 0$, $u_n \not\equiv 0$, $u_n \not\equiv 1$ for every n. One deduces by equation (1.1) that u_n is a bounded sequence in $H^2(\Omega)$ for instance, and one can extract a subsequence, which we also denote u_n, such that $u_n \to u$ strongly in $H^1(\Omega)$, and we get at the limit :

$$\Delta u = 0 \text{ in } \Omega$$

$$\frac{\partial u}{\partial n}\Big|_\Gamma = 0,$$

i.e. $u = c \in]0,1[$, which contradicts point (1).

Remark 2

A similar result hold for solutions not necessarily lying between 0 and 1 ; one has then to impose a growth condition in u to F.

III - In the case where (2.2) is not satisfied, there can exist one or several non trivial branches of solutions, bifurcating from the trivial curve $\lambda = 0$ at a point $(0,c)$, $c \in]0,1[$.

In the sequel, we shall therefore suppose that F satisfies :

(3.1) There exists $c \in]0,1[$ such that $\int_\Omega F(x,c)dx = 0$.

Remark 3

In example 1, if one suppose that f has no zeroes in $]0,1[$, condition (3.1) is equivalent to $\int_\Omega g(x)dx = 0$.

Similarly, in example 2, where $f(u) = u(1-u)$, condition (3.1) is equivalent to $\int_\Omega a(x)dx = c|\Omega|$ (a condition given in the one dimensional case by Peletier and Fife [5]).

We introduce now some notations :

We denote by A the unbounded operator in $L^2(\Omega)$ defined by :

$D(A) = \{u \in L^2(\Omega),\ \Delta u \in L^2(\Omega),\ \frac{\partial u}{\partial n}\big|_\Gamma = 0$ (1)$\}$, $Au = \Delta u$.

One can of course decompose $D(A)$ and $L^2(\Omega)$ as follows :

$D(A) = \text{Ker } A \oplus X_2$, $L^2(\Omega) = Y_1 \oplus R(A)$, with

$\text{Ker } A = 1\mathbb{R}$, $\dim Y_1 = 1$. We then denote Q the orthogonal projection on $R(A)$ associated to this decomposition ($Qu = u - \frac{1}{|\Omega|}\int_\Omega u\, dx$).

We introduce now the elements γ_i of X_2 defined by :

$$(3.2)\quad \begin{cases} \Delta\gamma_i = -QF_u^{(i)}(.,c) & i = 0,1,2 \\ \frac{\partial\gamma_i}{\partial n}\big|_\Gamma = 0 & \end{cases}$$

Note that this definition makes sense by (1.5).

The following three theorems describe completely the structure of S in the neighborhood of $(0,c)$, by examining the successive degeneracy cases of the bifurcation equation.

In what follows, we shall suppose, for technical reason, that $m \le 3$.

Theorem 2

We suppose that F verifies (1.3), (1.4), (1.5), (1.6) and (3.1).

(i) If $\int_\Omega F_u'(x,c)dx \neq 0$, *there exists a neighborhood* V *of* $(0,c)$ *in* $\mathbb{R}_+ \times H^2(\Omega)$ *such that*

$$(S \cap V) \setminus [\{0\} \times [0,1]] \cap V \text{ is a } C^k\text{-curve.}$$

(ii) If $\int_\Omega F_u'(x,c)dx = 0$, $\int_\Omega F_u'(x,c)\ \gamma_0(x)dx \neq 0$ *and*

(3.3) $\alpha \to \int_\Omega F(x,\alpha)dx$ *is not identically zero in the neighborhood of* $\alpha = c$ (2), *then the same conclusion as in (i) holds.*

Theorem 3

Let us suppose that F verifies (1.3), (1.4) (with $k \ge 3$*), (1.5), (1.6), (3.1). Suppose moreover that :*

---

(1) In the usual sense of traces, cf. Lions-Magenes [7].

(2) This hypothesis is to insure that the curve is not trivial.

(3.4) $$\int_\Omega F'_u(x,c)dx = \int_\Omega F'_u(x,c)\ \gamma_0(x)dx = 0$$

(3.5) $$\int_\Omega [F'_u(x,c)\ \gamma_1(x) + F''_u(x,c)\ \gamma_0(x)]dx \neq 0.$$

and

(3.6) $$\int_\Omega F''_u(x,c)dx = 0. \qquad (1).$$

Then there exists a neighborhood V *of* $(0,c)$ *in* $\mathbb{R}_+ \times H^2(\Omega)$ *such that* $S \cap V$ *consists in* $[\{0\} \times [0,1]] \cap V$ *and in two* C^{k-2} *- curves intersecting transversaly at* $(0,c)$. *One of these curves is always non trivial ; the other one is non trivial if and only if (3.3) holds.*

Theorem 4

Let us suppose that F satisfies (1.3), (1.4) (with $k = \infty$*), (1.5), (1.6) and (3.1).*

Suppose morever that

(3.7) $$\int_\Omega F'_u(x,c)dx = \int_\Omega F''_u(x,c)dx =$$

$$\int_\Omega [F'_u(x,c)\ \gamma_1(x) + F''_u(x,c)\ \gamma_0(x)]dx = 0.$$

and

(3.8) *a certain non degeneracy hypothesis (to be precised in the proof) is satisfied.*

Then there exists a neighborhood V *of* $(0,c)$ *in* $\mathbb{R}_+ \times H^2(\Omega)$, *such that* $S \cap V$ *consists in*

(i) Either $(\{0\} \times [0,1]) \cap V$

(ii) or $(\{0\} \times [0,1]) \cap V$ *and three* C^∞*- curves intersecting transversally at* $(0,c)$. *Two of these curves are always non trivial ; the third one is non trivial if and only if (3.3) holds.*

The alternative will be explicited in the proof.

Proof of theorem 2, 3 and 4

The principle of the proof is the following : we apply the Lyapounov-Schmidt method, then we examine the successive degeneracy cases of the bifurcation equation.

The idea is to reduce the study of the bifurcation equation to the study of

(1) This hypothesis is only used to simplify the statement of the theorem.

its Taylor polynomial at some order. To do this, one use implicit functions theorem, then Morse Lemma, and a 3-determination theorem for the jet of a certain application.

1) - Lyapouncv-Schmidt method :

Equation(1.1)is, in the neighborhood of (0,c), equivalent to :

$$\Delta u_2 + \lambda QF(.,\alpha+u_2) = 0 \tag{3.9}$$

$$\lambda(I-Q)\ F(.,\alpha+u_2) = 0 \tag{3.10}$$

where Q is the orthogonal projection on R(A) and

$$u = \alpha + u_2 \in \text{Ker } A \oplus X_2.$$

By implicit functions theorem, one can express u_2 as a function of α and λ, in a neighborhood of $\lambda = 0$, $\alpha = c$, $u_2 = 0$.

In fact, if one defines $A : \mathbb{R}_+ \times \mathbb{R} \times X_2 \to R(A)$ by $A(\lambda,\alpha,u_2) = \Delta u_2 + \lambda QF(.,\alpha+u_2)$, one sees that $D_{u_2} A(0,c,0) = \Delta$, and is therefore an isomorphism.

The resolution of (1.1) in the neighborhood of $\lambda = 0$, $u = c$ is then equivalent to that of :

$$\Phi(\alpha,\lambda) = \lambda(I-Q)\ F(.,\alpha+u_2(\alpha,\lambda)) = 0 \tag{3.11}$$

in the neighborhood of $\lambda = 0$, $\alpha = c$. Since we are only interested in non trivial solutions (i.e. $\lambda \neq 0$), one considers in fact the equation :

$$\psi(\alpha,\lambda) = (I-Q)\ F(.,\alpha+u_2(\alpha,\lambda)) = 0 \tag{3.12}$$

where ψ is a C^k-real function of two real variables.

Let us remark that :

$$\psi(c,0) = (I-Q)\ F(.,c+u_2(c,0)) = (I-Q)\ F(.,c) = 0 \tag{3.13}$$

by (3.1) and recalling that $(I-Q)v = \frac{1}{|\Omega|}\int_\Omega v\ dx$.

2) - Study of (3.12):

To study $\psi^{-1}(0)$ in the neighborhood of $\alpha = c$, $\lambda = 0$, we reduce, following Nirenberg [8] (cf. also [2]), by a suitable diffeomorphism, to the local study of the zero set of a polynomial.

This is equivalent to say that ψ is determined in the neighborhood of (c,0) by its Taylor polynomial at some order.

A) Order 1 degeneracy (Proof of theorem 2)

One has $\frac{\partial\psi}{\partial x}(\alpha,\lambda) = (I-Q)\ F'_u(.,\alpha+u_2)\left[1 + \frac{\partial u_2}{\partial\alpha}(\alpha,\lambda)\right]$

and $\frac{\partial\psi}{\partial\lambda}(\alpha,\lambda) = (I-Q)\ F'_u(.,\alpha+u_2)\ \frac{\partial u_2}{\partial\lambda}(\alpha,\lambda).$

One the other hand, by using (3.9), one finds, denoting by K the inverse of $\Delta|_{X_2}$:

$$\frac{\partial u_2}{\partial\alpha}(\alpha,\lambda) = -\ \lambda KQF'_u(.,u)\left[\frac{\partial u_2}{\partial\alpha}(\alpha,\lambda) + 1\right]$$

and $$\frac{\partial u_2}{\partial\alpha}(\alpha,\lambda) = -\ KQF(.,u) - KQF'_u(.,u)\ \frac{\partial u_2}{\partial\lambda}(\alpha,\lambda)$$

hence

(3.14) $\frac{\partial u_2}{\partial\alpha}(c,0) = 0$ and $\frac{\partial u_2}{\partial\lambda}(c,0) = -\ KQF(.,c).$

Finally, we get :

$$\frac{\partial\psi}{\partial\alpha}(c,0) = (I-Q)\ F'_u(.,c) = \int_\Omega F'_u(x,c)dx.$$

and $$\frac{\partial\psi}{\partial\lambda}(c,0) = -\ (I-Q)\ [F'_u(.,c)\ KQF\ (.,c)]$$
$$= \int_\Omega F'_u(x,c)\ \gamma_0(x)dx.$$

If $\int_\Omega F'_u(x,c)dx \neq 0$, there exists in the neighborhood of $(c,0)$ a unique C^k-curve $(\alpha(\lambda),\lambda)$, $|\lambda| < \varepsilon$, such that $\Psi(\alpha(\lambda),\lambda) = 0$, and we get the existence of a non trivial branch of solutions of (1) in a neighborhood of $u = c$, $\lambda = 0$.

If $\int_\Omega F'_u(x,c)dx = 0$ and $\int_\Omega F'_u(x,c)\ \gamma_0(x)dx \neq 0$, again by implicit functions theorem, there exists in a neighborhood of $\alpha = c$, $\lambda = 0$ a unique curve $(\alpha,\lambda(\alpha))$, $|\alpha - c| < \varepsilon$ such that $\psi(\alpha,\lambda(\alpha)) = 0$.

This curve will be non trivial if and only if :

$\psi(\alpha,0) \not\equiv 0$ for $0 < |\alpha - c| < \varepsilon$, i.e. if and only if

(3.15) $\alpha \to \int_\Omega F(x,\ +u_2(\alpha,0))dx \not\equiv 0$ for $0 < |\alpha - c| < \varepsilon$.

Hence, by (3.9), if and only if (3.3) holds.

Theorem 2 follows now readily, noting that a non trivial solution $u_2(\alpha,\lambda)$ of (3.11) corresponds bijectively to a non trivial solution $(u(\alpha,u_2(\alpha,\lambda)),\lambda)$ of (3.1).

One has clearly $\|u - c\|_{H^2(\Omega)} \leq \varepsilon(\lambda)$ where $\varepsilon(\lambda) \to 0$ with λ. For $m \leq 3$, one has the injection $H^2(\Omega) \subset C^o(\Omega)$ and so $\|u - c\|_{C^o(\Omega)} \leq \varepsilon'(\lambda)$, where $\varepsilon'(\lambda) \to 0$ with λ, proving that $0 \leq u(\alpha,u_2(\alpha,\lambda)) \leq 1$ for λ small enough, and then that $u \in S$ for λ small enough.

B) Order 2 degeneracy (Proof of theorem 3)

One easily computes by using (3.14) and by (3.12) :

$$\frac{\partial^2\Psi}{\partial\alpha^2}(c,0) = (I-Q)\,[F''_u(.,c) + F'_u(.,c)\,\frac{\partial^2 u_2}{\partial\alpha^2}(c,0)]$$

$$\frac{\partial^2\Psi}{\partial\lambda^2}(c,0) = (I-Q)\,[F'_u(.,c)\,\frac{\partial^2 u_2}{\partial\lambda^2}(c,0) + F'_u(.,c)\gamma_0^2]$$

$$\frac{\partial^2\Psi}{\partial\alpha\partial\lambda}(c,0) = (I-Q)\,[F''_u(.,c)\gamma_0 + F'_u(.,c)\,\frac{\partial^2 u_2}{\partial\alpha\partial\lambda}(c,0)].$$

By using (3.9) one finds :

$$\frac{\partial^2 u_2}{\partial\alpha^2}(c,0) = 0,\quad \frac{\partial^2 u_2}{\partial\alpha^2}(c,0) = 2KQF'(.,c)\,\gamma_0$$

$$\frac{\partial^2 u_2}{\partial\alpha\partial\lambda}(c,0) = -\,KQF'_u(.,c) = \gamma_1$$

$$\frac{\partial^2\Psi}{\partial\alpha^2}(c,0) = (I-Q)\,F''_u(.,c) = \int_\Omega F''_u(x,c)dx$$

$$\frac{\partial^2\Psi}{\partial\lambda^2}(x,0) = (I-Q)\,[2F'_u(.,c)\,KQF'_u(.,c)\,\gamma_0 + F''_u(.,c)\,\gamma_0^2]$$

$$\frac{\partial^2\Psi}{\partial\alpha\partial\lambda}(x,0) = (I-Q)\quad F''_u(.,c)\,\gamma_0 + F''_u(.,c)\,\gamma_1$$

$$= \int_\Omega [F''_u(x,c)\,\gamma_0(x) + F'_u(x,c)\,\gamma_1(x)]\,dx.$$

Let G_0 be the unique element in X_2 defined by :

$$\begin{cases} \Delta G_0 = - QF'_u(.,c)\ \gamma_0 \\ \frac{\partial G_0}{\partial n}\Big|_\Gamma = 0. \end{cases}$$

One can then write :

$$\frac{\partial^2\Psi}{\partial\lambda^2}(c,0) = 2\int_\Omega F'_u(x,c)\ G_0(x)dx + \int_\Omega F''_u(x,c)\ \gamma_0^2(x)dx.$$

Suppose $\int_\Omega F''_u(x,c)dx = 0$. Then :

$$\frac{\partial^2\Psi}{\partial\alpha^2}(c,0)\ \frac{\partial^2\Psi}{\partial\lambda^2}(c,0) - [\frac{\partial^2\Psi}{\partial\alpha\partial\lambda}(c,0)]^2 = - [\frac{\partial^2\Psi}{\partial\alpha\partial\lambda}(c,0)]^2$$

If $\frac{\partial^2\Psi}{\partial\alpha\partial\lambda}(c,0) \neq 0$ (which by the above computation is exactly condition (3.6)), the Hessian of Ψ at $(c,0)$ is non degenerate and indefinite.

By Morse Lemma (see for instance Nirenberg [8]), $\Psi^{-1}(0)$ consists in a neighborhood of $\alpha = 0$, $\lambda = 0$, in two C^{k-2}-curves intersecting transversaly.

At least one of these curves is therefore non trivial. By (3.15) the other one will be non trivial if and only if (3.3) holds.

Theorem 3 follows now readily, by the same argument as in case A).

If $\int_\Omega F''_u(x,c)dx \neq 0$, the condition for the critical point $(c,0)$ to be non degenerate is less easy to exprime in function of F. We let to the reader the care of stating the corresponding theorem.

C) <u>Order 3 degeneracy (Proof of theorem 4)</u>

One easily computes, using (3.7) :

$$A = \frac{\partial^3\Psi}{\partial\alpha^3}(c,0) = \int_\Omega F'''_u(x,c)dx$$

$$B = \frac{\partial^3\Psi}{\partial\alpha^2\partial\lambda}(c,0) = \int_\Omega F'_u(x,c)\ \gamma_2(x)dx + 2\int_\Omega F''_u(x,c)\ \gamma_1(x)dx + \int_\Omega F'''_u(x,c)\ \gamma_0(x)dx.$$

$$C = \frac{\partial^3\Psi}{\partial\alpha\partial\lambda^2}(c,0) = \int_\Omega F'_u(x,c)\ \frac{\partial^3 u_2}{\partial\alpha\partial\lambda^2}(c,0)dx + 2\int_\Omega F''_u(x,c)\ \gamma_1(x)\ \gamma_0(x)dx +$$

$$+ 2\int_\Omega F''_u(x,c)\ G_0(x)dx + 2\int_\Omega F''_u(x,c)\ G_0(x)dx.$$

$$D = \frac{\partial^3\Psi}{\partial\alpha\partial\lambda^3}(c,0) = \int_\Omega F'_u(x,c)\ \frac{\partial^3 u_2}{\partial\lambda^3}(c,0)\ dx +$$

$$+ 6\int_\Omega F''_u(x,0)\ G_0(x)[[1+\gamma_0(x)]\ dx + \int_\Omega F''_u(x,c)\ \gamma_0^3(x)dx.$$

On the other hand

$$\frac{\partial^3 u_2}{\partial\alpha\partial\lambda^2}(c,0) = -\ 2KQF'_u(.,c)\ \gamma_1 - 2KQF''_u(.,c)\ \gamma_0^2 \text{ and}$$

$$\frac{\partial^3 u_2}{\partial\lambda^3}(c,0) = -\ 6KQF'_u(.,c)\ G_0 - 3KQF''_u(.,c)\ \lambda_0^2.$$

So the third order Taylor expansion of Ψ in the neighborhood of $(c,0)$ writes :

$$\Psi(\alpha,\lambda) = A(\alpha-c)^3 + B\lambda(\alpha-c)^2 + C(\alpha-c)\lambda^2 + D\lambda^3 + O(|\lambda-c|^3, |\lambda|^3).$$

One wishes that the third order Taylor polynomial of Ψ (i.e. the 3-jet of Ψ) determines, up to a diffeomorphism, the structure of $\Psi^{-1}(0)$ in a neighborhood of $(c,0)$; that is to say, following the terminology used in the Thom-Mather theory (cf. for instance Zeeman's lecture in [11] p. 263), one looks for conditions under which Ψ is 3-determined in a neighborhood of $(c,0)$.

One knows (cf. Zeeman [11], or the appendix), that this will be the case if the cubic constituted by the 3-jet of Ψ is either an elliptic umbilic, or a hyperbolic umbilic. We suppose that $A = \int_\Omega F''_u(x,c)dx \neq 0$. (The case where $\int_\Omega F''_u(x,c)dx = 0$ is simpler and treated in [13]).

After trivial but lengthy computation, and setting $d = B^2C^2 - 4AC^3 - 27A^2D^2 + 18ABCD - 4DB^3$, one finds that the elliptic umbilic $\eta = x^3 - xy^2$ corresponds to $d > 0$, the hyperbolic umbilic $\eta = x^3 + y^3$ corresponds to $d < 0$; the case where $d = 0$ corresponds to the parabolic umbilic $\eta = x^2y$ which is not 3-determined (it would be necessary then to examine the 4th derivates !). So the degeneracy hypothesis (3.8) is simply $d \neq 0$.

One achieves the proof as in A) or B).

IV - We shall finish by a negative result concerning bifurcation from the trivial branches $u \equiv 0$, $u \equiv 1$.

Theorem 5

Let us denote μ_2 the first strictly positive eigenvalue of the Neumann problem

$- \Delta u = \mu u, \frac{\partial u}{\partial n}|_\Gamma = 0.$

Let $\varepsilon = 0$ *or* 1.

Then (λ, ε) *is not a bifurcation point for equation (1.1)*

if :

(i) $0 < \lambda < \dfrac{\mu_2}{\operatorname{Sup\,ess}_{x \in \Omega} F'_u(x,\varepsilon)}$ *and* $\int_\Omega F'_u(x,\varepsilon)dx \geqslant 0$

or if

$$0 < \lambda < \frac{\mu_2 \left| \int_\Omega F'_u(x,\varepsilon)dx \right|}{\int_\Omega F'_u(x,\varepsilon)^2 dx + \operatorname{Sup\,ess}_{x \in \Omega} F'_u(x,\varepsilon)] \left| \int_\Omega F'_u(x,\varepsilon)dx \right|}$$

and $\int_\Omega F'_u(x,\varepsilon)dx < 0.$

The proof of theorem 5 lies in fact on the study of the following spectral problem (1) :

For given Ω, a bounded open set in $\mathbb{R}^m$, of class C^2 and $r \in L^\infty(\Omega)$, $r \not\equiv 0$ (we don't make any sign hypothesis on r), to find $(\lambda,u) \in \mathbb{R} \times H^2(\Omega)$ verifying

$$(4.1) \quad \begin{cases} \Delta u + \lambda r u = 0 \\ \frac{\partial u}{\partial n}|_\Gamma = 0. \end{cases}$$

We note first by the classical results of Agmon-Douglis-Nirenberg [1] and Sobolev imbedding theorem, that $u \in W^{2,p}(\Omega)$, for every p, $1 < p < \infty$ if $m \leq 4$, and $u \in W^{2,2m/(m-4)}(\Omega)$ if $m \geq 5$.

We set $\Sigma = \{\lambda \in \mathbb{R}, \quad u \not\equiv 0 \ ; \ (\lambda,u) \text{ solution of } (4.1)\}$.

It is easy to see that Σ is discrete and denumerable : The operator A being defined as above, one defines

$T : L^2(\Omega) \to L^2(\Omega)$ by $Tu = Ru$.

Let us suppose $\Sigma \neq \mathbb{R}$. Then there exists $\lambda_0 \in \mathbb{R}$ such that $\operatorname{Ker}(A+\lambda_0 T) = \{0\}$.

The operator $A + \lambda_0 T$ is Fredholm of index 0 (one remarks that T is compact from $H^2(\Omega)$ to $L^2(\Omega)$: it suffices to use the compactness of the injection $H^2(\Omega) \subset L^2(\Omega)$).

By Banach theorem, $(A+ \lambda_0 T)^{-1}$, is therefore a bounded operator in $L^2(\Omega)$ and is

(1) We have not found any result on this problem in the literature.

compact (compactness of the injection $H^2(\Omega) \subset L^2(\Omega)$).

In particular, $\mathrm{Ker}(A+\lambda T) = \mathrm{Ker}\,[(A+\lambda_0 T)^{-1}(A+\lambda T)]$.

But $(A+\lambda_0 T)^{-1}(A+\lambda T) = I + (\lambda-\lambda_0)(A+\lambda_0 T)^{-1}$ and

$\mathrm{Ker}\ (A+\lambda_0 T)^{-1}(A+\lambda T) = \mathrm{Ker}\,[I + (\lambda-\lambda_0)(A+\lambda_0 T)^{-1}T]$.

We are therefore reduce to study the spectrum of $I + (\lambda-\lambda_0)(A+\lambda_0 T)^{-1}T$, which is the translated by λ_0 of the spectrum of $I + \lambda(A+\lambda_0 T)^{-1}T = I + \lambda K$.

But K is a compact operator by what precedes, and the announced assertion follows by the spectral theory of compact operators.

The following theorem shows that we always have $\Sigma \neq \mathbb{R}$ and gives an estimate on the first strictly positive eigenvalue of (4.1). We thank L. Tartar who suggested us an improvement of our initial estimate (we obtained $\|r\|_{L^\infty(\Omega)}$ in place of $\operatorname{Ess\,sup}_\Omega r$).

Let us recall that μ_2 is the first strictly positive eigenvalue of the Neumann problem :

$$\begin{cases} -\Delta u = \mu u & \text{in } \Omega \\ \left.\dfrac{\partial u}{\partial n}\right|_\Gamma = 0. \end{cases}$$

Theorem 6

Let λ_1 be the first strictly positive eigenvalue of (4.1). Then :

(4.2) $$\lambda_1 \geq \frac{\mu_2}{\operatorname{Ess\,sup}_\Omega r}, \quad \text{if } \int_\Omega r(x)dx \geq 0$$

(4.3) $$\lambda_1 \geq \frac{\mu_2 \left|\int_\Omega r\,dx\right|}{\int_\Omega r^2 dx + \operatorname{Ess\,sup}_\Omega r \left|\int_\Omega r\,dx\right|}$$

if $\int_\Omega r(x)dx < 0.$

Proof

1) If $\int_\Omega r(x)dx \neq 0$, let us introduce the operator

$Q_T : Y \to Y = L^2(\Omega)$ defined by :

$$Q_T(u) = u - \left(\frac{\int_\Omega u\,dx}{\int_\Omega r\,dx}\right)r.$$

It is easily verified that :

$\operatorname{Ker} Q_T = T(\operatorname{Ker} A)$

$R(Q_T) = R(A)$ (where we denote $R(B)$ the range of the operator B)

$Q_T \circ Q_T = Q_T$.

Thanks to projector Q_T, one can decompose $Y = L^2(\Omega)$ as $Y = R(A) \oplus Y_2$. One the other hand, the projector $u \to \frac{1}{|\Omega|}\int_\Omega u\,dx$ induces the decomposition $X = \operatorname{Ker} A \oplus X_2$. By projection, (4.1) can be then written on the equivalent form :

$$(4.4) \qquad \Delta u_2 + \lambda\, r u_2 - \left(\frac{\int_\Omega r u_2\,dx}{\int_\Omega r dx}\right) r = 0$$

$$(4.5) \qquad \lambda \frac{1}{\int_\Omega r dx} \int_\Omega r(\alpha + u_2) dx \quad r = 0$$

where $u = \alpha.1 + u_2 \in \operatorname{Ker} A \oplus X_2$.

Multiplying (4.4) by u_2 and integrating by parts, it follows :

$$(4.6) \qquad -\|\Delta u_2\|_0^2 + \lambda\left[\int_\Omega r u_2^2\,dx - \frac{1}{\int_\Omega r dx}\left(\int_\Omega r u_2 dx\right)^2\right] = 0.$$

where $\|.\|_0$ is the norm in $L^2(\Omega)$.

We distinguish now two cases :

α) $\int_\Omega r dx > 0$. Then it results from (4.6) :

$$(4.7) \quad \|\nabla u_2\|_0^2 \le \lambda \int_\Omega r u_2^2 dx \le \text{ (Ess sup r) } \|u_2\|_0^2.$$

But $\int_\Omega u_2 dx = 0$, so (4.7) implies by Poincaré inequality, (cf. Deny-Lions [4])

$$(4.8) \quad \|\nabla u_2\|_0^2 \le \frac{\lambda}{\mu_2}(\operatorname{Ess}\sup_\Omega r)\,\|\nabla u_2\|_0^2.$$

Therefore, for $\lambda \in\,]0, \lambda_*[$, where $\lambda_* = \frac{\mu_2}{\operatorname{Ess}\sup_\Omega})$, one has $\nabla u_2 = 0$ a.e. so $u_2 = 0$ a.e. (since $\in X_2$), which with (4.5) implies $\alpha = 0$, and $u \equiv 0$.

β) $\int_\Omega r dx < 0$. Then it results from (4.6) :

$$\|\nabla u_2\|_0^2 \le \lambda(\operatorname{Ess}\sup_\Omega r)\;\|u_2\|_0^2 + \frac{1}{\left|\int_\Omega r dx\right|}\|r\|_0^2\;\|u_2\|_0^2$$

and Poincaré inequality implies :

$$\|u_2\|_0^2 \le \frac{\lambda}{\mu_2}\left[(\text{Ess sup } r) + \frac{\|r\|_0^2}{\left|\int_\Omega r dx\right|}\right]\|\nabla u_2\|_0^2$$

and one concludes as in case a), but now :

$$\lambda_* = \mu_2(\operatorname*{Ess\,sup}_{\Omega} r + \frac{\|r\|_0^2}{|\int_\Omega r dx|})^{-1}.$$

2) When $\int_\Omega r(x)dx = 0$, one has to introduce another projection. In fact, one has now : $T(\text{Ker } A) \subset R(A)$ and one cannot decompose Y as in the preceding cases.

Now, by the projector $u \to u - \frac{1}{|\Omega|}\int_\Omega u\, dx$, we decompose Y as follows :

(4.9) $$Y = Y_1 \oplus R(A)$$

(Note that here, Y_1 is not equal to $T(\text{Ker } A)$).

We introduce now the operator : $Q_z : R(A) \to R(A)$ defined by

$$Q_z(u) = u - \frac{1}{|\int_\Omega r^2 dx|}[\int_\Omega ru\, dx]r.$$

It is easy to check that $\text{Ker } Q_z = T(\text{Ker } A)$, and

$$R(Q_z) = \{u \in R(A),\ \int_\Omega ru\, dx = 0\} \underset{\text{def}}{=} Z,$$

and $Q_z \circ Q_z = Q_z$.

This operator induces the decomposition :

(4.10) $$R(A) = T(\text{Ker } A) \oplus Z.$$

Using decomposition (4.9) and $\int_\Omega r dx = 0$, equation (4.1) is equivalent to :

(4.11) $$\Delta u_2 + \alpha\lambda r + \lambda r u_2 = 0$$

(4.12) $$\int_\Omega r u_2 dx = 0$$

(with $u = \alpha.1 + u_2$).

Using the decomposition (4.10), the equation (4.11) is equivalent to :

(4.13) $$\Delta u_2 - \frac{1}{\int_\Omega r^2 dx}(\int_\Omega r\, u_2 dx)\, r + \lambda[ru - (\frac{1}{\int_\Omega r^2 dx}\int_\Omega r^2 u dx)r] = 0$$

(4.14) $$\frac{1}{\int_\Omega r^2 dx}[r\int_\Omega r\Delta u_2 dx + \lambda r\int_\Omega r^2 u_2 dx] + \alpha\lambda r = 0.$$

We multiply (4.13) by u_2 and we integrate by parts ; one obtains, by using (4.12):

(4.15) $$-\|\nabla u_2\|_0^2 + \lambda\int_\Omega r u_2^2 dx = 0.$$

One can then conclude that $u_2 = 0$ a.e. as in the case $\int_\Omega r\, dx > 0$; (4.11) hence implies $\alpha = 0$, so $u \equiv 0$ when $\lambda \in]0,\lambda_*[$ where $\lambda_* = \frac{\mu_2}{\operatorname*{Ess\,sup}_{\Omega} r}$.

Remark 5

The estimates (4.2), (4.3) are not optimal. For instance, let $\Omega =]-1,2[$ and r be defined by :

$$r(x) = -1 \text{ if } -1 \le x \le 0$$

$$r(x) = \frac{1}{2} \text{ if } 0 \le x \le 2$$

(so $\int_\Omega r\, dx = 0$).

One has here $\mu_2 = \frac{\pi^2}{9}$ and $\underset{\Omega}{\text{Ess sup}}\, r = \frac{1}{2}$, so $\frac{\mu_2}{\underset{\Omega}{\text{Ess sup}}} = \frac{2\pi^2}{9}$.

But a simple computation shows that the eigenvalues of (4.1) are in this case solutions of the equation $\text{tg}\, \sqrt{2\lambda} = \sqrt{2}\, \text{th}\, \sqrt{\lambda}$.

One easily deduces from this that the first positive eigenvalue of (4.1) is $> \frac{\pi^2}{2}$. ∎

Theorem 5 follows readily from theorem 6.

APPENDIX

For the convenience of the non specialist reader, we recall here some notions used in the theory of singularities of differentiable mappings (See Zeeman's lecture in [11] for a nice introduction).

Let $\mathcal{E}_n$ be the vector space of germs in 0 of C^∞-functions : $\mathbb{R}^n \to \mathbb{R}$. We set $M = (x_1,\dots,x_n)$ the maximal ideal of $\mathcal{E}_n$ generated by the germs x_i, $i = 1,\dots,n$.

If $\eta \in \mathcal{E}_n$, the k-jet $j^k\eta$ is defined by $j^k\eta = \eta_0 + \dots \eta_k$ = Taylor polynomial of η at order k.

Two germs $\eta, \xi \in \mathcal{E}_n$ are said to be k-equivalent if they have same jet, i.e. $j^k\eta = j^k\xi$. We then note $\eta \overset{k}{\sim} \xi$.

A germ $\eta \in \mathcal{E}_n$ is said to be k-determinate if for every $\xi \in \mathcal{E}_n$, the relation $\eta \overset{k}{\sim} \xi$ implies $\eta \sim \xi$ (i.e. there exists a germ of C^∞-diffeomorphism γ such that $\eta = \xi\gamma$). Clearly, η k-determinate is equivalent to $j^k\eta$ k-determinate.

For $\eta \in \mathcal{E}_n$, one defines the jacobian ideal of η by $\Delta(\eta) = (\frac{\partial\eta}{\partial x_1},\dots,\frac{\partial\eta}{\partial x_n})$. We have then a criterion of k-determinacy, due to J. Mather.

Theorem (cf. Zeeman [11] p. 274).- *Let $\eta \in M$. Then $M^k \subset M\Delta(\eta)$ implies that η is k-determinate.* ∎

An easy computation then shows that the hyperbolic umbilic $\eta = x^3 + y^3$ and the elliptic umbilic $\eta = x^3 - xy^2$ are 3-determinate.

REFERENCES

[1] S. AGMON, A. DOUGLIS, L. NIRENBERG, *Estimates near the boundary for solutions of elliptic partial differential equations satisfying general boundary conditions*, I. Comm. Pure Appl. Math. 12, (1959), p. 623-727.

[2] S. CHOW, J.K. HALE, J. MALLET-PARRET, *Applications of generic bifurcation I*, Arch. Rat. Mech. Anal. 59, (1975) p. 159-188.

[3] M. CRANDALL, P.H. RABINOWITZ, *Bifurcation from simple eigenvalues*, J. Funct. Anal., 8, (1971), p. 321-340.

[4] J. DENY, J.L. LIONS, *Les espaces du type de Beppo Levi*, Ann. Inst. Fourier, 5, (1953-54), p. 305-370.

[5]$_a$ L.A. PELETIER, *On a nonlinear diffusion equation arising in population genetics*, Proc. 4th Conference on ordinary and partial differential equations at Dundee, Springer Lecture Notes (564), 1976.

[5]$_b$ L.A. PELETIER, *A non linear eigenvalue problem occuring in population genetics*, to appear.

[6] W.H. FLEMING, *A selection-migration model in population genetics*, J. Math. Biol. 2 (1975), p. 219-234.

[7] J.L. LIONS, E. MAGENES, *Problèmes aux limites non homogènes*, I. Dunod, Paris (1968).

[8] L. NIRENBERG, *Topics in non linear functional analysis*, Courant Institute, Lecture Notes, (1974).

[9] B. SCHEURER, *Exposé N° 8 dans le séminaire d'équations aux dérivées partielles non linéaires*, Orsay (1975-1976), Publications Mathématiques d'Orsay N° 77-76.

[10] M. SLATKIN, *Gene flow and selection in a cline*, Genetics 75, (1973), p.733-756.

[11] E.C. ZEEMAN, *The classification of elementary catastrophes of codimension* 5, Notes by D.J.A. Trotman, in Structural stability, the theory of catastrophes and applications in the sciences, edited by P. Hilton, Lecture Notes in Mathematics N° 525, Springer-Verlag, (1976).

[12] F.C. HOPPENSTEADT, *Analysis of a stable polymorphism arising in a selection-migration model in population genetics*, J. Math. Biol. 2, (1975) p. 235-240.

[13] J.C. SAUT, B. SCHEURER, *Remarks on a non linear equation arising in Population genetics*, Communications on P.D.E., 3 (10), 1978 - p.907-931.

TRIPLETS DE SOLUTIONS D'UNE EQUATION AUX DERIVEES PARTIELLES ELLIPTIQUE NON LINEAIRE.

par D. SERRE
Analyse Numérique et Fonctionnelle
C.N.R.S. et Université de Paris-Sud
91405 ORSAY (France)

Our purpose in this paper is to observe the existence of three solutions of a non-linear elliptic partial differential equation; more precisely, in at least one case, we cannot classify them by their spectral properties relatively to the equation.

Nous considérons l'équation aux dérivées partielles elliptique non linéaire suivante :

$$(E_L)\begin{cases}(L-q)u = u^2 \text{ dans } \Omega\\ u = 0 \quad \text{ sur } \partial\Omega\end{cases}$$

Le domaine Ω est un ouvert borné de $\mathbb{R}^n$, connexe et $\partial\Omega$ est une sous-variété fermée de codimension 1.

L'opérateur L est donné par la formule :

$$Lu = -\sum_{i,j=1}^{n} \frac{\partial}{\partial x_j}\left(a_{ij}\frac{\partial}{\partial x_j}\right)u$$

La matrice $A = (a_{ij})_{\substack{1\le i\le n\\ 1\le j\le n}}$ appartient à l'espace $USDP_n(L^\infty(\Omega))$ des matrices uniformément symétriques définies positives, à coefficients dans $L^\infty(\Omega)$. Il existe donc $\alpha > 0$, tel que pour tout vecteur ξ de $\mathbb{R}^n$:

$$\sum_{i=1}^{n}\sum_{j=1}^{n} a_{ij}\,\xi_i\,\xi_j \ge \alpha|\xi|^2$$

Le poids q est un élément de $L^3(\Omega)$. Le choix de cet espace sera motivé ultérieurement. Nous connaissons une solution de l'équation (E_L) :

$$u \equiv 0 \text{ dans } \Omega$$

Notre intérêt se portera en fait sur le problème :

$$(E_L')\begin{cases}(E_L)\\ u\not\equiv 0 \text{ dans } \Omega.\end{cases}$$

On montre dans [1] comment ce problème est lié au nombre de solutions du problème plus général :

$$(\mathcal{P}_L)\begin{cases}(L-q)u-\lambda u^2=f \quad \text{dans } \Omega\\ u=0 \quad \text{sur } \partial\Omega\end{cases}$$

où λ est un paramètre scalaire réel, et f une distribution sur Ω, suffisamment régulière.

Lorsque $f \equiv 0$ dans Ω, le problème $(\mathcal{P}_L)$ se ramène à l'équation (E_L) grâce au changement de fonction inconnue :

$$v = \lambda u.$$

Le rôle joué ici par une condition au bord homogène est donc crucial ; ceci apparaîtra dans le paragraphe II. Pour illustrer ce propos, J. Leray [2] a montré le comportement étonnant de l'équation :

$$\begin{cases} -\Delta u = \lambda u^k \text{ dans } \Omega \quad (k \geq 2) \\ u = 1 \quad \text{sur } \partial\Omega \end{cases}$$

dans le cas où Ω est une sphère de $\mathbb{R}^n$. Il obtient, lorsque n est assez petit, un paramètre critique λ^* pour lequel une infinité de solutions s'accumulent sur une fonction u_0 ne vérifiant pas la condition aux limites.

Des résultats identiques ont été retrouvés par C. Brauner et B. Nicolaenko [3], [4] sur l'équation singulière :

$$\begin{cases} -\Delta u = \lambda \left(\frac{1}{1-u}\right)^k \quad \text{dans } \Omega \quad k \geq 1 \\ u = 1 \quad \text{sur } \partial\Omega \end{cases}$$

Le choix d'une condition aux limites de Neumann ne semble pas simplifier la nature des résultats, comme le montre l'exemple suivant qui m'a été communiqué par F. Mignot et J.P. Puel :

$$\begin{cases} -\Delta u + u = \lambda e^u \text{ dans } \Omega \\ \frac{\partial u}{\partial n} = 0 \quad \text{sur } \partial\Omega \end{cases}$$

Ce problème admet des solutions évidentes :

$$\text{si } \alpha \in \mathbf{R} \quad \begin{cases} \lambda = \alpha e^{-\alpha} \\ u \equiv \alpha \quad \text{dans } \Omega \end{cases}$$

L'étude de la bifurcation est liée au spectre de l'opérateur :

$$-\Delta + 1 - \lambda e^u.$$

En dimension 1, si $\Omega =]a,b[$, pour chaque valeur :

$$\alpha_k = 1 + \left(\frac{k\pi}{b-a}\right)^2 \qquad k \in \mathbf{N}^*$$

cet opérateur est singulier, Fredholm de type (1,1) et la bifurcation apparaît ; ce qui donne une infinité de branches bifurquées.

Cependant, cette situation est due à la rapidité de croissance de e^x. Si on remplace e^x par un polynome dans cette équation, le nombre des bifurcations à partir de la branche triviale est fini.

L'équation (E_L) semble beaucoup moins pathologique ; en effet le changement d'inconnue :

$$v = \lambda u$$

montre qu'il n'y a pas de bifurcation par rapport à λ dans l'équation :

$$\begin{cases} (L-q)u = \lambda u^2 & \text{dans } \Omega \\ u = 0 & \text{sur } \partial\Omega \end{cases}$$

Les résultats démontrés ici utiliserons en fait la dépendance par rapport au poids q de l'équation (E_L). Notons :

$$\lambda_1(L-q) < \lambda_2(L-q) \leq \cdots \leq \lambda_k(L-q) \leq \cdots$$

les valeurs propres de l'opérateur L-q avec la condition de Dirichlet :

$$\begin{cases} (L-q)u_k = \lambda_k\ u_k & \text{dans } \Omega \\ u_k = 0 & \text{sur } \partial\Omega \end{cases}$$

Cet article a pour objet les résultats suivants :

<u>THEOREME 1</u> : Si $n \leq 5$, pour tout $k \geq 2$, et pour toute matrice

$$A \in USDP_n(L^\infty(\Omega))$$

il existe $q \in L^3(\Omega)$ tel que :

1. $$\lambda_1(L-q) > 0$$
2. L'équation (E_L) a (au moins) deux solutions non triviales u et v, vérifiant :

$$\lambda_1(L-q-2u) < 0 \leq \lambda_2(L-q-2u)$$

$$\lambda_k(L-q-2v) \leq 0 \leq \lambda_{k+1}(L-q-2v)$$

□

Une amélioration de la démonstration donne aussi :

<u>THEOREME 2</u> : Si $n \leq 5$, et si $A \in USDP_n(L^\infty(\Omega))$, il existe

$q \in L^3(\Omega)$ tel que :

1. $$\lambda_1(L-q) > 0$$
2. L'équation (E_L) a (au moins) deux solutions <u>distinctes</u> non triviales u et v, vérifiant :

$$\lambda_1(L-q-2u) < 0 \leq \lambda_2(L-q-2u)$$

$$\lambda_1(L-q-2v) < 0 \leq \lambda_2(L-q-2v)$$

□

Il s'agit donc du cas k=1, pour lequel les inégalités ne suffisent pas pour prouver que u et v sont distinctes.

Nous verrons, au cours des démonstrations, qu'il est correct de parler des valeurs propres des opérateurs

L-q, L-q-2u, L-q-2v.

Dans ce qui suit, nous supposerons que $n \leq 5$.

I - CADRE FONCTIONNEL

L'espace

$$H^1(\Omega) = \{u \quad L^2(\Omega) \ ; \ \frac{\partial u}{\partial x_i} \quad L^2(\Omega) \quad \forall \ i \leq n\}$$

est un espace de Hilbert pour son produit scalaire usuel. Il contient $\mathcal{C}^1(\bar{\Omega})$ comme sous - espace dense. L'application "trace" :

$$\begin{array}{ccc} \mathcal{C}^1(\bar{\Omega}) & \rightarrow & \mathcal{C}(\Gamma) \\ u & \mapsto & u|_\Gamma \end{array}$$

se prolonge de manière unique en une application linéaire continue surjective :

$$H^1(\Omega) \xrightarrow{\gamma_o} H^{1/2}(\Gamma)$$

où $H^{1/2}(\Gamma)$ est l'interpollé $[H^1(\Gamma), L^2(\Gamma)]_{1/2}$ (Lions-Magenès [5]).

Les solutions cherchées seront des éléments du noyau :

$$H_o^1(\Omega) = \text{Ker } \gamma_o$$

et c'est de cette façon que nous interprèterons la condition :

$$u = 0 \quad \text{sur } \partial\Omega.$$

L'espace $H_o^1(\Omega)$ à une structure naturelle d'espace de Hilbert pour le produit scalaire :

$$((u,v)) = \sum_{i=1}^{n} \int_\Omega \frac{\partial u}{\partial x_i} \frac{\partial u}{\partial x_j} dx$$

avec la norme correspondante :

$$\|u\| = ((u,u))^{1/2}.$$

Lorsque $A \in USDP_n(L^\infty(\Omega))$, on munit $H_o^1(\Omega)$ de la structure d'espace de Hilbert associée au produit scalaire :

$$((u,v))_A = \sum_{i=1}^{n} \sum_{j=1}^{n} \int_\Omega a_{ij}(x) \frac{\partial u}{\partial x_i} \frac{\partial u}{\partial x_j} dx$$

La norme

$$\|u\|_A = ((u,u))_A{}^{1/2}$$

est équivalente à la précédente.

Puisque l'ouvert Ω est borné, que $n \leq 5$, le théorème de Sobolev affirme que :

$$H_o^1(\Omega) \hookrightarrow L^3(\Omega)$$

et que cette injection est compacte.

Lorsque $L^3(\Omega)$ est muni de sa norme usuelle

$$|u|_3 = \{\int_\Omega |u(x)|^3 \ dx\}^{1/3}$$

et que $H_o^1(\Omega)$ est muni de la norme $\| \ \|_A$, nous notons

C_A la norme de cette injection. On pose $C = C_I$.

II - L'EQUATION (E_L) A L-q FIXE

Nous supposons dans cette section que A et q sont fixés :

$$A \in USDP_n(L^\infty(\Omega))$$
$$q \in L^3(\Omega) \quad .$$

Alors, la multiplication par q est un endomorphisme compact de $H_o^1(\Omega)$; il s'ensuit que l'opérateur L-q est Fredholm d'indice nul de $H_o^1(\Omega)$ dans son dual $H^{-1}(\Omega)$, et que son spectre est bien de la forme

$$\lambda_1(L-q) < \lambda_2(L-q) \leq \cdots \leq \lambda_k(L-q) \leq \cdots \xrightarrow{k\to+\infty} +\infty \quad .$$

Supposons $\lambda_1(L-q) > 0$.

Soit

$$\|u\|_1 = \{\|u\|_A^2 - \int_\Omega q(x)\, u^2(x)\, dx\}^{1/2} \qquad \forall u \in H_o^1(\Omega)$$

C'est une norme équivalente à la norme $\|\cdot\|$ sur $H_o^1(\Omega)$.

Proposition 1 : Si $\lambda_1(L-q) > 0$, il existe une solution non triviale u de (E_L), vérifiant :

$$\lambda_1(L-q-2u) < 0 \leq \lambda_2(L-q-2u)$$ □

Démonstration : Soit $S(u) = \dfrac{\int_\Omega u^3(x)dx}{\|u\|_1^3} \qquad \forall u \in H_o^1(\Omega)\backslash\{o\}$

L'application $u \longmapsto \|u\|_1^3$ est convexe continue, donc faiblement semi-continue-inférieure sur $H_o^1(\Omega)$. D'autre part $u \longmapsto \int_\Omega u^3(x)\, dx$ est une application faiblement continue sur $H_o^1(\Omega)$. Il en résulte que $u \longmapsto S(u)$ est faiblement semi-continue-supérieurement sur $H_o^1(\Omega)\backslash\{o\}$.

Comme $H_o^1(\Omega) \hookrightarrow L^3(\Omega)$, on a :

$$S(u) \leq C^3 \left(\frac{\|u\|}{\|u\|_1}\right)^3 \qquad \forall u \in H_o^1(\Omega)\backslash\{o\}$$

et, vu l'équivalence des normes, on obtient

$$\text{Sup } S(u) < +\infty \tag{1}$$

Soit u_n une suite d'éléments de $H_o^1(\Omega)\backslash\{o\}$ maximisant S :

$$\lim_{n\to\infty} S(u_n) = \text{Sup } S(u).$$

Comme S est homogène d'ordre 0, on peut supposer que :

$$\|u_n\|_1 = 1 \qquad \forall\, n \in \mathbb{N} . \tag{2}$$

Quitte à extraire une sous-suite de la suite u_n, nous pouvons donc supposer également que :

$$u_n \xrightarrow{n \to \infty} u_\infty \qquad \text{dans } H_0^1(\Omega) \text{ faible} \tag{3}$$

Comme l'injection $H_0^1(\Omega) \hookrightarrow L^3(\Omega)$ est compacte, et vu (2) et (3) :

$$\int_\Omega u_\infty^3(x)\, dx = \lim_{n\to\infty} \int_\Omega u_n^3(x)\, dx = \lim_{n\to\infty} S(u_n) = \operatorname{Sup} S(u) > 0$$

ce qui prouve :

$$u_\infty \not\equiv 0 \tag{4}$$

Alors, la semi-continuité-supérieure faible de S donne :

$$S(u_\infty) \geq \overline{\lim_{n\to\infty}}\, S(u_n) = \operatorname{Sup} S(u)$$

C'est-à-dire :

$$S(u_\infty) = \operatorname{Sup} S(u) \tag{5}$$

Lemme 1 : L'application S atteint sa borne supérieure sur $H_0^1(\Omega)\backslash\{o\}$ □

Soit alors $u \in H_0^1(\Omega)\backslash\{o\}$ qui maximise S. Comme S est homogène d'ordre 0, on peut supposer que :

$$\|u\|_1^2 = \int_\Omega u^3(x)\, dx \quad . \tag{6}$$

On a, car S est de classe $\mathcal{C}^\infty$ sur $H_0^1(\Omega)\backslash\{o\}$:

$$\begin{cases} S'(u) = 0 & (7) \\ S''(u)(h,h) \leq 0 \qquad \forall\, h \in H_0^1(\Omega) & (8) \end{cases}$$

Et

$$S'(u) = \frac{3}{\|u\|_1^5} \{ \|u\|_1^2\, u^2 - (\int_\Omega u^3(x)\, dx)\ (L-q)u \}$$

Les égalités (6) et (7) nous donnent :

$$(L-q)u = u^2 \tag{9}$$

et on a bien :

$$u \in H_0^1(\Omega)\backslash\{o\} .$$

Il reste à prouver les inégalités de la proposition 1. D'après le principe du maximum, appliqué à l'opérateur L-q, on a :

$$(u^2 \gtrless 0) \Leftrightarrow (L-q)u \gtrless 0) \Rightarrow (u > 0 \text{ pp dans } \Omega) \quad . \tag{10}$$

Or :

$$\begin{cases} (L-q-u)u = 0 & \text{dans } \Omega \\ u \in H_0^1(\Omega)\backslash\{o\} \end{cases}$$

Il existe donc $K \in \mathbb{N}^*$ tel que :

$$\lambda_k(L-q-u) = 0.$$

Le vecteur propre associé, u, étant positif, on a en fait :

$$k = 1.$$

Utilisant (10), l'inégalité :

$$L-q-2u < L-q-u$$

implique :

$$\lambda_1(L-q-2u) < \lambda_1(L-q-u) = 0 \tag{11}$$

Enfin

$$\frac{1}{3}\|u\|_1^7 S''(u)(h,h) = 2\|u\|_1^4 \int_\Omega u\, h^2 dx - 6\|u\|_1^2 \int_\Omega h[(L-q)u]dx \int_\Omega u^2 h\, dx$$
$$+ 5\int_\Omega u^3 dx \left(\int_\Omega h[(L-q)u]\, dx\right)^2 - \|u\|_1^2 \int_\Omega u^3\, dx \int_\Omega h[(L-q)h]\, dx$$

Utilisons (6) et (9) pour simplifier :

$$\frac{1}{3}\|u\|_1^5 S''(u)(h,h) = 2\|u\|_1^2 \int_\Omega uh^2 dx - \left(\int_\Omega u^2 h\, dx\right)^2 - \|u\|_1^2 \|h\|_1^2 \tag{12}$$

Soit F l'hyperplan fermé de $H_0^1(\Omega)$ défini par :

$$F = \{h \in H_0^1(\Omega) \ ; \int_\Omega u^2\, h\, dx = 0\}$$

Pour tout $h \in F$, on a, compte-tenu de (8) et (12) :

$$\|h\|_1^2 - 2\int_\Omega u\, h^2\, dx \geq 0 \tag{13}$$

où :

$$<(L-q-2u)h,h> \quad \geq 0 \qquad \forall h \in F \tag{14}$$

Puisque codim F = 1, (14) est équivalent à :

$$\lambda_2(L-q-2u) \geq 0 \tag{15}$$

Ce qui termine la preuve de la proposition 1.

On précise ce résultat de la manière suivante :

<u>Proposition 2</u> : Si $S(u) = \text{Max } S(v)$, $u \in H_0^1(\Omega)\setminus\{o\}$; si u est normalisé par (6) (c'est -à-dire que (9) a lieu), et si (15) est une égalité :

$$\lambda_2(L-q-2u) = 0.$$

Alors, $(h,k,l) \in (\text{Ker}(L-q-2u))^3$, on a

$$\int_\Omega h(x)\ k(x)\ l(x)\ dx = 0 \qquad (16)$$ □

<u>Démonstration</u> : Sous les hypothèses de la proposition 2, on a :

$$\begin{cases} S'(u) = 0 \\ S''(u)(h,h) = 0 \end{cases} \qquad \forall h \in \text{Ker}(L-q-2u) \qquad (17)$$

Alors $S(u) = \text{Max } S(v)$ implique, avec (17) :

$$\forall h \in \text{Ker}(L-q-2u) \qquad \begin{cases} S'''(u)(h,h,h) = 0 & (18) \\ S^{IV}(u)(h,h,h,h) \le 0 & (19) \end{cases}$$

Or, en tenant compte de (6), (9), on a :

$$\forall h \in \text{Ker}(L-q-2u) \qquad \begin{cases} S'''(u)(h,h,h) = \dfrac{6}{\|u_1\|_1^3} \displaystyle\int_\Omega h^3(x)\ dx & (20) \\ S^{IV}(u)(h,h,h,h) = -342\ \dfrac{\int_\Omega u(x)\ h^2(x)\ dx}{\|u\|_1^5} & (21) \end{cases}$$

L'inégalité (19) est donc certainement réalisée,et l'égalité (18) est :

$$\int_\Omega h^3(x)\ dx = 0 \qquad (22)$$

Si $(h,k,l) \in \text{Ker}(L-q-2u)$, formons le polynome :

$$P(\lambda,\mu) = \int_\Omega (h+\lambda k+\mu l)^3(x)dx$$

Il est identiquement nul d'après (22), puisque :

$$h+\lambda k+\mu l \in \text{Ker}(L-q-2u)$$

La nullité du coefficient dans P du monome $\lambda\mu$ est l'égalité demandée. □

<u>III - DEMONSTRATION DU THEOREME I</u>

Supposons les hypothèses du théorème 1 vérifiées.

Soit $p \in H_0^1(\Omega) \cap \mathcal{C}^2(\bar{\Omega})$ une fonction strictement positive sur Ω. Le problème spectral :

$$\begin{cases} L\Psi = \mu p\Psi \\ \Psi \in H_0^1(\Omega)\backslash\{o\} \end{cases}$$

admet le système de valeurs propres :

$$0 < \mu_1^A(p) < \mu_2^A(p) \leq \cdots \leq \mu_k^A(p) \leq \cdots$$

associées aux vecteurs propres $\Psi_1^A(p),\ldots,\Psi_k^A(p),\ldots$

Ayant fixé p ainsi, fixons $\mu \in [\mu_k^A(p), \mu_{k+1}^A(p)]$. L'opérateur $L - \mu p$ est symétrique et il est aisé de voir que :

$$\lambda_k(L - \mu p) \leq 0 \leq \lambda_{k+1}(L - \mu p).$$

Ce qui prouve qu'on pouvait choisir au départ p tel que :

$$\lambda_k(L-qp) \leq 0 \leq \lambda_{k+1}(L-p) \qquad (1)$$

Nous supposerons dans cette section que p vérifie (1).

Soit g la solution du problème de Dirichlet :

$$\begin{cases} Lg = Lp - \frac{1}{2}p^2 & \text{dans } \Omega \\ g \in H_0^1(\Omega) \end{cases}$$

Posons :

$$r = \operatorname{Max}(0, -\operatorname*{Inf}_{x\in\Omega} \frac{p+g}{2}(x), -\operatorname*{Inf}_{x\in\Omega} g(x))$$

Comme $p \in \mathcal{C}^2(\bar{\Omega})$, $Lp - \frac{1}{2}p^2 \in L^\infty(\Omega)$, et donc :

$$g \in L^\infty(\Omega)$$

Ce qui prouve : $\quad r < +\infty$

Nous définissons une suite récurrente g_n par :

$$\begin{cases} \forall n \in \mathbb{N} & \begin{cases} (L+r)g_{n+1} = rg_n + \frac{1}{2}g_n^2 + Lp - \frac{1}{2}p^2 & \text{dans } \Omega \\ g_{n+1} \in H_0^1(\Omega) \end{cases} \\ g_0 = g \end{cases}$$

<u>Lemme 2</u> : La suite g_n est croissante, et vérifie, $\forall n \in \mathbb{N}$:

(i) $Lg_n - \frac{1}{2}g_n^2 \leq Lp - \frac{1}{2}p^2$ dans Ω

(ii) $g_n < p$ pp dans Ω

(iii) $\lambda_1(L-g_n) > 0$

□

La démonstration procède par récurrence :

D'après le principe du maximum, appliqué à L :

$$g_o = g < p \qquad \text{pp dans } \Omega.$$

Et $$(L-g)(p-g) = \frac{1}{2}(p-g)^2 + \frac{1}{2}g^2 > 0 \text{ dans } \Omega.$$

A l'aide du Lemme 3, nous obtenons, puisque p-g > 0 :

$$\lambda_1(L-g_o) > 0.$$

<u>Lemme 3</u> : Soit $h \in L^3(\Omega)$. Supposons qu'il existe $\phi \in H_0^1(\Omega)$ tel que

$$\begin{cases}(L-h)\phi \gneqq 0 & \Omega \\ \phi \geq 0 & \Omega\end{cases}$$

alors $\lambda_1(L-h) > 0$. □

Les assertions i) ii) iii) étant vérifiées si n = 0, supposons qu'elles le sont pour $n \leq m$, et que :

$$g_o \leq g_1 \leq \cdots \leq g_m.$$

Alors
$$(L+r)(g_{m+1}-g_m) = (r + \frac{g_m+g_{m-1}}{2})\ (g_m - g_{m-1})$$

Or
$$r + \frac{g_m+g_{m-1}}{2} \geq r + g_o \geq 0 \quad \text{pp dans } \Omega$$

grâce à la définition de r.

Ainsi, $(L+r)(g_{m+1} - g_m) \geq 0$ car $g_m - g_{m-1} \geq 0$, et le principe du maximum, appliqué à L+r prouve :

$$g_{m+1} - g_m \geq 0 \tag{2}$$

Ensuite :
$$(L-g_{m+1})(p-g_{m+1}) \geq \frac{1}{2}\ (p-g_{m+1})^2 \geq 0 \tag{3}$$

Et $$(L+r)(p-g_{m+1}) = (r + \frac{p+g_m}{2})\ (p-g_m) \tag{4}$$

Comme :
$$r + \frac{p+g_m}{2} \geq 0 \text{ dans } \Omega$$

l'égalité (4) implique, avec le principe du maximum :

$$p - g_{m+1} > 0\ . \tag{5}$$

On applique alors le Lemme 3 aux inégalités (3) et (5), et on obtient :

$$\lambda_1(L-g_{m+1}) > 0$$

D'où le Lemme 2. □

Démonstration du Lemme 3 : Soit Ψ le premier vecteur propre, positif, de L-h, et λ_1 la première valeur propre :

$$\begin{cases} (L-h)\ \Psi = \lambda_1\ \Psi & \text{dans } \Omega \\ \Psi > 0 & \text{pp dans } \Omega \\ \Psi = 0 & \text{sur } \partial\Omega \end{cases}$$

On a :

$$\lambda_1 \int_\Omega \Psi(x)\ \phi(x)\ dx = \int_\Omega [(L-h)\ \Psi]\ (x)\ \phi(x)\ dx$$

$$= \int_\Omega [(L-h)\ \phi]\ (x)\ \Psi(x)\ dx > 0$$

et

$$\int_\Omega \phi(x)\ \Psi(x)\ dx > 0$$

donc

$$\lambda_1 > 0$$

□

La suite g_n, croissante et majorée, converge simplement vers sa borne supérieure q. D'après ii), on a

$$q \le p \qquad \text{pp dans } \Omega$$

Vu la formule de récurrence, on en déduit que g_n est une suite uniformément équicontinue. Ainsi, elle converge uniformément sur $\bar\Omega$ vers q. La suite Lg_n est donc elle aussi, uniformément convergente. Donc q vérifie :

$$Lq - \frac{1}{2}\ q^2 = Lp - \frac{1}{2}\ p^2 \tag{6}$$

Passons à la limite dans iii) :

$$\lambda_1(L-q) \ge 0 \tag{7}$$

ce qui montre que $q \neq p$, puisque $\lambda_1\ (L-p) < 0$. Posons :

$$v = \frac{1}{2}\ (p-q).$$

On a

$$(L-q)v = \frac{1}{2}(Lp-Lq) - \frac{1}{2}\ q(p-q) = \frac{1}{4}\ (p-q)^2 = v^2 \tag{8}$$

et $v \in H^1_0(\Omega)$.

Enfin

$$L-q-2v = L-p$$

donc

$$\lambda_k(L-q-2v) \le 0 < \lambda_{k+1}(L-q-2v)$$

Et (8), joint à $v \geq 0$, montre (Lemme 3) :

$$\lambda_1(L-q) > 0 \tag{9}$$

Nous appliquons alors la proposition 1 à L-q, et nous obtenons une solution u de (E_L) vérifiant :

$$\lambda_1(L-q-2u) < 0 \leq \lambda_2(L-q-2u)$$

Ce qui prouve le théorème 1. □

IV - DEMONSTRATION DU THEOREME 2

Avec les notations du paragraphe précédent, nous prenons

$$\mu = \mu_2^A(p)$$

Si bien que :

$$\lambda_2(L-\mu p) = 0$$

De même que précédemment, nous normalisons :

$$\mu = 1 \qquad\qquad \lambda_2(L-p) = 0$$

Supposons que $\phi \in \mathrm{Ker}(L-p)\setminus\{o\}$. Vu la régularité de p, ϕ est contenu sur Ω, et comme cette fonction n'est pas de signe constant (cela contredirait (1)), il existe un domaine $\omega \subset\subset \Omega$ et une constante $\varepsilon > 0$ tels que :

$$\phi(x) \geq \varepsilon \quad \text{dans } \omega.$$

Soit alors h_o un vecteur propre associé à la première valeur propre de L sur ω :

$$\begin{cases} L\, h_o = \alpha\, h_o \quad \text{dans } \omega \\ h_o \in H_o^1(\omega),\ h_o \not\equiv 0 \end{cases}$$

Convenons de prolonger h par 0 à Ω tout entier. Alors :

$$h_o \in H_o^1(\omega) \Rightarrow h_o \in H_o^1(\Omega)$$

Définissons, si $r \in \mathbb{R}$:

$$\begin{cases} p_r(x) = \dfrac{L(\phi+rh_o)(x)}{\phi(x)+rh_o(x)} & \text{si } x \in \omega \\ p_r(x) = p(x) & \text{si } x \notin \omega \end{cases}$$

Si $r \in \left]-\dfrac{\varepsilon}{|h_o|_\infty}, \dfrac{\varepsilon}{|h_o|_\infty}\right[$ on a :

$$p_r \in \mathcal{C}^1 \cap H_o^1(\Omega)$$

D'autre part :

$$\begin{cases} (L-p_r)(\phi+rh_o) = 0 \\ \phi+rh_o \in H_o^1(\Omega)\setminus\{o\} \end{cases}$$

Donc, il existe $j \in \mathbb{N}^*$ tel que :

$$\lambda_j(L-p_r) = 0.$$

Mais il est immédiat, vu le choix de r, que $(\phi+rh_0)^{-1}(\mathbb{R}^*)$ a le même nombre de composantes connexes que $\phi^{-1}(\mathbb{R}^*)$, c'est-à-dire deux puisque ϕ est le deuxième vecteur propre de L-p. Cela implique j = 2.

I.e :

$$\lambda_2(L-p_r) = 0 \qquad \forall r \in \left]-\frac{\varepsilon}{|h_0|_\infty}, \frac{\varepsilon}{|h_0|_\infty}\right[$$

La définition de h_0 montre que (quitte à changer h par -h) :

$$h_0 > 0 \quad \text{dans } \omega$$

D'où

$$\int_\Omega \phi^2(x)\, h_0(x)\, dx > 0$$

C'est-à-dire que l'application :

$$r \longmapsto \int_\Omega (\phi+rh_0)^3(x)\, dx$$

n'est pas identiquement nulle au voisinage de 0, et qu'on peut choisir r de façon que :

$$\begin{cases} p_r \in \mathcal{C}^1(\bar{\Omega}) \cap H_0^1(\Omega) \\ \lambda_2(L-p_r) = 0 \\ \int_\Omega (\phi+rh_0)^3(x)\, dx \neq 0 \end{cases}$$

La méthode du paragraphe précédent donne une solution q du problème :

$$\begin{cases} Lq - \frac{1}{2} q^2 = Lp_r - \frac{1}{2} p_r^2 \\ q \in H_0^1(\Omega) \\ \lambda_1(L-q) > 0 \end{cases}$$

En prenant $v = \frac{1}{2}(p_r-q)$, on a une solution de (E_L) vérifiant :

$$\lambda_2(L-q-2v) = 0 \tag{2}$$

La méthode du paragraphe II fournit elle aussi une solution u de (E_L), vérifiant

$$\lambda_1(L-q-2u) < 0 \leq \lambda_2(L-q-2u) \tag{3}$$

Les solutions u et v sont distinctes ; sinon, en prenant $h=k=l=\phi+rh_0$, on contredirait la proposition 2. D'où le théorème □

Remarque : Une étude fine de l'équation (E_L) lorsqu'on fait varier q à partir de la valeur obtenue ci-dessus montre qu'on peut rendre stricte les inégalités du théorème 2, et que cela se réalise lorsque q est dans un ouvert non vide de $L^3(\Omega)$.

BIBLIOGRAPHIE

[1] SERRE D. : Sur l'image de l'opérateur $u \to -\Delta u - u^2$, avec $u \in H_0^1(\Omega)$. (A paraître).

[2] LERAY J. : Etude de diverses équations intégrales non linéaires. J. Math. Pures et Appl. 12, (1933), p. 20-21.

[3] BRAUNER C.M. et NICOLAENKO B. : Sur une classe de problèmes elliptiques non linéaires. Comp. Rend. Sean. Acad. Scie. 286, série A, (1978), p.1007.

[4] BRAUNER C.M. et NICOLAENKO B. : Voir la présente publication.

[5] LIONS J.L. et MAGENES E. : Problèmes aux limites non homogènes et applications. Paris, Dunod, 1968.